The Solar System 1

SCIENCES

Universe, Field Director – Fabienne Casoli

Solar System, Subject Head – Thérèse Encrenaz

The Solar System 1

Telluric and Giant Planets, Interplanetary Medium and Exoplanets

Coordinated by
Thérèse Encrenaz
James Lequeux

iSTE

WILEY

First published 2021 in Great Britain and the United States by ISTE Ltd and John Wiley & Sons, Inc.

ISTE Ltd
27-37 St George's Road
London SW19 4EU
UK

www.iste.co.uk

John Wiley & Sons, Inc.
111 River Street
Hoboken, NJ 07030
USA

www.wiley.com

Library of Congress Control Number: 2021940349

British Library Cataloguing-in-Publication Data
A CIP record for this book is available from the British Library
ISBN 978-1-78945-033-0

ERC code:
PE9 Universe Sciences
PE9_1 Solar and interplanetary physics
PE9_4 Formation of stars and planets

MIX
Paper from responsible sources
FSC www.fsc.org FSC® C013604

Contents

Preface

Thérèse ENCRENAZ[1] **and James LEQUEUX**[2]
[1] *LESIA, Paris Observatory, PSL University, Paris, France*
[2] *LERMA, Paris Observatory, PSL University, Paris, France*

The aim of this book is to present a global and synthetic vision of planetology to the reader, in other words, the study of the objects of the Solar System. This is an ambitious objective, because planetology has undergone considerable development over the last decades and, today, it presents interfaces with multiple disciplines. In our approach, we have chosen to prioritize the study of physico-chemical processes, in order to shed light on the mechanisms that are at the origin of the formation of the objects of the Solar System, or that are responsible for their evolution.

This work is a continuation of the book *Le Système solaire* (T. Encrenaz, J.-P. Bibring, M. Blanc, M.-A. Barucci, F. Roques, P. Zarka), published in 2003 by EDP-Sciences and CNRS Editions, in the "Savoirs Actuels" collection. This work was itself the third edition of *Le Système solaire* (T. Encrenaz, J.-P. Bibring, M. Blanc), first published in 1987 in co-edition with InterEditions and CNRS Editions. More than 30 years after this first edition, a complete revision of the work was essential: during this period, planetology has undergone several revolutions. The first ones, both realized thanks to ground-based telescopes, were the discovery of the first trans-Neptunian objects other than Pluto in 1992, and then, in 1995, the discovery of the first extrasolar planets around solar-type stars. The 1990, 2000 and 2010 decades saw the deep exploration of the planets Jupiter and Saturn and their system, with the Galileo and Cassini space missions. The beginning of the 21st century saw the resumption of the Mars exploration program with the launch of numerous probes, both in orbit and on the surface of the planet. Venus and Mercury were also visited by space probes, as well as several asteroids including Ceres and

The Solar System 1,
coordinated by Thérèse ENCRENAZ and James LEQUEUX.

Vesta; lastly, the comet Churyumov-Gerasimenko was the subject of extended exploration thanks to the spectacular European Rosetta mission. Meanwhile, the first detections of exoplanets paved the way to a new field of research in full development, that of exoplanetology. The extraordinary variety of objects discovered around other stars, both in terms of their orbits and their physical parameters, has raised new questions about the formation scenarios of planetary systems; the discovery of the phenomenon of migration in these systems has encouraged new research on the origin and evolution of our own Solar System. The discovery of a large proportion of rocky exoplanets – the "super-Earths" – has also stimulated work in exobiology with the ultimate goal of finding life on one of these exoplanets. Over the last 30 years, planetary scientists have forged increasingly close links with other astronomers (who are at the origin of the discovery of exoplanets), but also with geophysicists, chemists and biologists interested in the problem of the emergence of life, on Earth or elsewhere.

The plan of the work follows the main course of the previous works. The first volume opens with three general chapters. Chapter 1 presents a general description of the Solar System. Chapter 2 describes the environment of the Solar System, the formation of stars and planets within the disks as well as the extrasolar planets. Chapter 3 deals with the interaction of the Solar System with the interplanetary medium. Chapters 4 and 5 present a description of telluric and giant planets. The second volume presents the satellites and rings of giant planets (Chapter 1), comets, asteroids and dwarf planets (Chapter 2), meteorites and cosmochemistry (Chapter 3), the formation and dynamics of the Solar System (Chapter 4), the origin of life and extraterrestrial life (Chapter 5), and lastly, the methods of study of the Solar System (Chapter 6). Almost all of these chapters are composed of original texts. In the few cases where we have relied on texts from the last edition of the Solar System (2003), we have explicitly mentioned it. We thank J.-P. Bibring, M. Blanc, M.-A. Barucci and P. Zarka for their contributions in the previous book.

This book is intended primarily for students of planetology, astrophysics and geophysics at the undergraduate and graduate level, but we hope that it can be read with benefit by all scientists, researchers and engineers wishing to deepen their knowledge of planetology.

June 2021

1

General Presentation of the Solar System

Thérèse ENCRENAZ, Françoise ROQUES and Laurent LAMY
LESIA, Paris Observatory, PSL University, Paris, France

1.1. Introduction

The Solar System is defined as the set of objects subjected to the gravitational field of the Sun. According to this definition, it extends up to about two light-years (about 2 10^{13} km), halfway from the nearest star, Proxima Centauri. The outermost boundary of the Solar System is the Oort cloud, a vast spherical reservoir of comets located at about 20,000 to 40,000 astronomical units (au) and whose outer edge could extend to a few hundred thousand au, or a third to half a light-year. Apart from comets, all the objects in the Solar System that we know are located at distances not exceeding a few thousand au. Most of them are located close to the plane of the equator of the Sun, itself very close to the plane of the Earth's orbit; it is this plane, called "ecliptic", which is adopted as a reference. The Solar System thus has the structure of a disk (resulting from its formation scenario – see Chapters 2 and 4 of Volume 2), in which the planets, their satellites, asteroids and comets evolve at great distances from each other. The mass of this disk is very low (about 0.15%) with respect to that of the Sun: it is essentially distributed in the planets.

There are eight planets in the Solar System and they are divided into two categories: the telluric planets (Mercury, Venus, Earth and Mars), close to the Sun; and the giant planets (Jupiter, Saturn, Uranus and Neptune), that are further away. The brightest (Mercury, Venus, Mars, Jupiter and Saturn), observable with the naked eye, have been known since antiquity. Their number has varied over the

The Solar System 1,
coordinated by Thérèse ENCRENAZ and James LEQUEUX.

centuries, depending on successive discoveries. Following the discovery of Pluto in 1930, this number reached 9, until in 2006, the International Astronomical Union (IAU) reviewed the definition of a planet (see Box 1.1) and reduced it to 8, Pluto now belonging to the category of Trans-Neptunian objects (TNO). With the exception of Mercury, they all have an atmosphere. The telluric planets are relatively small (the largest, Earth, has a radius of 6,400 km) and dense, and have a small number of satellites. On the other hand, the four giant planets are large (with a radius greater than 20,000 km) and not very dense; all of them have a system of rings (formed of particles whose size can vary from a micrometer to a meter) and a procession of satellites (also of very variable size, the largest being comparable to Mercury); we will see further on that these characteristics are the consequence of their formation scenario. With a few rare exceptions (including Titan, Saturn's largest satellite), satellites have no atmosphere.

The list of the planets of the Solar System is, however, not definitively closed. In 2016, astronomers K. Batygin and M. Brown (Batygin and Brown 2016) suggested, based on numerical simulations, the existence of a ninth planet of about 10 Earth masses that would have acted as a massive disturbance of the outer Solar System, and whose existence would explain the very asymmetrical distribution of the orbits of the most distant TNO. This planet, formed in the current environment of the giant planets, would have been ejected towards the outside by the play of the gravitational disturbances, for which they are responsible. If it was not definitively ejected from the Solar System, it would evolve between some 200 and 1,200 au, on an inclined orbit of about 30° on the ecliptic plane. Given its distance, this planet will be extremely difficult to observe. Other studies suggest that the particular configuration of the distant TNO orbits could also be caused by the combined gravitational forces of several small trans-Neptunian objects.

Historically, the planets of the Solar System have been defined as the "wandering objects" (πλανετεσ in Greek) of the sky. In the 16th century, the Earth was added to this list, then Uranus and Neptune. The asteroids, discovered in the 19th century, were classified rather quickly in a category apart because of their ever-increasing number. Pluto, discovered in 1930, was first considered as a planet. In 2003, the discovery of the TNO Eris, of a size comparable to that of Pluto, needed a new reflection on the specificity of a planet. After much debate, the IAU adopted the following definition in 2006: a planet is a celestial body that (1) is in orbit around the Sun, (2) has sufficient mass to be in hydrostatic equilibrium (in other words, spherical in shape), and (3) has eliminated the neighborhood around its orbit. According to this definition, there are eight confirmed planets of the Solar System; the four telluric ones and the four giants. There is perhaps a ninth planet at the confines of the Solar System, whose existence remains to be confirmed. The IAU has also introduced the concept of the "dwarf planet" which includes the first two criteria, but not the third. There are currently five objects in this category, one asteroid (Ceres) and four TNOs (Pluto, Haumea, Makemake and Eris) (see section 2.4, Volume 2).

The discovery of exoplanets, since 1995, has forced astronomers to redefine the border between planet and star. Unlike planets, stars have the ability to trigger nuclear reactions within them. Brown dwarfs are intermediate objects between stars and planets, which are the seat of the first nuclear reaction cycle causing the fusion of deuterium, but not of the next, responsible for the fusion of helium. The brown dwarfs have a mass ranging between 13 M_J (mass of Jupiter) and 0.07 M_S (mass of the Sun), in other words, 75 M_J. This definition was adopted by the IAU in 2003.

Box 1.1. *How do you define a planet?*

The following books present general studies on the physics of Solar System objects: (Kivelson and Russell 1995; Lewis 1995; Encrenaz *et al.* 2003; Hanel *et al.* 2003; Lissauer and de Pater 2013; Spohn *et al.* 2014; Catlin and Kasting 2017). Figure 1.1 presents a simplified diagram of the Solar System.

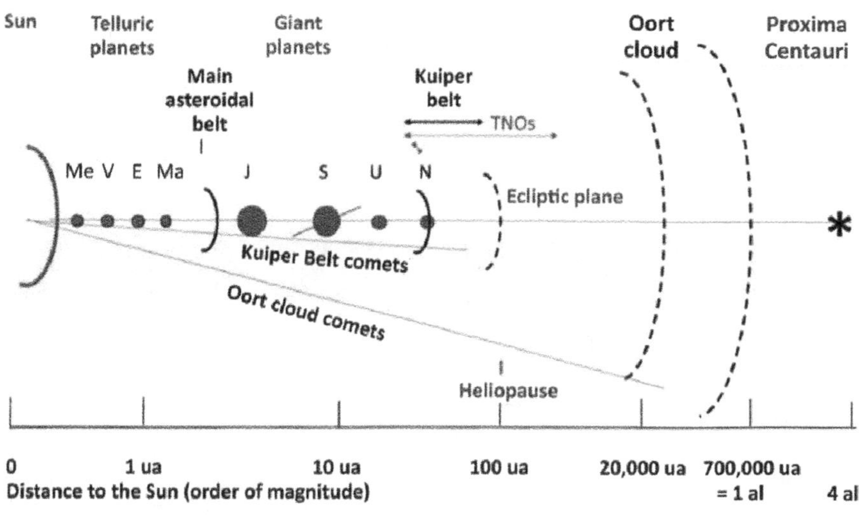

Figure 1.1. *Simplified diagram of the structure of the Solar System. The dimensions are not respected on the figure and the scale of the distances to the Sun indicates an order of magnitude. The outer limits of the Kuiper belt, the trans-Neptunian object region (TNO) and the Oort cloud are very imprecise. For a color version of this figure, see www.iste.co.uk/encrenaz/solar1.zip*

Asteroids are small objects (with a radius of less than 500 km), also devoid of atmosphere. Most of them are located in a torus called the "main asteroid belt", between the orbits of Mars and Jupiter. Beyond the orbit of Neptune extends the "Kuiper Belt", occupied by the trans-Neptunian objects. These objects, discovered since 1992, have a diameter of a few hundred kilometers, on highly different orbits; Pluto, discovered in 1930, is one of the largest objects of this family. Lastly, comets are even smaller objects (less than 10 kilometers). Unlike all the preceding objects, they evolve in very elliptical orbits that can take them from the farthest reaches of the Solar System to near the Sun. The extraterrestrial matter received on Earth mainly comes from asteroids, in the case of meteorites, and from comets, in the case of shooting stars and micrometeorites.

Planet	Semi-major axis (au)	Eccentricity	Inclination on the ecliptic (°)	Period of sidereal revolution (year)
Mercury	0.387	0.206	7.00	0.24
Venus	0.723	0.007	3.39	0.62
Earth	1.000	0.017	0.00	1.00
Mars	1.524	0.093	1.85	1.88
Jupiter	5.203	0.048	1.30	11.86
Saturn	9.537	0.054	2.48	29.42
Uranus	19.218	0.046	0.77	84.02
Neptune	30.110	0.009	1.77	164.9

Table 1.1. *Orbital characteristics of the planets of the Solar System*

Planet	Mass (M_T)	Radius (R_T)	Density (g/cm^3)	Rotation period	Obliquity (°)	V_{escape} (km/s)	Atm. Comp.
Mercury	0.055	0.382	5.43	58.65 d	0.04	4.43	-
Venus	0.815	0.949	5.20	243.0 d	177.33	10.36	CO_2, N_2
Earth	1.00	1.000	5.52	23.93 h	23.45	11.19	N_2, O_2
Mars	0.107	0.532	3.93	24.62 h	25.19	5.03	CO_2, N_2
Jupiter	317.9	11.21	1.33	9.92 h	3.08	59.54	H_2, He, CH_4, NH_3
Saturn	95.16	9.45	0.69	10.65 h	26.73	35.49	H_2, He, CH_4, NH_3
Uranus	14.53	4.00	1.32	17.24 h	97.92	21.33	H_2, He, CH_4
Neptune	17.14	3.88	1.64	16.11 h	28.80	23.61	H_2, He, CH_4

Table 1.2. *Physical characteristics of the planets of the Solar System*

Planet	Satellite	a (R_P)	e	i (°)	P (days)	R (km)	D (g/cm^3)
Earth	Moon	60.1	0.0554	5.16	27.32	1,737	3.34
Mars	Phobos	2.76	0.0151	1.075	0.319	11.1	1.87
	Deimos	6.90	0.0002	1.793	1.262	6.2	2.25
Jupiter	Io	5.90	0.0041	0.036	1.769	1,821.6	3.528
	Europa	9.39	0.0094	0.469	3.551	1,560.8	3.014
	Ganymede	14.97	0.0011	0.170	7.155	2,631.2	1.942
	Callisto	26.33	0.0074	0.187	16.69	2,410.3	1.834
	Amalthea	2.54	0.0031	0.388	0.498	83.5	3.1
	Thebe	3.10	0.0177	1.070	0.675	49.3	3.0
	Adrasteia	1.80	0.0018	0.054	0.298	8.2	3.0
	Metis	1.79	0.0012	0.019	0.295	21.5	3.0
Saturn	Mimas	3.08	0.0206	1.566	0.942	198.6	1.14
	Enceladus	3.95	0.0001	0.010	1.370	249.4	1.00
	Tethys	4.89	0.0001	0.168	1.888	529.8	1.00
	Dione	6.26	0.0002	0.002	2.737	559	1.50
	Rhea	8.75	0.0009	0.327	4.518	764	1.24
	Titan	20.27	0.0288	1.634	15.95	2,575	1.88
	Hyperion	24.29	0.0175	0.568	21.28	133	1.1
	Iapetus	59.08	0.0284	7.570	79.33	718	1.02
Uranus	Ariel	7.47	0.0012	0.041	2.520	578.9	1.66
	Umbriel	10.40	0.0039	0.128	4.144	584.7	1.40
	Titania	17.07	0.0011	0.079	8.706	788.9	1.71
	Oberon	22.83	0.0014	0.068	13.46	761.4	1.63
	Miranda	5.08	0.0013	4.338	1.413	235.8	1.20
Neptune	Triton	33	0.0000	156.834	5.877	1,353	2.06
	Nereid	222.6	0.7512	7.232	360.14	170	1.5

Table 1.3. *The main natural satellites of the planets of the Solar System: a is the semi-major axis of the orbit, expressed in planetary radii; e is the eccentricity, i is the inclination with respect to the equatorial plane of the planet (except in the case of the Moon, where it is the plane of the ecliptic); P is the period of revolution, R is the radius in kilometers and D is the density*

All objects in the Solar System are immersed in the interplanetary medium, close to absolute vacuum but still filled with matter (dust, gas) and ionized particles of the solar wind, produced by the continuous radial expansion of the solar corona. This plasma carries the solar magnetic field and flows into the heliosphere, a large-scale cavity (110 to 120 au wide) hollowed out in the interstellar medium. During its propagation in the heliosphere, the solar wind interacts directly with the objects of the Solar System in different ways, according to the nature of their envelope and the possible presence of a large-scale magnetic field. In the case of planets, this interaction generates magnetospheres, magnetic cavities dug in the interplanetary medium (Chapter 3). Figure 1.2 gives some illustrations of these interactions.

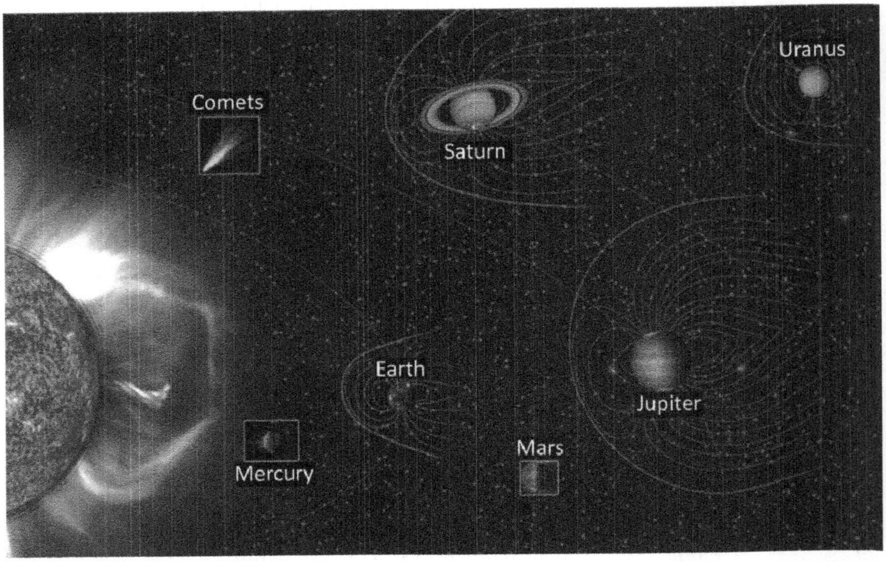

Figure 1.2. *The solar wind interacts in various ways with objects in the Solar System, giving rise to planetary magnetospheres (source: adaptation of ESA/NASA images; credits: Sigal, APIS, Lesia, Paris Obs.). For a color version of this figure, see www.iste.co.uk/encrenaz/solar1.zip*

1.2. Mechanics and dynamics of the Solar System[1]

The observation of objects in the Solar System orbiting the Sun has made it possible to discover the laws governing their mutual motions. It was Johannes Kepler (1571–1630) who first discovered the laws describing the movement of the planets around the Sun, which has since then borne his name. Sometime later, Isaac Newton (1643–1727) set out the general laws of gravitation that allow us to account

1 This section is taken up and adapted from the text written by Françoise Roques in the book *The Solar System* (Encrenaz *et al.* 2003, Chapter 1).

for the relative movements of bodies in the Universe. The next step was the theory of relativity, stated by Albert Einstein (1879–1955) in 1905. Today, large computers make it possible to study the dynamics of the Solar System and its evolution on long time scales using the equations defining the N-body problem.

1.2.1. *Newton's law of gravitation*

The law of gravitation governs the motion of bodies on all length scales, from the millimeter range to the dimensions of the Solar System. It is shown on the Earth by the gravitational attraction (gravity) exerted by the planet from its center, as well as in the motion of the Moon relative to the Earth, and it governs the motion of all bodies in the Solar System in orbit around the Sun.

Newton's universal law of gravitation is expressed as follows: two bodies of masses m_1 and m_2 separated by a distance d undergo a force of attraction F whose intensity is:

$$F = \mathrm{G} m_1 m_2 / d^2 \qquad [1.1]$$

where G is the universal constant of gravitation. In the MKSA system, G takes the value 6.672 10^{-11} m^3 kg^{-1} s^{-2}. In the vicinity of large masses such as that of the Sun, it is necessary to take into account the equations of general relativity; this is particularly the case for the calculation of the orbit of Mercury.

1.2.2. *Kepler's laws*

Based on the observations of Tycho Brahe (1546–1601), Kepler stated three laws in 1609 to describe the motion of the planets, that have since borne his name (see section 4.1.1, Volume 2):

1) the trajectory of a planet is an ellipse, of which the Sun is one of the focuses. Its motion obeys the following equation:

$$r = a(1\text{-}e^2)/(1+e \cos \vartheta) \qquad [1.2]$$

where r is the heliocentric distance, a is the semi-major axis of the ellipse, e its eccentricity and ϑ is the polar angle (called "true anomaly") with origin at the perihelion (the point closest to the Sun). The farthest point is the aphelion;

2) the vector ray joining the center of the Sun to the planet describes equal surfaces in equal times (law of equal areas); this law illustrates the fact that the planet accelerates near the Sun and slows down when it moves away from the Sun;

3) the ratio of the cube of the semi-major axes a to the square of the periods of revolution P is the same for all planets:

$$a^3/P^2 = GM/4\pi^2 = C \qquad [1.3]$$

M being the mass of the Sun. It is an approximation of the exact formula $a^3/P^2 = G(M+m)/4\pi^2$. This approximation is valid when m, the mass of the planet, is negligible in front of M.

The quantity C is proportional to the mass of the Sun, or, in the case of an extrasolar planetary system, to the mass of the central star. In the case of the Solar System, C is 1 when a is expressed in au and P in year.

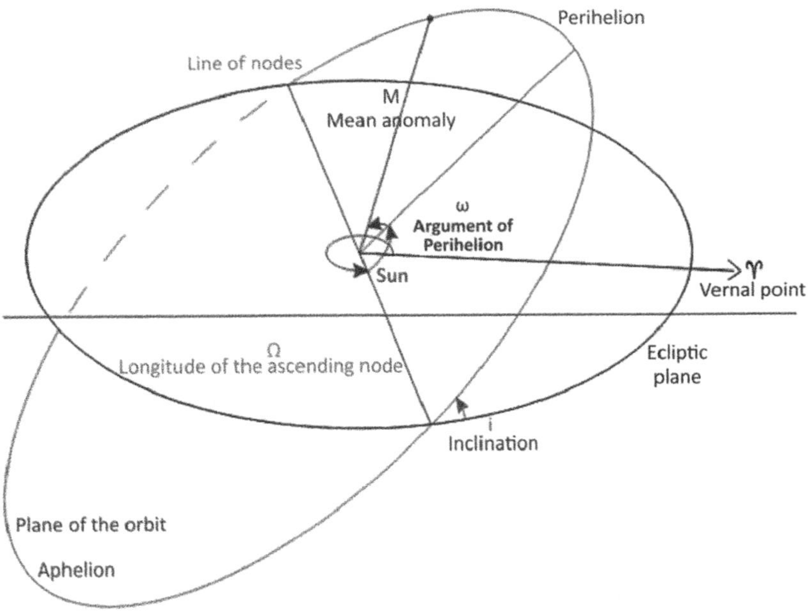

Figure 1.3. *Diagram representing the parameters of a Keplerian orbit. The mean anomaly represents the position (measured with respect to the perihelion) of a (hypothetical) point animated by a uniform "average" velocity 2π/P, P being the revolution period. This quantity is proportional to the area swept by the line connecting the Sun and the planet (source: Wikimedia Commons). For a color version of this figure, see www.iste.co.uk/encrenaz/solar1.zip*

To determine the motion of an object around the Sun, five parameters are necessary, the reference plane being the ecliptic: the semi-major axis a, the eccentricity e of the ellipse, the inclination i of the plane of the orbit with respect to the ecliptic, the longitude Ω of the ascending node, the argument of the perihelion ω, and the instant t of passage at the perihelion. Ω is measured from the direction of the vernal point, which is defined by the position of the Earth at the autumnal equinox. Two other parameters are also used in the case of comets: the perihelic distance $a(1\text{-}e)$ and the aphelic distance $a(1+e)$. Figure 1.3 shows the parameters representing a Keplerian orbit. They also apply in the case of a satellite in orbit around a planet; in this case, the reference plane is the equatorial plane of the planet.

The above equations describe the motions of a two-body system, one having a negligible mass relative to the other. In the case of the Solar System, the objects (planets and satellites) are subject to interactions with other neighboring bodies, which induces a variation of the orbit parameters. One of these perturbations is the precession of the nodes, whose period is several tens of thousands of years. Another perturbation is the advance of the perihelion. In Mercury's case, Newton's law predicts an advance of 488 seconds of degree per century, while the measured advance is 531 seconds of degree; it is necessary to take into account the equations of general relativity to explain the measured value.

1.2.3. *Mean motion resonances*

When several objects gravitate around the same star, their mutual interaction can present a particular configuration called "mean motion resonance" (see section 4.5, Volume 2). This phenomenon occurs when the periods of revolution of the two objects are in a commensurable ratio, in other words, the ratio of two integers that are generally lower than 10.

Two objects are in mean motion resonance if their mean motion: $n_1 = 2\pi/P_1$ and $n_2 = 2\pi/P_2$ (P_1 and P_2 being their periods of revolution) obey the following type of equation:

$$n_1 N_1 - n_2 N_2 = 0 \qquad [1.4]$$

N_1 and N_2 being integers. Mean motion resonances can affect two (or even more) satellites around a planet (this is the case of the three Galilean satellites closest to Jupiter), one satellite and the ring to which it is close, the Jupiter planet and the main asteroid belt. Near a ring, the presence of a satellite has the effect of confining this ring: it is called "shepherd satellite". In the case of the main asteroid belt, Jupiter's gravity field induces empty zones (Kirkwood gaps) and accumulation zones through resonances. The effect of the mean motion resonances is therefore to structure the perturbation of the orbits of the various bodies orbiting around a central star. Thus, at the beginning of the history of the Solar System, the four giant planets evolved through several phases of mutual resonances (see Chapter 4, Volume 2). Table 1.4 presents some examples of mean motion resonances in the Solar System.

	Objects	Type of resonance
Planets	Jupiter-Saturn	Close to 5:2
Planet-Asteroids	Jupiter-Asteroids	Kirkwood gaps 4:1, 3:1, 5:2, 2:1, etc. Accumulations 1:1, 3:2
Jupiter System	Io-Europe-Ganymede	4:2:1
Saturn System	Mimas-Thetis	4:2
	Enceladus-Dione	2:1
	Titan-Hyperion	4:3
	Janus-Epimetheus	1:1
	Mimas-Cassini Division	2:1
Neptune Kuiper Belt	Neptune, TNO resonators	3:2

Table 1.4. *Examples of mean motion resonances in the Solar System*

1.2.4. *The N-body problem*

In order to calculate the orbit of the objects of the Solar System with accuracy, it is necessary to be able to take into account all the gravitational perturbations (described by Newton's law) induced by other massive objects, near or far: this is the N-body problem. In most cases, there is no exact solution. Mathematicians have used the perturbation method, which allows finding an approximate solution from serial developments. Since the advent of large computers, it has been possible to find approximate numerical solutions to calculate the evolution and stability of orbits over very long periods of time, and to trace the dynamic history of the Solar System. This work highlights the importance of the initial conditions and the impossibility to predict the behavior of certain orbits because of their chaotic nature.

An important special case is the three-body problem (see section 4.4, Volume 2), which makes it possible to account for the motion of an asteroid (subject to the gravity fields of the Sun and Jupiter) or the motion of a satellite (affected by a neighboring satellite). Examined by many mathematicians including Joseph-Louis Lagrange (1736–1813) and Henri Poincaré (1854–1912), the problem has many solutions. There are five points of equilibrium, called "Lagrange points", three of which are located on the Sun-planet (or planet-satellite) axis, and the two others on the orbit of the planet (or satellite), 60° upstream or downstream. The last two (L_4 and L_5) are stable: this is the configuration of the Trojan and Greek asteroids on the orbit of Jupiter. The Lagrange points L_1 and L_2 of the Sun–Earth system (see Figure 1.4) are used in space astronomy to send probes intended for the observation of the Sun (L_1) or the Universe (L_2), which orbit close to these points.

Mean motion resonances lead to a structuring of planetary orbits that some astronomers have tried to translate into a geometrical law. Following the first works of Christian Wolf (1679–1754), Johann-Daniel Titius (1729–1796) and Johann-Elert Bode (1747–1826) proposed the following equation:

$$a = 0.4 + 0.3 \text{ x } 2^n$$

in which a is the average heliocentric distance of the planet. n takes the value $-\infty$ for Mercury, 0 for Venus, 1 for Earth, 2 for Mars, 4 for Jupiter, 5 for Saturn.

This equation, accurate to within 5%, was validated by the later discoveries of Uranus ($n = 6$), discovered in 1781, and of the asteroid Ceres ($n = 3$) in 1801, but then invalidated by the discoveries of Neptune in 1946 (located at 30 au) and Pluto in 1930 (located at 39 au); indeed, the law of Titius–Bode, with $n = 7$ and 8, would place them at 39 and 77 au respectively. Two other recent discoveries have contributed to invalidate this law: the discovery of trans-Neptunian objects populating the Kuiper belt between 40 and 100 au, and the study of the past migration of giant planets, which shows that their distance from the Sun has considerably changed over time.

It appears today that Titius–Bode's "law" has no physical foundation. The fact that there are simple ratios between the orbits of certain planets is rather the consequence of the mean motion resonances. These resonances have evolved during the history of the Solar System, so one cannot describe the distances of the orbits with a perennial geometrical relation.

Box 1.2. *The Titius–Bode's law*

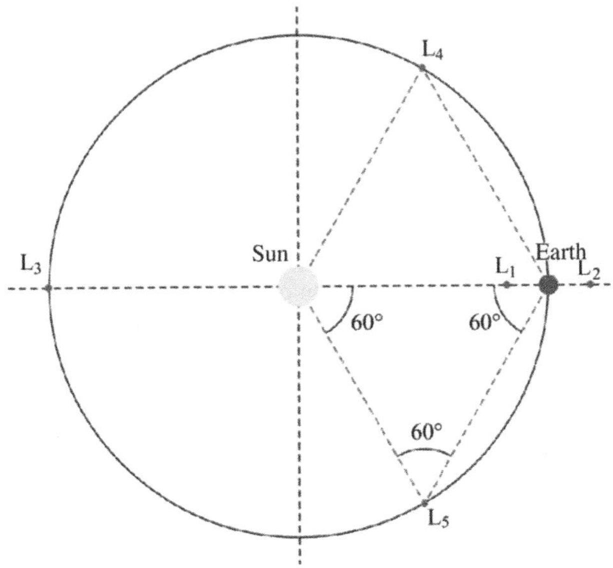

Figure 1.4. *Diagram representing the five Lagrange points of the Sun–Earth system (source: Wikimedia Commons)*

1.2.5. *The role of collisions*

Although the Solar System is essentially made of vacuum, on the scale of its lifetime (4.6 billion years), one must take into account collisions, as well as resonances and migration phenomena, in order to understand its dynamics. The effect of collisions is generally to modify the trajectory of the two bodies; it can also lead to one of them being dislocated, the creation of a disk, followed by the coalescence of a satellite in orbit around the more massive body; this is the scenario used to explain the formation of the Earth–Moon system (see section 4.2.5.2). The collision between asteroids can also be at the origin of a family of asteroids, resulting from a collision with the parent body.

The role of collisions was particularly important during the era of the Late Heavy Bombardment (LHB). According to the dynamic models, this event occurred about 800 million years after the birth of the Solar System, following the passage of the Jupiter-Saturn couple at the 2:1 resonance, leading to a strong instability of the orbits of small bodies (see below and section 4.10, Volume 2); the cratered surfaces of asteroids and satellites devoid of atmosphere, like the Moon, bear the trace of this event.

Nowadays, collisions are rarer, but they exist nevertheless: this is the case of the comet Shoemaker–Levy 9's collision with Jupiter in 1994, an object whose diameter before rupture was of the order of one kilometer. The Earth has also suffered major impacts, such as that caused by an object ten kilometers in size that formed the Chicxulub crater in Mexico 65 million years ago, causing a mass extinction of animal species.

1.2.6. *Migrations in the Solar System*

At the beginning of the history of the Solar System, another phenomenon influenced the dynamic of the stars. This phenomenon was the migration of the planets, due to the gravitational interaction between the giant planets and the protoplanetary disk in which the planets were formed (see section 4.10, Volume 2).

Indeed, stellar formation models (Lequeux 2013) show that they are born following the collapse of a rotating interstellar cloud into a disk, within which planets are formed (see Chapter 2). This disk, composed of gas and dust, has a lifetime of about ten million years at most. This period is very short compared to the age of the Solar System, estimated, based on the lifetime of long-lived radioactive elements, at 4.56 billion years (see Box 1.3; Chapter 3, Volume 2). It is essential in the history of the dynamics of the Solar System, because the interaction between the giant planets and the protoplanetary disk leads, according to numerical simulation models, to an inward migration of the most massive giant protoplanets, Jupiter and Saturn. According to the Nice model (Gomes *et al.* 2005; Morbidelli *et al.* 2005, Tsiganis *et al.* 2005) that is generally accepted today in the scientific community, the simultaneous presence of two giant planets led to a halt of their inward progression around the orbit of Mars, then an outward migration, the orbits of Jupiter and Saturn being blocked in 3:2 resonance. This outward migration also brought Uranus and Neptune in their wake (see Figure 1.5). During its external migration, the Jupiter-Saturn couple passed the 2:1 resonance, which caused (still according to numerical simulations) a total disruption of the orbits of the small bodies, with a dispersion towards the outside of the disk of the planetoids towards the Kuiper belt and the Oort cloud. It was the time of the Late Heavy Bombardment, mentioned above. Note that the simulations of the Nice model are not intended as a demonstration, but they do present a possible scenario that gives a good account of all the observations (see section 4.10, Volume 2).

In the case of the Solar System, the migration of the giant planets was moderate, and did not disperse the inner Solar System. The discovery of exoplanets, for about twenty years, has shown that the phenomenon of internal migration is very common in exoplanetary systems (see Chapter 2).

Numerical simulations have enabled us to retrace the dynamic history of the Solar System; can they also predict its future evolution, using longer and longer series to improve the accuracy of the calculations? The answer is no. Indeed, as the mathematician Henri Poincaré (1854–1912) showed, the series used in celestial mechanics can diverge beyond a certain order, making the evolution of phenomena unpredictable (see section 4.6, Volume 2). We know several chaotic configurations in the Solar System; this is the case, in particular, of the axis of rotation of Hyperion, Saturn's satellite, on a scale inferior to a month, or of the orbit of Pluto on a scale of about 20 million years.

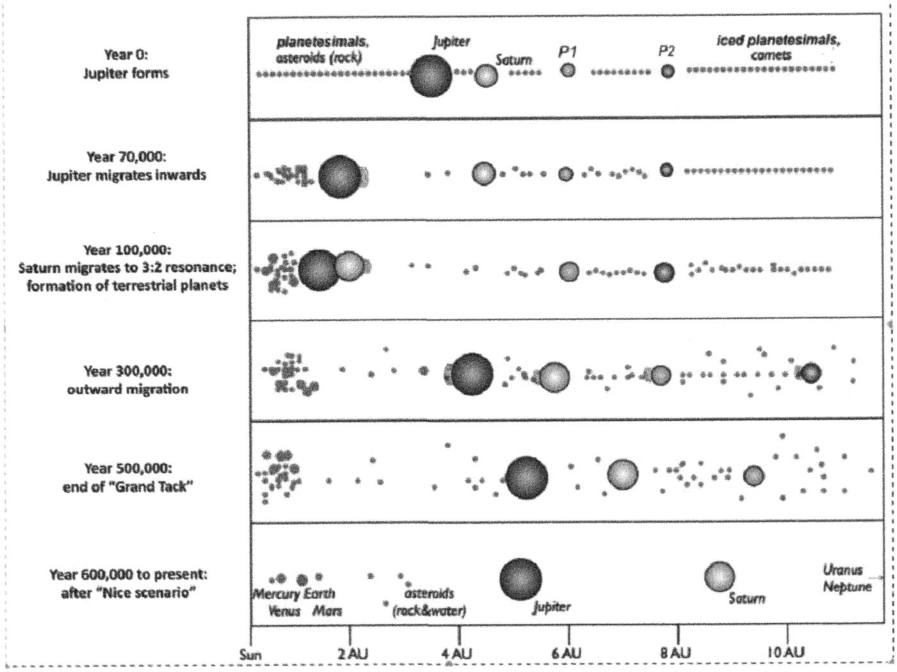

Figure 1.5. *Migration of the giant planets during the first million years of the history of the Solar System, according to the Nice model. For a color version of this figure, see www.iste.co.uk/encrenaz/solar1.zip*

COMMENT ON FIGURE 1.5. – *The masses of the planets are represented by the size of their circle. The rocky planetesimals are represented in red and the icy planetesimals (formed further from the Sun, therefore at lower temperature), in blue. The passage of Jupiter and Saturn at the 2:1 resonance occurred a little later, about 800,000 years after the beginning of the formation of the planets (source: the diagram is taken from (Lequeux* et al. *2020)).*

The most accurate dating method uses the radioactive decay of long-lived radioactive elements contained in the most primitive meteorites. By measuring the abundances of the elements produced, compared to those of the original elements, it is possible to assess the age of the sample from which they were taken. In particular, rubidium 87 and its decay product strontium 87, which has a lifetime of 4.7 10^{10} years, is used; strontium 86 (stable) is used as a reference. The age of the Solar System thus determined is 4.6 billion years.

Another method consists of measuring the rate of craters present on the surface of the Solar System's atmosphere-free bodies (asteroids or outer satellites). The higher the cratering rate, the older the surface. Since 1969, the collection of lunar samples, whose age has been assessed in the laboratory, has made it possible to establish an absolute scale.

Box 1.3. *How old is the Solar System?*

1.2.7. *The role of gravity in a solid body*

In the preceding sections, in order to describe the movement of a celestial body, we have considered it as a material point, whose mass is concentrated in its center. In reality, they have a certain dimension and the distribution of masses inside the body depends on gravity. The shape that the object takes depends on the nature of the elements that compose it. In the case of a gaseous object (star or giant planet), the interior of the object is in hydrostatic equilibrium, the force of gravity being balanced by the internal pressure gradient, and the object has a more or less flattened spherical shape. Indeed, the formation of a body, star, planet, galaxy, always results in the rotation of this body by conservation of angular momentum. In the case of a solid object, if the forces of cohesion are lower than the forces of gravitation, the object also takes a more or less flattened spherical shape: this is the case of telluric planets, external satellites, asteroids and trans-Neptunian objects whose diameter is greater than a few hundred kilometers (asteroids and TNOs which have this property are grouped in the class of dwarf planets, see Box 1.1). Objects of smaller size have irregular shapes resulting from their formation processes; this is particularly the case for comets or small asteroids.

Another effect of gravitation is to lead to the internal differentiation of solid bodies, the heaviest elements migrating towards the center of the object, if the internal temperature is sufficient. This differentiation can occur in the formation phase of the object, or under the effect of the energy released by the radioactive elements present inside. In the case of a rotating planet, the coupling of centrifugal force and gravity causes its flattening into an ellipsoid, the equatorial diameter being greater than the diameter measured on the axis of rotation.

1.2.7.1. *Tidal effects*

In the case of a system consisting of a planet and its satellite, each body is in orbit around the center of gravity of the system (usually very close to the center of the planet). Since the satellite is not a point, the attraction exerted by the planet is not the same at all parts (see section 4.8, Volume 2). This differential gravitational force, called the tidal effect, is responsible for the deformation of the satellite. In the case of a rotating satellite, this deformation propagates to follow the direction of the planet, creating a braking or acceleration effect that tends to align this deformation with the direction of the planet. This is the reason why large satellites close to their planet are, for the most part, rotating synchronously with their planet; it is the 1:1 spin-orbit resonance, in which the satellite always presents the same face to the planet. This is the case, in particular, of the Moon with respect to the Earth and the Galilean satellites of Jupiter. The planet also undergoes a braking effect that tends to slow down its rotation, but on a much longer time scale. The ultimate stage of evolution is the double synchronous rotation, a configuration in which the two bodies are permanently facing each other. This is the case for Pluto and its satellite Charon. In the case of the Earth–Moon system, the time needed to reach this equilibrium would be several tens of billions of years, a time much longer than the lifetime of the Solar System (see section 4.8.1 and Box 4.1, Volume 2).

In the case of the Earth–Moon system, the tidal effects induced by the Moon on the Earth are particularly important because of the fluid envelope surrounding the planet. On the Earth–Moon axis, water is more attracted to the Moon and to the opposite side, which induces a deformation of the sea level, with an average amplitude of about 20 centimeters in a day. Terrestrial tides are made very complex due to the presence of continents. The influence of the Sun is also added to that of the Moon, whose gravitational effect is about half as strong. As the plane of the lunar orbit is close to the ecliptic, the contributions due to the Moon and the Sun can add (during the full Moon and the new Moon), or subtract (when the Moon and the Sun are in quadrature).

1.2.7.2. *The Roche limit*

If a satellite is very close to its planet, the difference in the gravity field between the closest point on the planet and the farthest point can induce a differential force greater than the cohesion forces of the satellite, resulting in its rupture. The Roche limit (see section 4.8.3, Volume 2) is the distance from the planet below which rupture occurs.

Two particles of mass m, in contact with each other, located respectively at distances R and $R+d$ from the center of the planet, are subjected to a force of attraction:

$$F_{att} = Gm^2/d^2 \quad [1.5]$$

and to a tidal force due to the differential gravitational attraction of the planet:

$$F_{mar} = GMm/(R+d)^2 - GMm/R^2 \cong -2GMmd/R^3 \quad [1.6]$$

in which G is the gravitational constant and M the mass of the planet. M and m can be expressed as follows:

$$M = 4/3\ \rho_P\ \pi R_P{}^3 \quad [1.7]$$

and:

$$m = 4/3\ \rho\ \pi(d/2)^3 \quad [1.8]$$

ρ_P and ρ being the respective densities of the planet and the particle, and R_P being the radius of the planet.

At equilibrium, $F_{att} = F_{mar}$, hence:

$$R = 2^{4/3}\ R_P\ (\rho/\rho_P)^{1/3} \quad [1.9]$$

We see that the Roche limit is located at about 2.5 times the radius of the planet. This is the distance within which most of the rings of the giant planets are located. It should be noted that the above calculation does not take into account the cohesive forces between the two particles. If they exist, they can allow the presence of small satellites inside the Roche limit; this is what we observe in the case of the rings of Saturn.

1.2.8. *Special configurations of the Sun–Earth–Moon system*

The alignment of the Sun, the Earth and the Moon gives rise to a well-known phenomenon: eclipses. If the Moon was located in the plane of the ecliptic, eclipses would occur at each conjunction (new Moon, the Moon being between the Sun and the Earth: solar eclipse) and at each opposition (full Moon, the Earth being between the Sun and the Moon: lunar eclipse). In reality, the Moon's orbit being inclined by 5° on the ecliptic, eclipses occur when the Moon, at the time of alignment, is at one of the nodes of its orbit. Since the Moon and the Sun, seen from the Earth, have very similar apparent diameters (about 0.5°), the eclipse of the Sun can be total or annular, depending on the geometry of the event. The solar eclipse is total in the shadow zone and partial in the penumbra zone. The duration of a total solar eclipse is in the range of several minutes.

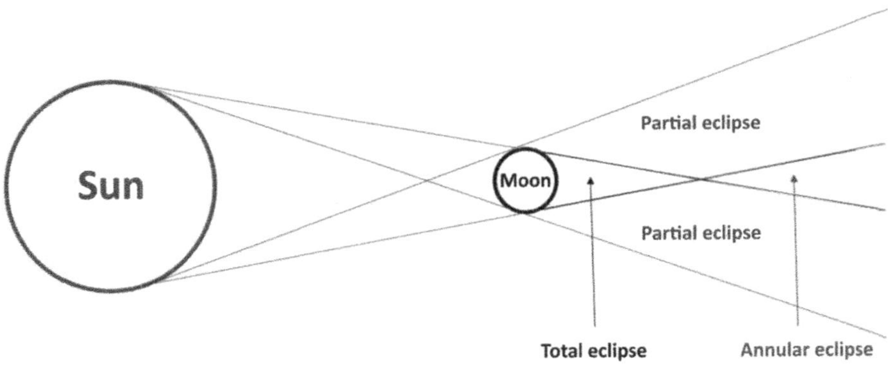

Figure 1.6. *Geometrical configuration of a solar eclipse; for the points of the Earth located in the shadow cone, the eclipse is total; in the areas of penumbra, it is partial; if the Earth is too far from the Moon, the eclipse is annular. For a color version of this figure, see www.iste.co.uk/encrenaz/solar1.zip*

1.3. Physics of the Solar System

From a physical point of view, the main difference between the Sun and the objects of the Solar System is the fact that the latter have no – or very little – radiation of their own, whereas the Sun, like all stars, produces intrinsic radiation due to the nuclear reactions responsible, within it, for the fusion of hydrogen into helium. Objects in the Solar System, whose mass is, in most cases, well below one-thousandth of the solar mass, have insufficient internal temperatures to generate such reactions (see Box 1.1). In the absence of thermonuclear reactions, the internal energy of Solar System objects is generally very low compared to the energy received from the Sun.

However, it must be taken into account; it is a matter of:

– the contraction (or even the internal differentiation) following the formation phase of objects (this is particularly the case of giant planets);

– the energy provided by the radioactive decay of the radioelements present in the interior (especially for external satellites and small bodies of the Solar System).

In most cases, the main source of energy received by Solar System objects is solar energy. This energy can be absorbed by the object or reflected back to the outside, either by simple reflection (in the case of objects without an atmosphere) or by diffusion (in the case of objects with an atmosphere rich in clouds or aerosols).

1.3.1. *Equilibrium temperature of an object in the Solar System*

As a first approximation, the proper radiation (or thermal radiation) of an object in the Solar System is that of a blackbody, whose temperature is that of its surface. This is given by Planck's law:

$$B(\nu) = 2h\nu^3/c^2\ (e^{\frac{h\nu}{kT}} - 1)^{-1} \qquad [1.10]$$

ν being the frequency; h being Planck's constant; c being the speed of light; k being Boltzmann's constant; and T, the body's temperature.

This formula can be written as:

$$B(\lambda) = 2hc^2/\lambda^5\ (e^{\frac{hc}{\lambda kT}} - 1)^{-1} \qquad [1.11]$$

λ being the wavelength. According to Stefan's law, the integral of the radiation is related to the temperature T by the relation:

$$\int_0^\infty B(\nu)d\nu = \int_0^\infty B(\lambda)d\lambda = \sigma T^4 \qquad [1.12]$$

σ being Stefan's constant.

Two units are commonly used in astrophysics: the wavenumber (expressed in cm^{-1}), unit of frequency, inverse of the wavelength expressed in centimeters, and the micrometer, or micron (μm), unit of wavelength. A wavelength of 100 μm corresponds to a frequency of 100 cm^{-1}.

Using these units, one can deduce a simple relationship between blackbody temperature and the frequency (or wavelength) of the blackbody emission maximum:

$$T/\nu_{max} = 0.5099 \text{ cm K} \qquad [1.13]$$

$$\lambda_{max}\ T = 2880\ \mu\text{m K} \qquad [1.14]$$

These equations show that the colder an object is, the more it radiates at long wavelengths.

Another component is added to this thermal radiation: the solar flux received by the object. This is, as a first approximation, that of a blackbody at the temperature of the solar photosphere, in other words, 5,770 K. A solar photon received by the object can be absorbed by the object (it then contributes to its thermal radiation), or be reflected or diffused towards the outside.

Every object in the Solar System emits radiation with two components:

– the reflected solar component, whose maximum is located at 0.5 μm and whose intensity decreases as one moves away from the Sun;

– the thermal component, which essentially comes from the absorption of solar radiation converted into heat; the maximum of this component moves towards the far infrared as one moves away from the Sun.

For each object of the Solar System, the intensity of these two components depends on the albedo A, in other words, the fraction of the solar flux received by the object that is reflected outward. In the Solar System, albedos range from a few percent (for dark comets and asteroids) to about 0.9 (for the brightest outer satellites). The lower the albedo, the greater the thermal radiation. The equilibrium temperature T_e of the object is defined as the temperature that the blackbody would have when absorbing the same amount of solar energy.

Using Planck's law, the surface temperature of a body of radius R, located at the Sun's distance D, is written as follows:

$$\Phi/D^2\ \pi R^2\ (1-A) = \sigma T_e^4 \times 4\pi R^2 \qquad [1.15]$$

if the object is in fast rotation (the solar flux Φ, received on a single hemisphere, is then distributed over the entire planet; this is the case of giant planets), or:

$$\Phi/D^2\ \pi R^2\ (1-A) = \sigma T_e^4 \times 2\pi R^2 \qquad [1.16]$$

if the object is in slow rotation (this is the case of planet Venus).

Expressing D in au and T in K, equations [1.15] and [1.16] are expressed as follows:

$$T_e = (1-A)^{1/4} \times 279/D^{1/2} \qquad [1.17]$$

in the case of an object in fast rotation, and:

$$T_e = (1-A)^{1/4} \times 332/D^{1/2} \quad [1.18]$$

in the case of an object in slow rotation.

These equations make it possible to determine an order of magnitude of the equilibrium temperature of an object that is located at a given distance from the Sun; indeed, even if the albedo is unknown, it intervenes only at the power ¼ in these equations.

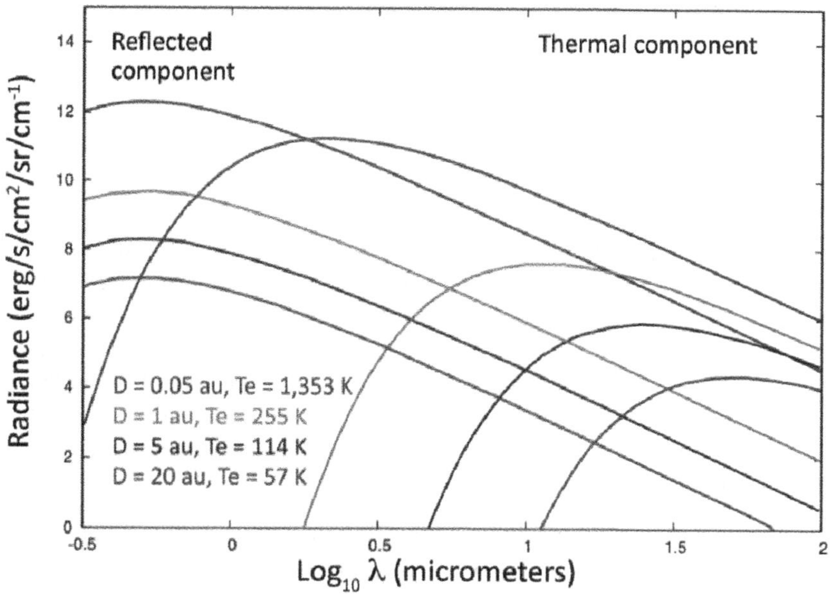

Figure 1.7. *Reflected solar component and thermal component of radiation from objects at different distances from the Sun. For a color version of this figure, see www.iste.co.uk/encrenaz/solar1.zip*

COMMENT ON FIGURE 1.7. – *Equilibrium temperatures are calculated according to equations [1.17] and [1.18] using an albedo of 0.3. Case D = 0.05 au corresponds to a fictitious planet in synchronous rotation, as found in exoplanetary systems. The temperature of 255 K corresponding to D = 1 au is the* equilibrium *temperature that the Earth would have in the absence of the greenhouse effect (source: based on (Tinetti* et al. *2013)).*

It should be noted that the surface temperature of an object can be higher than the equilibrium temperature for several reasons:

– there are non-negligible internal energy sources (it is the case of Jupiter and Neptune);

– it is in the presence of a greenhouse effect (as on Venus and, to a lesser extent, the Earth).

Figure 1.7 illustrates the relative contributions of the reflected solar component and the thermal component for objects at different distances from the Sun.

1.3.2. *Planets*

1.3.2.1. *Telluric planets and giant planets*

The planets of the Solar System are divided into two distinct classes (see Tables 1.1 to 1.3):

– the telluric planets (in increasing order of heliocentric distance: Mercury, Venus, Earth and Mars) are relatively small and of high density; they have no, or very few, satellites and are devoid of ring systems. They all have a rocky surface and, with the exception of Mercury, all have a stable neutral atmosphere. That of Venus and Mars is composed mainly of CO_2 with a few percent of N_2. In the case of the Earth, the primitive atmosphere, probably also rich in CO_2 and N_2, evolved with the trapping of CO_2 at the bottom of the oceans and the appearance of oxygen following the development of life;

– the giant planets (in ascending order of heliocentric distance: Jupiter, Saturn, Uranus and Neptune), farther from the Sun, are more voluminous and less dense: they all have a procession of satellites and a system of rings. They do not have a surface and the pressure in their center exceeds one million bars. Their gaseous envelope consists essentially of hydrogen and helium, hence their low density. Jupiter and Saturn are the only giant planets observable with the naked eye, and thus known since Antiquity, like the telluric planets; Uranus was discovered in 1781 by William Herschel (1738–1822) thanks to the improvements brought to the implementation of telescopes; Neptune was discovered in 1846 by Urbain Le Verrier (1811–1877), from calculations of celestial mechanics aiming to account for the anomalies of the orbit of Uranus.

The separation of the planets into two distinct classes can be explained in its outline from the Solar System formation scenario that is commonly accepted today. In this scenario, the Sun (like stars) is formed from the collapse into a disk of a rotating nebula of interstellar matter (see section 2.4). The proto-Sun forms from the central part of the disk, while the planets form by accretion of solid matter – the planetesimals – within the disk. The nature of the accreted object thus depends on the composition of the solid matter available at the time of accretion, and this depends on the temperature of the medium, which decreases as one moves away from the Sun. There are therefore two possible scenarios:

– near the Sun, at a temperature above about 200 K, the only solid elements are rocks and metals. However, as we can see from the table of cosmic abundances of elements, the lightest elements (starting with H and He) are the most abundant (see Table 1.5); the other elements, whose synthesis within stars requires more energy the heavier they are, are less and less abundant. There are therefore relatively few heavy materials to make up the telluric planets, hence their relatively low mass and high density;

– away from the Sun, the temperature is low enough to allow the most abundant small molecules – H_2O, CH_4, NH_3, CO_2, etc. – to be in ice form (whereas they were in gaseous form in the previous case). They can thus constitute solid nuclei that can reach ten Earth masses. At this stage, the models show that their gravity field is sufficient to accrete what remains of the surrounding protosolar nebula around them, consisting mainly of hydrogen and helium. This scenario explains the formation of the giant planets, rich in H and He, and therefore voluminous and not very dense. It also explains the presence of rings and numerous satellites (called "regular") in the equatorial plane of the giant planets, following the possible collapse of the sub-nebula into a disk around the protoplanet.

According to this scenario, there is a limit corresponding to the condensation of small molecules, whose mass contribution is dominant: it is called the *ice line* (see section 2.4).

This limit corresponds, in fact, to the condensation of H_2O, and this is for two reasons:

– water is particularly abundant due to the high abundance of H and O (see Table 1.5);

– compared to other small molecules, H_2O is by far the first molecule that condenses as the temperature decreases.

According to the formation models of the Solar System, the ice line was located at around 3 au when the planets were formed.

1.3.2.2. *Stability of an atmosphere*

Among the planets of the Solar System, only the one closest to the Sun, Mercury, is devoid of atmosphere. To understand the reason for this, we must use the notion of escape velocity, V_{esc}.

For a molecule of mass m, subjected to the gravity field of a planet of mass M and radius R, we have the relation:

$$\frac{1}{2} m V_{esc}^2 = mMG/R \qquad [1.19]$$

G being the universal gravitational constant, from which we deduce:

$$V_{esc} = [2GM/R]^{1/2} \qquad [1.20]$$

The escape velocities of the planets are shown in Table 1.2. The thermal agitation velocity of a molecule at temperature T is given by the Maxwell distribution:

$$V_{th} = [2kT/m]^{1/2} \qquad [1.21]$$

k being Boltzmann's constant. The comparison between V_{esc} and V_{th} gives the probability of stability of an atmosphere. It is all the more stable the heavier its components are, the more massive the planet is, and the lower its temperature is. In the case of Mercury, the escape velocity is low and, moreover, the surface temperature on the day side reaches 700 K, which explains the absence of a stable atmosphere. In the case of Mars, whose escape velocity is also low, it is the low surface temperature (always below 300 K) that allows the existence of a stable atmosphere.

1.3.2.3. *Chemical composition of planetary atmospheres*

In the case of giant planets, the escape velocity is high enough for their atmosphere to be able to retain all the elements, even the lightest, hydrogen and helium, which are by far the most abundant (see Table 1.5). This explains the low density of these planets, whose atmosphere is called "primary" because it was formed in the early stages of planetary accretion. On the other hand, the atmospheres of telluric planets could not retain hydrogen and helium, and are made up of heavier molecules (see Table 1.2). These atmospheres are said to be "secondary" because they were probably formed after the initial accretion phase of the nuclei, either by outgassing from the interior, by meteoritic impact, or by chemical or (in the case of the Earth) biological evolution.

What chemical composition can we expect for planetary atmospheres? Let us consider a mixture (H, C, N, O) conforming to the cosmic abundances (Table 1.5) cooling within the protosolar disk, while respecting the reactions of thermodynamic equilibrium. The condensation sequence, described in (Lewis 1995) begins with the condensation of water, the partial conversion of water into ammoniac NH_4OH, then the conversion of the residual water ice into methane clathrates (CH_4, $6H_2O$), and lastly the condensation of methane CH_4.

Two balance reactions are particularly important:

$$CH_4 + H_2O \leftrightarrow CO + 3\ H_2 \quad [1.22]$$

$$2\ NH_3 \leftrightarrow N_2 + 3\ H_2 \quad [1.23]$$

Element	Log_{10} of solar abundance	Log_{10} of meteorite abundance
H	12	–
He	10.97 +/- 0.01	–
C	8.43 +/- 0.05	7.39 +/- 0.04
N	7.83 +/- 0.05	6.26 +/- 0.06
O	8.69 +/- 0.05	8.40 +/- 0.04
Na	6.24 +/- 0.04	6.27 +/- 0.02
Mg	7.60 +/- 0.04	7.53 +/- 0.01
Al	6.45 +/- 0.03	6.43 +/- 0.01
Si	7.51 +/- 0.01	7.51 +/- 0.01
S	7.12 +/- 0.03	7.15 +/- 0.02
K	5.03 +/- 0.09	5.08 +/- 0.02
Ca	6.34 +/- 0.04	6.29 +/- 0.02
Ti	4.95 +/- 0.05	4.91 +/- 0.03
Cr	5.64 +/- 0.04	5.64 +/- 0.01
Mn	5.43 +/- 0.04	5.48 +/- 0.01
Fe	7.50 +/- 0.04	7.45 +/- 0.01
Co	4.99 +/- 0.07	4.87 +/- 0.01
Ni	6.22 +/- 0.04	6.20 +/- 0.01
Pb	1.75 +/- 0.10	2.04 +/- 0.03
U	–	-0.54 +/- 0.03

Table 1.5. *Abundances of the main elements measured in the solar photosphere (left) and in carbonaceous chondrites (the most primitive meteorites), normalized to silicon (right). In the case of the elements C, N and O, the discrepancies come from the fact that these elements are only partially condensed in meteorites (source: from (Asplund* et al. *2009))*

Figure 1.8 shows the regions (P, T) in which CH_4 and NH_3 are expected to dominate over CO and N_2. We see that CH_4 and NH_3 dominate at high pressure and low temperature (which corresponds to the environment of the giant planets), while CO and N_2 dominate under the opposite conditions (which are those of the telluric planets). The majority of CO_2 in the telluric planets is explained by the reaction:

$$CO + H_2O \leftrightarrow CO_2 + H_2 \qquad [1.24]$$

followed by the escape of hydrogen.

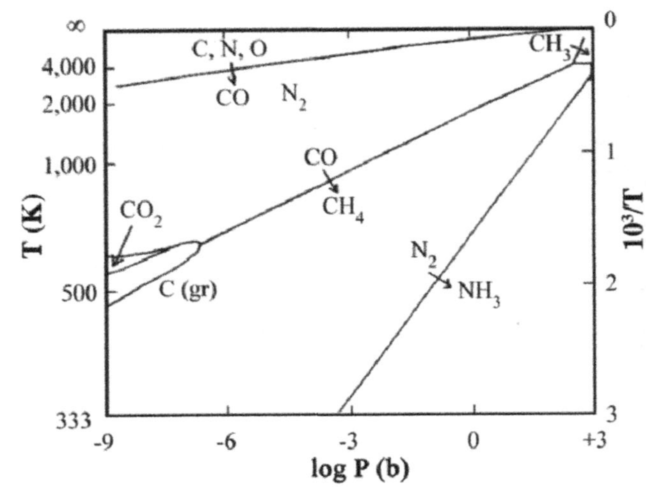

Figure 1.8. *Limits of CO/CH_4 and N_2/NH_3 equilibrium as a function of pressure and temperature under thermodynamic equilibrium conditions (source: based on (Lewis 1995)). For a color version of this figure, see www.iste.co.uk/encrenaz/solar1.zip*

However, it should be noted that the hypothesis of thermochemical equilibrium has its limits, as shown by the spectroscopic study of giant planets and exoplanets. Species out of equilibrium appear under the effect of two mechanisms:

– photochemistry, due to the irradiation by the solar (or stellar) UV flux, which especially explains the presence of several hydrocarbons in the giant planets;

– the vertical transport, which can bring compounds out of equilibrium coming from the deep layers to observable levels, thanks to a transfer time shorter than that of their chemical destruction; this is the case, in particular, of PH_3 on Jupiter and Saturn.

1.3.2.4. *Structure of a planetary atmosphere*

In a planetary atmosphere, the variation of pressure P as a function of altitude z is defined by the balance between hydrostatic equilibrium and the ideal gas law. The hydrostatic equilibrium is written as:

$$dP = -\rho g dz \quad [1.25]$$

ρ being the density and g the gravity. The ideal gas law is written as:

$$P = R\rho T/\mu \quad [1.26]$$

R being the perfect gas constant, T being the temperature and μ the average molecular weight. By combining these two equations, we obtain the following:

$$P(z) = P(z_0)e^{-(z - z_0)/H)} \quad [1.27]$$

H being the scale height:

$$H = RT/\mu g \quad [1.28]$$

If we approximate the atmosphere as isothermal, we see that the pressure decreases exponentially as the altitude increases. The scale height is about 8 km on the Earth's surface, 16 km on Venus, 11 km on Mars, and 20 km and 30 km on Jupiter and Saturn at a pressure level of 0.5 bar. The approximation of the isothermal atmosphere is only justified in the first order, since the variations in pressure are much faster than those in temperature.

A planetary atmosphere consists of a lower part, the homosphere, in which all the gaseous constituents are mixed, surmounted, above the homopause, by a heterosphere, in which the gases separate according to their own scale height, the number of collisions becoming insufficient to maintain the mixing of the gases. The homosphere is divided into several regions, depending on the mode of energy transfer, and the temperature is not constant as a function of the altitude.

The troposphere is the region in which the transfer of energy takes place by convection. It is the region located just above the surface, for planets that have it; in the case of giant planets, it extends from the deep layers to the pressure level of 0.1 bar. Heat transfer from the surface (or deep atmospheric layers) to the higher layers is achieved by convection. On the assumption of an adiabatic exchange, in which the ratio P/ρ^γ is constant (γ being the ratio of specific heat at constant pressure and volume, C_P/C_V), and taking into account the ideal gas law, we obtain the following:

$$dT/dz = -(\gamma-1)g\mu/k \quad [1.29]$$

At the Earth's surface, the expected adiabatic gradient is 13 K/km. The observed gradient is closer to 8 K/km or even less. The difference is due to the condensation of water vapor which contributes to heat transfer. In general, the moist adiabatic gradient is lower than the adiabatic gradient in a dry atmosphere.

The tropopause marks the upper limit of the convective region of the atmosphere. Above this limit, energy is transferred by radiation. The thermal profile depends on the amount of energy absorbed at each level, which, in turn, depends on the atmospheric composition. In the case of the Earth and the giant planets, a temperature inversion occurs, due to the absorption of solar ultraviolet radiation which causes the photodissociation of certain molecules (O_2 for the Earth, CH_4 for the giant planets) and the formation of by-products (O_3 for the Earth, various hydrocarbons for the giant planets): this is the stratosphere. In the case of Mars and Venus, in the absence of these constituents, the temperature remains isothermal above the tropopause: this is the mesosphere.

Above the homosphere, solar ultraviolet radiation interacts directly with atoms, radicals and molecules to ionize them, which induces heating of the atmosphere; the temperature increasing rapidly with altitude. Another source of heating comes from the interaction with charged particles in the solar wind (see Figure 1.9).

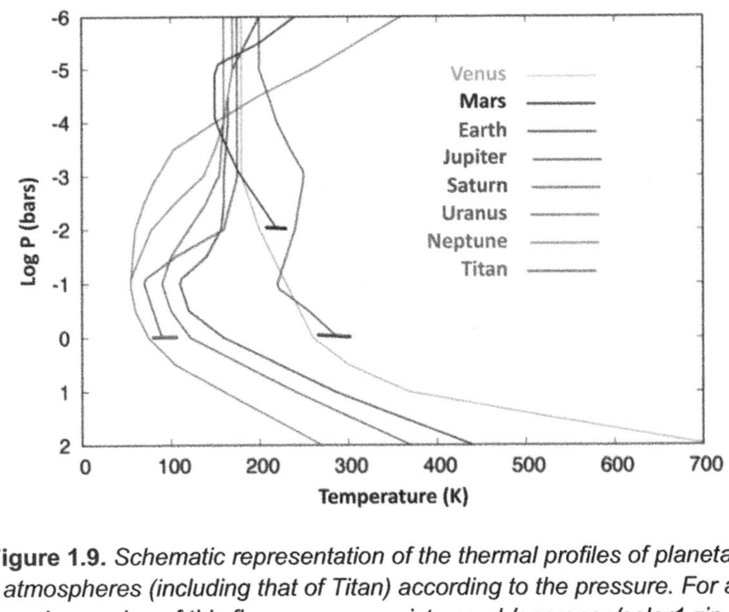

Figure 1.9. *Schematic representation of the thermal profiles of planetary atmospheres (including that of Titan) according to the pressure. For a color version of this figure, see www.iste.co.uk/encrenaz/solar1.zip*

COMMENT ON FIGURE 1.9. – *The giant planets, Titan and Earth have a stratosphere, while Mars and Venus have none. The tropospheres of the giant planets present a remarkable similarity, with an adiabatic gradient defined by hydrogen and helium and a tropopause of 0.1 bar.*

1.3.3. *Satellites*

Most of the satellites in the Solar System orbit around giant planets for two reasons:

– in the case of giant planets, the mode of formation by accretion around an ice core favors the formation, in the equatorial plane of the planet, of accreted objects in the disk, formed by the collapse of the protoplanetary nebula surrounding the giant planets: these are the regular satellites;

– the strong gravitational field of the giant planets favors the capture of the small surrounding objects: these are the irregular satellites.

In the case of telluric planets, their formation model does not lead to the existence of satellites *a priori*. Mercury and Venus are devoid of them. The three existing inner satellites – the Moon around the Earth, Phobos and Deimos around Mars – seem to be the result of collisions between the planet and a body passing nearby at the beginning of the planet's history.

1.3.3.1. *Telluric planet satellites*

The Earth–Moon pair presents a special case in the Solar System, because of the high radius (Moon/Earth) ratio (0.27); we could almost talk about a binary system. It should be noted that the Moon does not orbit in the equatorial plane of the Earth, but close to the plane of the ecliptic (see Table 1.3). Current theories on its origin agree on an impact with an object called Theia, of a size comparable to that of Mars, some 100 million years after the formation of the Earth (see section 4.2.5.2). A cloud of debris (composed of the materials of the impactor and the Earth's mantle) would have formed in orbit around the Earth and would have re-agglomerated to form the Moon. It remains to be understood why the Moon and the Earth have the same isotopic composition for the elements O, Ti, Cr, W and K, whereas the chemical composition of Theia should *a priori* have been different from that of the Earth; several theories have been developed to try to account for this constraint.

Like many satellites relatively close to their planet, the Moon is in synchronous rotation around the Earth, that is to say that it always presents the same face to the Earth. As the Moon is located well beyond the Earth's geostationary orbit, the tidal effects of the Earth–Moon system have progressively moved the Moon away from the Earth and caused a slowing down of its rotation period; at the time of its formation, it is estimated that its distance from the Earth was half of its current value, the Earth's rotation period then being six hours. Today, as a result of tidal effects, the Earth–Moon distance is increasing by 3.78 cm/year (see section 4.2.5.3). There is a remarkable dichotomy between the visible and hidden sides of the Moon, the latter being more rugged, extremely cratered and with a thicker crust (about 80 km) (see Chapter 4). The Moon is not massive enough to have a stable atmosphere. Following the meteorite bombardment that it has been subject to during its history, the surface of the Moon is covered with a layer of fine dust, regolith, that is several meters thick.

The Moon is the only satellite to have been the subject of human exploration, with the American Apollo program (see section 4.1.2). Between 1969 and 1972, a dozen astronauts set foot on the Moon and brought back about 380 kg of lunar samples; simultaneously, the Soviet robotic missions of the Luna program brought back a few hundred grams of samples. These brought decisive information on the dating of the main metamorphic events and provided an absolute scale for the dating of objects in the Solar System, defined in a relative way by the counting of surface craters (these being all the more numerous the older the surface is).

Discovered by the American astronomer Asaph Hall (1829–1907) in 1877, two very small satellites Phobos and Deimos (see Table 1.3) orbit Mars, also in synchronous rotation, in close proximity to the equatorial plane of the planet. Unlike the Moon, they have a low density, which makes them more akin to the asteroid family. However, it is difficult to explain their origin by a capture. Just as is the case with the Moon, current scenarios favor a collision with a body smaller than Mars, within a few hundred million years after its formation. The debris ejected in orbit would have re-accreted into several small objects, some of which would have fallen back on Mars due to tidal forces. Located below the geostationary orbit of Mars, Phobos is also destined to crash on Mars in a few tens of millions of years, while Deimos, located beyond this orbit, is expected to move away indefinitely from its planet, like the Moon.

1.3.3.2. *Giant planet satellites*

More than 600 natural satellites are known in the outer Solar System, including more than 200 around Saturn, most of them discovered by the Cassini-Huygens mission (see Chapter 6). Satellites of giant planets form two families:

– the regular satellites, located in the equatorial plane of the planet, formed by accretion within the disk resulting from the collapse on the protoplanetary nucleus of the surrounding protosolar sub-nebula;

– irregular satellites, whose inclination on the equatorial plane of the planet is random, being the result of a capture by the gravitational field of the giant planet.

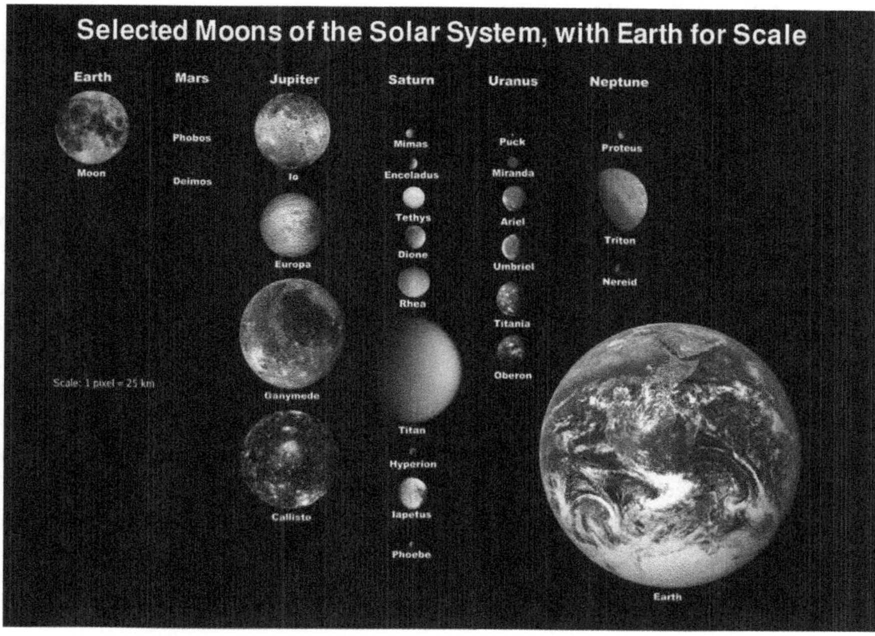

Figure 1.10. *Schematic representation of the largest natural satellites of the planets and Pluto. The Earth is represented to fix the scale of the diameters. We also know satellites around some asteroids and some trans-Neptunian objects, like Pluto (source: from a NASA image). For a color version of this figure, see www.iste.co.uk/encrenaz/solar1.zip*

The satellite systems of the four giant planets have various configurations (see Figure 1.10). Jupiter and Uranus have a few large regular satellites that are relatively close to the planet (in the case of Jupiter, these are the four Galilean satellites, see section 1.2.1, Volume 2), while Saturn has a procession of small regular satellites followed by a large regular satellite, Titan (see section 1.3.1, Volume 2), the only one with a stable and dense atmosphere; all are in synchronous rotation around their planet. As for Neptune, it has a large irregular satellite, Triton, probably a trans-Neptunian object captured by Neptune, also surrounded by a very tenuous stable atmosphere (see section 1.4.2, Volume 2).

Numerical simulations can be used to account for the distribution of regular satellites according to their distance from their planet. They involve the formation of an initial planetary ring, formed in the accretion phase. Satellites would form beyond the outer edge of this ring, outside the rock boundary, as soon as the force of gravity becomes greater than the tidal forces, and would then move away from the planet. This model seems to account well for the current distribution of the satellites of Uranus and the small satellites of Saturn.

1.3.3.3. *Asteroid and TNO satellites*

There are also satellites around asteroids and trans-Neptunian objects. The first object in this category is Charon, a satellite of Pluto discovered in 1978 by James Christy, through the examination of photographic plates. The second is Dactyl, a satellite of the asteroid Ida of the main asteroid belt, discovered during the flyby of the asteroid by the *Galileo* probe in 1994. Several hundreds of binary asteroids are known today, with a very variable size ratio between the two bodies. These systems could result from the capture of the smaller of the two bodies by the main object (asteroid or TNO), or be a result of the collision between two bodies, followed by the ejection of material in orbit of the large object and reconstitution of the debris (according to the process favored for satellites of telluric planets), or, lastly, aggregation of debris ejected by an asteroid in rapid rotation on itself: this is the YORP effect (see section 4.9.4, Volume 2). Note that the binary systems of asteroids and TNOs are of particular interest because their observation allows the mass of the system to be deduced by application of Kepler's laws.

1.3.4. *Small bodies of the Solar System*

In 2006, the IAU adopted a resolution defining small bodies as all objects in the Solar System that are neither planets, satellites nor dwarf planets (see Box 1.1). In practice, this category includes all objects in solar orbit that have not acquired sufficient mass to have a spherical shape and to have cleared their orbit. This therefore includes asteroids, trans-Neptunian objects, comets and rings. This definition is not totally satisfactory since there is a strong interconnection between satellites and rings; furthermore, the transition between comet and asteroid is not well defined, as a comet behaves like an asteroid at the end of its life.

1.3.4.1. *Rings*

As illustrated in the previous section, the distinction between rings and satellites is mostly historical: rings (see section 1.5, Volume 2) are made up of small particles which, under certain conditions, agglomerate to form small satellites. In the case of

Saturn, there is a strong interaction between certain rings and the small shepherd satellites which ensure their confinement, whose life span can be short.

It was Galileo who, at the beginning of the 17th century, first noticed the unusual and variable appearance of Saturn; Huygens, sometime later, understood that it was a disk surrounding the planet, seen from the Earth under a variable incidence. Saturn's rings were explored in depth by the Cassini-Huygens space mission between 2005 and 2017.

Since the 1980s, it has been known that the four giant planets have a ring system, but apart from Saturn, these systems are very tenuous. The rings of Jupiter (section 1.5.3, Volume 2), dark and close to the planet, were discovered by the Voyager 1 probe in 1979. Those of Uranus (section 1.5.5, Volume 2) and Neptune (section 1.5.6, Volume 2), also narrow, tenuous and dark, were discovered from the Earth, in 1977 and 1984, by stellar occultation measurements, then confirmed during flybys by the Voyager 2 probe of Uranus in 1986 and Neptune in 1989. The rings of Neptune have the peculiarity of being composed of continuous diffuse rings and denser arcs, which were discovered from the Earth by stellar occultations. This configuration is due to a particular resonance between the shepherd satellites, which confine the rings.

Observations of stellar occultations, which become more and more efficient as the precision of the ephemerides increases, allowed, in 2014, the discovery of a two-ring system around the asteroid Chariklo (Braga-Ribas *et al.* 2014). This small object, 250 km in diameter, belongs to the Centaurs, a family of asteroids that cross the orbit of giant planets. This discovery was completely unexpected, as no model predicted the existence of stable rings around such a small body. Another system of rings was then discovered around the TNO Haumea in 2017, still by the stellar occultation method. These discoveries raise new questions about the origin and evolution of the rings. Indeed, the mechanisms and containment time scales are different around a giant planet and a body of a few hundred kilometers.

1.3.4.2. *Asteroids*

Asteroids (or minor planets, see section 2.2, Volume 2) are objects in orbit around the Sun whose mass is insufficient to have cleared their orbit (see Box 1.1). The diameter of the largest of them is several hundred kilometers, that of the smallest currently detectable ones is of the order of a meter. The first of them, Ceres (and also the largest, with a diameter approaching 1,000 km), was discovered in 1801 by Giuseppe Piazzi (1746–1826). This discovery was followed by that of three other asteroids located in neighboring orbits – Pallas, Juno and Vesta – highlighting the existence of the main asteroid belt between Mars and Jupiter. Today, several hundreds of thousands of asteroids are known, divided into several large families:

– the main asteroid belt, located at about 2.7 au from the Sun, regroups the planetesimals that could not accrete into a planet because of Jupiter's very strong gravity field. Its structure, according to the distance from the Sun, presents gaps (called Kirkwood gaps) and accumulation regions that correspond to particular resonances with Jupiter (see Table 1.4);

– Near Earth Objects (NEOs) have relatively close orbits to the Earth; they are subject to special surveillance because of the potential risk of collision of one of them with the Earth (see section 2.2.2, Volume 2);

– the Trojan and Greek asteroids are two families located on the orbit of Jupiter, at 60° upstream and downstream of the planet, at Lagrange points L_4 and L_5 (see Figure 1.3). Other families of asteroids, also called "Trojans" by extension, are located at points L_4 and L_5 of the orbits of the Earth, Mars and the giant planets (see section 2.2.3, Volume 2);

– centaurs are asteroids evolving between the orbits of giant planets. A particular one counted in this family is Chiron, discovered in 1977, and Chariklo, discovered in 1997 and endowed with a system of rings (see section 1.3.4.1). Some of them show transient signs of cometary activity. They could be TNOs ejected from the Kuiper belt (see section 2.3.1, Volume 2).

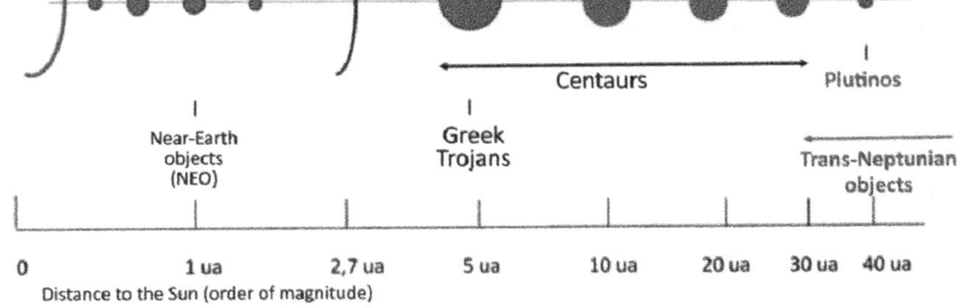

Figure 1.11. *Schematic representation of the different classes of asteroids. For a color version of this figure, see www.iste.co.uk/encrenaz/solar1.zip*

The surface composition of asteroids is known to us mainly by visible and infrared spectrophotometry from the Earth. According to the slope of the spectra and their spectral signatures, asteroids have been classified into different categories including carbonaceous (C), silicate (S) and metallic (M). Thanks to various flybys by space probes, and then dedicated probes, our knowledge of asteroids has considerably progressed during the last 20 years. Two missions have returned samples, Hayabusa 1 and 2, and another one, OSIRIS-REx, is currently in operation. Figure 1.11 shows the distribution of the different classes of asteroids. For a review on the physics and dynamics of asteroids, see (Michel *et al.* 2015).

1.3.4.3. *Trans-Neptunian objects*

Pluto was, in 1930, the first trans-Neptunian object discovered, but it was not until 1992, with the discovery of the object 1992QB1 by David Jewitt and Jane Luu, that the existence of this new family of objects was recognized (see section 2.3.2, Volume 2). It includes the Kuiper-Edgeworth Belt, between 30 and 55 au, and, at greater heliocentric distance, the scattered disk and extreme trans-Neptunian objects (ETNO). The existence of the Kuiper belt had been suggested for several decades on the basis of celestial mechanics arguments, notably by Kenneth Edgeworth (1880–1972) in 1943.

The Kuiper Belt resembles some of the debris disks observed around other stars; it is much less massive because its constituents were widely dispersed during the last stages of the migration of giant planets. For a recent review of the trans-Neptunian system, see (Prianik *et al.* 2019).

Kuiper objects can be divided into two main categories:

– objects in resonance with Neptune, including Plutinos in resonance 3:2;

– classical objects (the most numerous), of low eccentricity and moderate inclination (< 30°), with a semi-major axis between 42 and 47 au. Numerical simulations show that the Kuiper belt is the reservoir for comets of short period and of low inclination (see section 1.3.4.4). Scattered objects, of strong eccentricity, have their perihelion located beyond the orbit of Neptune; their distance from the Sun can reach several hundred au, or even more. It is from the orbital distribution of the scattered objects that the hypothesis of the ninth planet was put forward (see section 1.1).

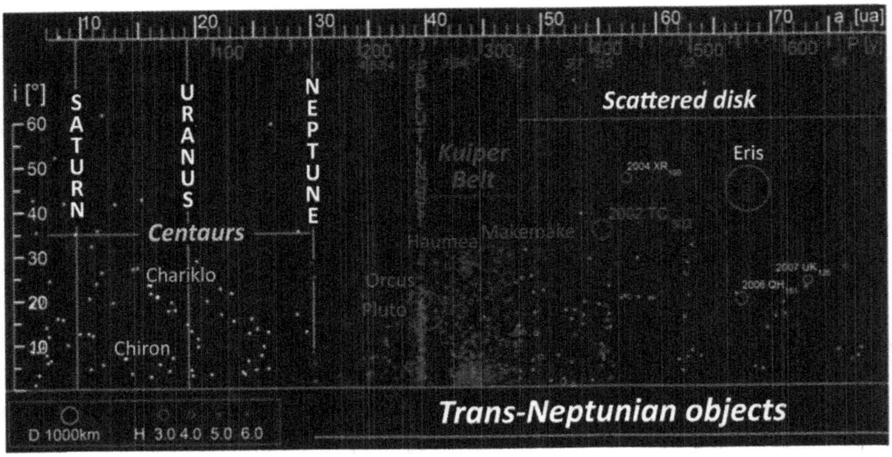

Figure 1.12. *TNO distribution according to the distance from the Sun (source: Wikimedia Commons). For a color version of this figure, see www.iste.co.uk/encrenaz/solar1.zip*

1.3.4.4. *Comets*

Comets (see section 2.1, Volume 2) have been known since Antiquity, and their unpredictable appearance has always fascinated mankind. It was Edmond Halley who, in 1705, was the first to highlight the periodic character of comets, based on the observation of previous appearances of the comet that has since borne his name.

Comets are very small bodies, generally less than 10 kilometers in diameter, in a very elliptical orbit around the Sun, consisting of a silicate and carbonaceous core of dust and ice. When they approach the Sun, the ice on their surface sublimates, causing the ejection of gas and dust that reflects sunlight in the form of a coma and a tail of dust.

Most of the time, having evolved in a very cold environment and too small to be affected by internal differentiation processes, comets are considered as privileged witnesses of the formation and evolutionary processes at the beginning of the Solar System's history (Festou *et al.* 2004). The water from comets may have contributed to the constitution of the Earth's oceans and may have also deposited prebiotic molecules, perhaps at the origin of life on Earth.

The exploration of comets has experienced several important stages during the last thirty years, particularly with the appearance of Halley's comet in 1986, accompanied by the flight of five space probes, in addition to the appearance of two particularly brilliant new comets, Hyakutake in 1996 and Hale-Bopp in 1997, the return of the cometary sample by the Stardust mission (1999–2011) and, lastly, the Rosetta mission and the follow-up of the comet Churyumov-Gerasimenko (67P/CG) between 2014 and 2016.

According to their origin, comets have been classified into two main categories:

– comets from the Kuiper belt have a low inclination on the ecliptic and a period limited to few years; having made multiple passages near the Sun, they are generally not very active. This is the case for 67 P/CG;

– comets coming from the Oort cloud, a vast shell located between 40,000 and 100,000 au. This reservoir would contain some 10^{11} comets, ejected following the gravitational disturbances due to the giant planets. Some comets, following perturbations by neighboring stars or Galactic gravity waves, are injected towards the interior of the Solar System in the form of new comets, sometimes very bright and very inclined on the ecliptic; this is the case of comet Hale-Bopp. Others, following other perturbations by giant planets, settle on orbits of shorter period: this is the case of comet Halley, whose period is 76 years.

1.3.5. *The interplanetary medium*[2]

The medium in which the bodies we have described evolve has a density that is not zero, but much lower than that of the best laboratory vacuums. The matter of the interplanetary medium is made up of two components:

– a solid component, formed of gas, dust or small fragments, the interplanetary matter;

– a flow of ionized particles from the Sun, the solar wind. In addition, the interplanetary medium is bathed in a radiation field, essentially of solar origin.

1.3.5.1. *Interplanetary matter*

The existence of interplanetary matter manifests itself in the form of meteorite falls (when a solid object reaches the Earth's surface) or meteors (when there is disintegration in the Earth's atmosphere: these are the "shooting stars" that we are familiar with). Some of these bombardments take place periodically: in this case, the meteors come from the dust cloud along the orbit of a comet, or from a remnant of a comet; this is the case of the Leonids, associated with comet Tempel-Tuttle, or the

2 This section has been taken and adapted from the text written by P. Zarka in the book *The Solar System* (Encrenaz *et al.* 2003, Chapter 1).

Orionids, associated with Halley's comet. Meteors, originating from grains of 200 μm to a few centimeters in diameter, are probably of comet origin.

Bodies that reach the Earth's surface without being totally destroyed are meteorites, with an initial mass greater than one kilogram (Lauretta and McSween 2006). They lose a fraction of their mass as they pass through the Earth's atmosphere, and arrive on the ground with a frequency of 2 to 10 falls per day. They probably come from asteroids (see Chapter 3, Volume 2).

Micrometeorites, with a diameter of less than 10 microns and a mass of less than 10^{-7} g, can enter the Earth's atmosphere without being destroyed. Like meteors, these particles are of cometary origin; they constitute the bulk of extraterrestrial matter received on Earth.

The existence of interplanetary dust is also proven by the observation of the zodiacal cloud: the solar radiation reflected by the dust present in the Solar System. It is mainly observed at the ecliptic, hence its name. In the anti-solar direction, this radiation is called "Gegenschein". As shown in 1983 by the infrared satellite IRAS, dust is not confined in the plane of the ecliptic, but appears at certain well-defined latitudes.

1.3.5.2. *The solar wind and the heliosphere*

The discovery of the solar wind dates back to the 1950s, when Ludwig Biermann (1907–1986) demonstrated the existence of a direct current of ionized particles coming from the Sun, through the systematic action of this current on the ionized tail of comets. It was observed for the first time *in situ* by space probes in the mid-1960s, which revealed a composition of ions (about 95% protons, 5% He^{2+} with traces of other heavy ions) and electrons (see Chapter 3). In 1958, Eugene Parker developed a theory to describe the flow of this hot and magnetized plasma, directly linked to the solar corona and its high thermal conductivity. The temperature of the corona (of the order of a million degrees) is such that a stable stationary state requires a continuous expansion of the corona, forming a monokinetic "wind". Above a few solar radii, the plasma velocity becomes supersonic.

The average velocity measured at 1 au is about 400 km s^{-1}, with a flux of about 2 10^8 particles cm^{-2} s^{-1} corresponding to an average density of 5 ions cm^{-3}. In periods of high solar activity, the velocity and density can increase by more than a factor of 2–3 and 4–5, respectively. As one moves away from the Sun, the ion density decreases as the square of the heliocentric distance. The solar wind is a hot plasma (its temperature at the level of the Earth's orbit is about 400,000 K) and therefore highly conductive. It carries with it the solar magnetic field. At 1 au, the interplanetary magnetic field has an amplitude of about 1 to 10 gammas (1 gamma = 1 nanotesla = 10^{-5} gauss). This field can increase by a factor of 10 during the passage of perturbations related to large solar flares, which are at the origin of the emission of solar cosmic radiation. The parameters indicated above are those corresponding to the "slow solar wind", located in the plane of the ecliptic during the solar minimum period, as opposed to the "fast wind" (a few hundred km s^{-1} faster) detected by the Ulysses probe at higher latitude.

The interaction of the solar wind with the bodies of the Solar System is of a different nature, depending on whether or not the obstacle has a gas envelope and/or an intrinsic magnetic field. The main cases are described in Chapter 3. The solar wind itself does not "relax" in an absolute vacuum, but in the local interstellar medium, forming a plasma cavity called the heliosphere. The properties of the local interstellar medium, measured by various methods, indicate a temperature of about 7,000 K, a density of 0.1 cm^{-3}, a relative speed to the Sun of 21 to 26 km/s and a magnetic field of about 0.3 nT. The total pressure of the interstellar medium, dominated by that of its magnetic field, thus reaches ~ 10^{-13} Pa.

A similar pressure, dominated by the *dynamic pressure*, is reached by the solar wind at an average heliocentric distance of 110 to 120 au, which varies during the solar cycle. It is in this region that the heliopause, the boundary between the heliosphere and the interstellar medium, is located, which the Voyager 1 and 2 probes crossed in 2012 and 2018 at distances of 122 (in the north) and 119 au (in the south). Prior to that, the two probes had crossed a "shock zone" in 2004 and 2007 at respective distances of 94 and 84 au. This internal shock in the heliosphere, known as the terminal shock, corresponds to the place where the solar wind speed becomes subsonic again. Each of the probes crossed these boundaries several times over several months/years, reflecting their highly dynamic nature. A possible external shock, expected for a supersonic flow of the interstellar medium around the heliosphere, has not been confirmed.

1.3.5.3. *The interaction of the planets with the solar wind*

The Solar System offers a large variety of types of interaction with the planets (obstacles with/without magnetic field, with/without atmosphere, near/far from the sun), as well as interaction geometries, due to the various inclinations between the magnetic axis and the planetary rotation axis, and between the latter and the ecliptic plane (see Tables 1.1, 1.2 and 1.6; Chapters 3, 4 and 5).

1.3.5.3.1. Large-scale magnetospheres

The solar wind interacts in a particular way with planets that have an intrinsic magnetic field (the Earth, the four giant planets and Mercury). In this case, the solar wind flux is deflected by the Lorentz force:

$$\boldsymbol{F} = q\boldsymbol{V} \times \boldsymbol{B} \qquad [1.30]$$

undergone by a particle of charge q and velocity $\boldsymbol{V}$ in a magnetic field $\boldsymbol{B}$, and distantly bypasses the planetary obstacle.

The region where the planetary magnetic field dominates the dynamics of charged particles is called the "magnetosphere". The planetary magnetic shield thus carves a cavity in the solar wind, with an almost hemispherical shape on the solar side, and is stretched out in a wake called a "magnetospheric tail" on the opposite side. The boundary of this cavity is the "magnetopause". It is not totally impermeable, and a fraction of the mass, energy and momentum of the interplanetary plasma can be transmitted to the magnetosphere, notably through a partial reconnection of the planetary and interplanetary magnetic fields. Other possible sources of magnetospheric plasma are the planet's ionosphere (ionized upper atmosphere), its rings and satellites, including satellites with active volcanism such as Io, Enceladus or Triton.

As the solar wind propagates at a much higher speed than the propagation speeds of the "mechanical" perturbations of the environment, the magnetopause is preceded, on the solar side, by a stationary shockwave. The region between this bow shock and the magnetopause is called the "magnetosheath", where the heated, slowed down and deflected solar wind flows and where the interplanetary magnetic field stacks up (see Figure 1.13).

The characteristic radius of the subsolar magnetopause (R_{MP}) can be estimated by the equilibrium of the dominant pressures on either side of this boundary: inside, the pressure of the planetary magnetic field ($B^2/2\mu_o$), and outside, the dynamic pressure of the solar wind plasma flow (NmV^2, where N is the ion concentration, m the mean ion mass, close to the proton mass, and V the overall speed of the solar wind).

By approximating a dipolar planetary field, where:

$$B(r) = B_e/r^3 \qquad [1.31]$$

at a distance r, in planetary radii R_P, in the plane of the equator, B_e being the equatorial surface magnetic field, we obtain:

$$R_{MP} \approx (B_e/2\mu_o\, NmV^2)^{1/6} \qquad [1.32]$$

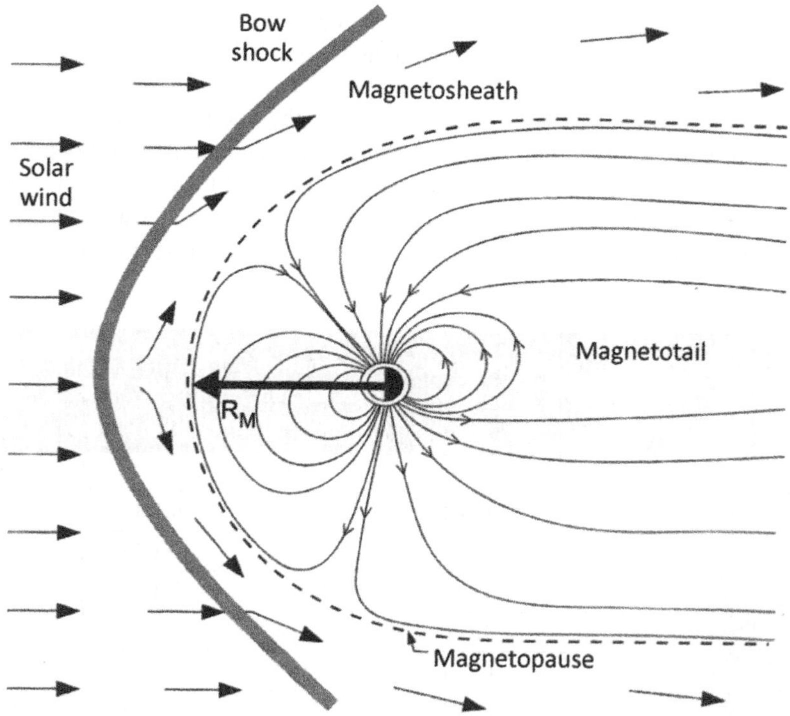

Figure 1.13. *Schematic view of a planetary magnetosphere (source: based on (Bagenal 2001))*

The decrease in $1/d^2$ of the solar wind concentration with distance d from the Sun, and the different intensities of the planetary magnetic fields, thus lead to magnetospheres of very different sizes (see Table 1.6), from the "dwarf" magnetosphere of Mercury crushed against the planet (see Chapter 4) to the giant magnetosphere of Jupiter, which permanently encompasses all its satellites (see Chapter 5). The thickness of the magnetosheath is about 0.3 to 0.5 R_{MP}. On the anti-solar side, the magnetospheric tail can extend up to ~ 100 R_{MP}.

1.3.5.3.2. Induced magnetospheres

In the different case of non-magnetized obstacles, but having an atmosphere and thus an ionosphere (Mars, Venus, as well as comets), the high electrical conductivity of the ionosphere allows the existence of currents J, resulting from the electric field $\boldsymbol{E} = -\boldsymbol{V} \times \boldsymbol{B}_{IP}$ ($\boldsymbol{B}_{IP}$ being the interplanetary magnetic field) which appears in the reference frame of the obstacle. The interaction of these currents with $\boldsymbol{B}_{IP}$ slows down the incident solar wind flow through the Lorentz force $\boldsymbol{F} = \boldsymbol{J} \times \boldsymbol{B}_{IP}$.

We can also consider that it is the pressure of the interplanetary magnetic field, stacked upstream of the obstacle, that provides the magnetic pressure opposing the dynamic pressure of the solar wind, hence the qualification of an "induced magnetosphere" for this type of interaction. The solar wind is always deflected, a shock is formed upstream of the obstacle, and a wake downstream. The equivalent of magnetopause here is called "ionopause".

Planet	Dipole moment ($G.km^3$)	Field at equator (G)	Inclination [B, Ω] and direction (°)	Typical radius of the magnetopause (km)	(R_P)
Mercury	$3\ 10^7$	0.002	< 0.8	3,700	1.5
Venus	$< 3\ 10^7$	< 0.0003	N/A	6,100	1
Earth	$8\ 10^{10}$	0.299	+ 9.7	64,000	10
Mars	$< 2\ 10^6$	< 0.00005	N/A	3,400	1
Jupiter	$1.5\ 10^{15}$	4.170	– 10.3	6,430,000	90
Saturn	$4.6\ 10^{13}$	0.211	0	1,200 000	20
Uranus	$4.1\ 10^{12}$	0.225	– 59.8	470,000	18
Neptune	$2.1\ 10^{12}$	0.142	– 46.9	570,000	23

Table 1.6. *Planetary magnetic fields and magnetospheres (see also Table 3.1)*

1.4. References

Asplund, M., Grevesse, N., Sauval, A.J., Scott, P. (2009). The chemical composition of the Sun. *Annual Review of Astronomy and Astrophysics*, 47(1), 481–522.

Bagenal, F. (2001). Planetary Magnetospheres. In *Encyclopedia of Astronomy and Astrophysics*, Murdin, P. (ed.). IOP Publishing, Bristol.

Batygin, K. and Brown, M.E. (2016). Evidence for a distant giant planet in the Solar System. *Astronomical Journal*, 151(22).

Braga-Ribas, F., Ortiz, J.L., Sicardy, B., Snodgrass, C., Rogues, F., Vieira-Martins, R., Camargo, J.I.B., Assafin, M., Duffard, R., Jehin, E., Pollock, J. *et al.* (2014). A ring system detected around Chariklo. *Nature*, 508, 72–75.

Catlin, D.C. and Kasting, J.F. (2017). *Atmospheric Evolution of Inhabited and Lifeless Worlds*. Cambridge University Press, Cambridge.

Encrenaz, T., Bibring, J.-P., Blanc, M., Barucci, M.-A., Roques, F., Zarka, P. (2003). *Le Système solaire*. CNRS Éditions/EDP-Sciences, Paris.

Encrenaz, T., Bibring, J.-P., Blanc, M., Barucci, M.-A., Roques, F., Zarka, P. (2004). *The Solar System*, Springer-Verlag, Berlin.

Festou, M.C., Keller, U., Weaver, H.A. (2004). *Comets 2*. University of Arizona Press, Tucson.

Gomes, R., Levison, H.F., Tsiganis, K., Morbidelli, A. (2005). Origin of the cataclysmic late heavy bombardment of the terrestrial planets. *Nature*, 435, 466–769.

Hanel, R.A., Conrath, B.J., Jennings, R.E., Samuelson, R.E. (2003). *Exploration of the Solar System by Infrared Remote Sensing*. Cambridge University Press, Cambridge.

Kivelson, M.G. and Russell, C.T. (1995). *Introduction to Space Physics*. Cambridge University Press, Cambridge.

Lauretta, D.S. and McSween, H.Y. (2006). *Meteorites and the Early Solar System*. University of Arizona, Tucson.

Lequeux, J. (2013). *Birth, Evolution and Death of Stars*. World Scientific Publishing Co., Singapore.

Lequeux, J., Casoli, F., Encrenaz, T., (2020). *The Exoplanet Revolution*. EDP Sciences, Les Ulis.

Lewis, J.S. (1995). *Physics and Chemistry of the Solar System*. Academic Press-Elsevier, Cambridge.

Lissauer, J.J. and de Pater, I. (2013). *Fundamental Planetary Science*. Cambridge University Press, Cambridge.

Michel, P., Bottke, W.F., Deleo, F.E. (2015). *Asteroids IV*. University of Arizona Press, Tucson.

Morbidelli, A., Levison, H.F., Tsiganis, K., Gomes, R. (2005). Chaotic capture of Jupiter's Trojan asteroids in the early Solar System. *Nature*, 435(7041), 462–465.

Prianik, D., Barucci, M.-A., Young, L. (2019). *The Transneptunian Solar System*. Elsevier, Saint-Louis.

Spohn, T., Breuer, D., Johnson, T.V. (2014). *Encyclopedia of the Solar System*, 3rd edition. Elsevier, Saint-Louis.

Tinetti, G., Encrenaz, T., Coustenis, A. (2013). Spectroscopy of planetary atmospheres in our Galaxy. *Astronomy & Astrophysics Reviews*, 21.

Tsiganis, K., Gomes, R., Morbidelli, A., Levison, H.F. (2005). Origin of the orbital architecture of the giant planets of the Solar System. *Nature*, 435, 459–461.

2

Solar and Planetary Systems

James LEQUEUX
LERMA, Paris Observatory, PSL University, Paris, France

2.1. The Sun in the Galaxy

The Sun is just one of 100 billion or so stars that make up the Milky Way, our Galaxy. It is relatively isolated, with its closest companion (Proxima Centauri) at a distance of 1.30 parsec. The parsec (pc), which is the unit of length that we will use throughout this book, along with the astronomical unit (au), has historically been defined as the distance at which the semi-major axis of the Earth's orbit (1 au) subtends an angle of one second of a degree. Since 2015, to simplify calculations, it is defined as being exactly (648,000/π) au, or 3.0857 10^{16} m. It is equal to 3.2616 light-years.

The distances of stars located less than 100 pc from the Sun are known to 10% accuracy thanks to the Hipparcos satellite. Its successor, the European satellite Gaia, is providing much more accurate distances for more than a billion stars, located up to the confines of the Milky Way. These distances are obtained by the *geometric parallax* method. This method consists of observing the position of the star in relation to very distant celestial objects that can be considered as fixed, the quasars, several times in a year, in order to locate its displacement. This displacement is composed of a linear *proper motion*, and an annual elliptical motion due to the orbital motion of the Earth, which is a *parallax* effect. The inverse of the parallax of the star, which is the length of the semi-major axis of the ellipse thus described, expressed in arc seconds, is the distance of the star in parsecs.

The Solar System 1,

The Galaxy is a disk with a radius of about 15 kpc. The Sun is located 8 kpc from the center, inside the disk, near its plane of symmetry. As a result, the Galaxy seen from the Earth appears as a light trail along a large circle of the celestial vault: it is the Milky Way, which appears more or less divided in two by a dark band, due to the extinction of starlight by interstellar dust: the latter, like the interstellar gas with which it is mixed, occupies a disk that is thinner than most of the stars. The present situation of the Sun does not imply anything about the conditions of its birth, which could have taken place either in isolation or within a cluster of stars born all at about the same time and which then disintegrated. On the other hand, it is almost certain that, today, the Sun is not at the same distance from the galactic center as it was where it was formed. Indeed, it contains more heavy elements (relative to hydrogen) than the interstellar matter and the young stars that surround it, whereas one would expect it to contain less since it was born 4.6 billion years ago and since nucleosynthesis in stars continuously enriches the Galaxy with heavy elements. However, observation shows that the abundance of heavy elements increases as one gets closer to the galactic center: this strongly suggests that the Sun was born closer to the galactic center than it is now. Its displacement comes from a radial migration of some stars, which is due to their gravitational interaction with the mass of the disk and its irregularities, in particular the spiral arms which are density waves. These migration phenomena make it very difficult to study the temporal evolution of the Galaxy, an evolution that is currently very poorly understood and which we will therefore not talk about; fortunately, Gaia's data are revolutionizing the subject.

In the Galaxy, there are stars of all masses, from 0.08 $M_\odot$ ($M_\odot$ designates the mass of the Sun) to several hundred $M_\odot$; for an introduction to stellar physics, see (Lequeux 2011). The more massive the star is, the more luminous it is: the luminosity varies approximately as the power 2.5 of the mass. Moreover, since massive stars consume their nuclear energy much faster than low mass stars, they live for a short time (a few million years for the most massive stars), while a 1 $M_\odot$ star has a lifetime of about 13 billion years: the Sun, which is 4.6 billion years old, has therefore only lived through a third of its life so far. As long as stars burn hydrogen in their central regions, their luminosity and temperature vary little, but nevertheless significantly: for example, the Sun's luminosity has increased by 30% since its birth.

Thus, stars of different masses form a sequence where luminosity increases with the mass, as indicated above, as well as the surface temperature: this sequence is called *main sequence* and the stars in it are called *dwarf stars*: we will learn why. Indeed, when the hydrogen is exhausted in the core of the star, it starts to gravitationally collapse on itself, which causes the temperature to rise again: helium is then transformed into carbon, and the effect of this new input of energy makes the star swell and become a *giant star*. Here, we will only focus on the dwarf stars of the main sequence, whose main characteristics are given in Table 2.1.

Spectral type	Radius (R_{Sun})	Mass (M_{Sun})	Luminosity (L_{Sun})	Effective temperature (K)
A5V	2.0	4.0	54	9,520
F0V	1.4	1.7	6.5	7,200
F5V	1.2	1.29	3.2	6,440
G0V	1.05	1.10	1.5	6,030
G2V (Sun)	1	1	1	5,770
G5V	0.93	0.93	0.79	5,500
K0V	0.85	0.78	0.42	5,250
K5V	0.74	0.69	0.15	4,350
M0V	0.63	0.47	0.077	3,850
M5V	0.32	0.21	0.011	3,240
M8V	0.12	0.08	0.0005	2,560

Table 2.1. *Properties of dwarf stars of different spectral types (the Roman numeral V shows that they are dwarf stars, whereas giant stars are labelled by the numeral III). With the exception of the Sun, these data are approximate due to the evolution of stars on the main sequence (for more details, see (Schaller* et al. *1992))*

It should be known that the number of stars in the Galaxy is all the greater the less massive they are: high mass stars are very rare, and the Galaxy is therefore dominated by dwarf stars of type K and, especially, M, with low mass and low luminosity, that have a red color because of their relatively low surface temperature.

2.2. Planetary systems in the Galaxy

Our anthropocentrism has long led us to believe that our Solar System, of which the Earth was to be the center, was unique in the Universe. The Copernican revolution gradually placed the Sun, and not the Earth, at the center of the Universe, and some have even imagined that other planetary systems could well exist, around stars similar to the Sun. This was already the case of Epicurus (c. 342–270 BCE), then of Lucretia (c. 94–54 BCE), who assert that the atoms that formed the Earth and the living species could not fail to generate similar systems elsewhere. For Lucretia, "the sky, the earth, the sun, the moon, the sea, and all bodies are not unique, but rather infinite in number".

It was not possible for Christians to adopt such unorthodox ideas, and it was not until the Renaissance that a few daring minds returned to them: first, Nicholas of Cues (1401–1464), in his 1440 work *De docte ignorantia*, and then Giordano Bruno (1548–1600), who went much further in *The Ash Wednesday Supper*, published in 1584. A convinced Copernican, he believed that the planets revolve around the Sun, and even that others must surround the stars, which he considered to be as many suns; they might even have inhabitants.

Then, René Descartes (1596–1650), perhaps inspired by Epicurus, built a new system of the world, based on whirlpools of subtle matter. One of these whirlpools is centered on the Sun and drags the planets, others surround the stars and move their own planetary systems. Secondary whirlpools drive the satellites of Jupiter. Descartes' theories had a great influence in the Netherlands for a few decades, and longer in France, until they were dethroned by Newton's universal gravitation and its consequences on the motion of the planets.

For lack of data, things remained as they were until the first exoplanet, orbiting a solar-type star, was discovered in 1995. Systematic research, which we will discuss later, has today identified more than 4,400 exoplanets, including 770 systems with several planets. They are extremely varied, and the Solar System seems more the exception than the rule. Today, it appears that the vast majority of stars in our Galaxy are surrounded by one or more planets, while there are also many isolated planets, ejected from planetary systems. There must therefore be more than 100 billion planets in the Galaxy.

While we still have very few observations of planets in formation, the birth of stars and their planetary procession is fairly well understood. In his *Allgemeine Naturgeschichte und Theorie der Himmels* (Universal History of Nature and Theory of the Heavens) in 1755, Immanuel Kant (1724–1804) had already proposed that all the matter that forms the Solar System was originally in the form of a large nebulous mass; it would have contracted into a flat rotating disk under the effect of its own gravitation, then distributed between the Sun and the planets, born from inhomogeneities in the disk. The idea of the rotating disk obviously comes from the fact that all the planets rotate in the same direction on orbits located in approximately the same plane. Perhaps without knowing Kant, Pierre-Simon Laplace (1749–1813) put forward a very similar theory in his *Exposition du système du monde*, the first edition of which dates from 1796. He was familiar with the observations of William Herschel (1738–1822), who saw a central star in many nebulae. Both thought that they were primitive nebulae within which a sun had already formed, which supported Laplace's theory. Yet they were wrong: we now know that the vast majority of these objects, planetary nebulae, are made of a star at the end of its life that has ejected some of its matter. However, the ideas of Kant and Laplace remain valid in their broad outlines. But, at that time, the original matter of planetary systems, *interstellar matter*, remained to be discovered.

2.3. Interstellar matter[1]

2.3.1. *History*

As early as 1890, photographs by Edward E. Barnard (1857–1923) and Max Wolf (1863–1932) revealed the existence of regions of the Milky Way apparently devoid of distant stars, which was correctly attributed in the 1910s to the presence of opaque dust clouds. Then, in 1912, the American Vesto Slipher (1875–1969) showed that the cloudiness surrounding Pleiades stars was made of dust illuminated by these stars, and suggested that this dust may well attenuate the light of the stars behind them. We then began to interpret the dark band that seems to divide the Milky Way in two, not by an absence of stars, but by dust extinction.

Dust is not the only thing found between the stars of the Milky Way. There is also gas. For a long time, bright nebulae had been spotted, the most famous of which is the Orion Nebula. However, their nature was unknown until the first spectra were obtained by William Huggins (1824–1910) in 1864 and by Fr. Angelo Secchi (1818–1878) the following year. They saw only three lines, one of which was emitted by hydrogen and the other two were unknown. It was not until much later, in 1927, that the American astronomer Ira S. Bowen (1898–1973) realized that they were actually lines of a very common element, doubly ionized oxygen. They are only observable in the very rarefied gas of gaseous nebulae, which is ionized by the ultraviolet radiation of the very hot stars within them.

There is gas outside these nebulae. In 1904, in Potsdam, Johannes Hartmann (1865–1936) studied the spectrum of the double star δ Orionis, for which the spectral lines continuously change wavelength because its components periodically move away from and towards us during their orbital motion, and noticed that a line of ionized calcium remained fixed: it could not come from the star, but from an interposed gas cloud. Fifteen years later, and in the same way, the American Mary Lea Heger (1897–1983) discovered fixed sodium lines in the spectrum of the same star, as well as broad spectral bands whose wavelength did not vary either: they were called "diffuse interstellar bands", and today, more than 500 of them are known.

1 For a complete study, see (Lequeux 2002).

After that, little progress was made for about 30 years. More and more interstellar absorption lines were discovered, suggesting that the medium consists of a large number of clouds containing ionized calcium and titanium, sodium, potassium and neutral lithium: these are the only elements that can be detected by their absorption lines in the visible range. In the 1940s, three interstellar molecules were discovered by the Belgian Polydore Swings (1906–1983) and the Canadian Andrew McKellar (1910–1960), also by their absorption lines in the visible range: cyanogen CN, as well as CH and CH^+. These molecules are highly reactive, and their mere presence indicates that they are in a very dilute medium, where they very rarely encounter an atom or molecule with which they could combine. However, these elements and molecules are relatively rare: most of the interstellar matter must be hydrogen, which is by far the most abundant element in the Universe (see Table 1.5). As early as 1945, the Dutch scientist Hendrick van de Hulst (1918–2000) had put forward the idea to detect it by a line emitted by the hydrogen atom in the radio domain, at 21 cm wavelength; this was done in 1951 in the United States by Harold I. Ewen (1922–2015) and Edward M. Purcell (1912–1997).

Thanks to this, we were able to determine the quantity of interstellar gas present in our Galaxy – a few hundredths of its mass – and its temperature, which turned out to be very low, less than a hundred K. Then came the observation in ultraviolet light by several artificial satellites, in particular NASA's *Copernicus* satellite, launched in 1972, then the HST (Hubble Space Telescope) and the Fuse satellite (Far-Ultraviolet Spectroscopic Explorer). It revealed the absorption lines of many elements that were inaccessible to visual observation.

After the discovery of hydrogen in its atomic form, the discovery of the hydrogen molecule, which produces important absorptions in the far ultraviolet, could have been expected. This discovery was made in 1970 by George R. Carruthers, with a spectrograph carried by a rocket.

2.3.2. *Chemical composition*

The abundance of elements in the interstellar medium is generally estimated in relation to atomic hydrogen, which produces a very strong line at 121.6 nanometers wavelength, the Lyman α line. Molecular hydrogen is also present in variable quantities. We can see that many elements are very under-abundant in the gas, compared to what is found in the Sun. This is because they are mainly found in the form of dust, and not in the gas. Thus, the link is made between the two components of the interstellar medium: gas and dust. It is unfortunately difficult to directly obtain the abundances of the elements in the very dense gas of the regions where stars form, which are opaque at the ultraviolet wavelengths where the measurements are made.

Hydrogen is mainly in the form of molecules, and there are also a large number of other molecules, of which the main ones were discovered in the 1960s and 1970s by their radio emission. One of the most abundant is carbon monoxide (CO), which has proven to be an excellent tracer of molecular gas in general. It is in these dark and molecular clouds that stars are born.

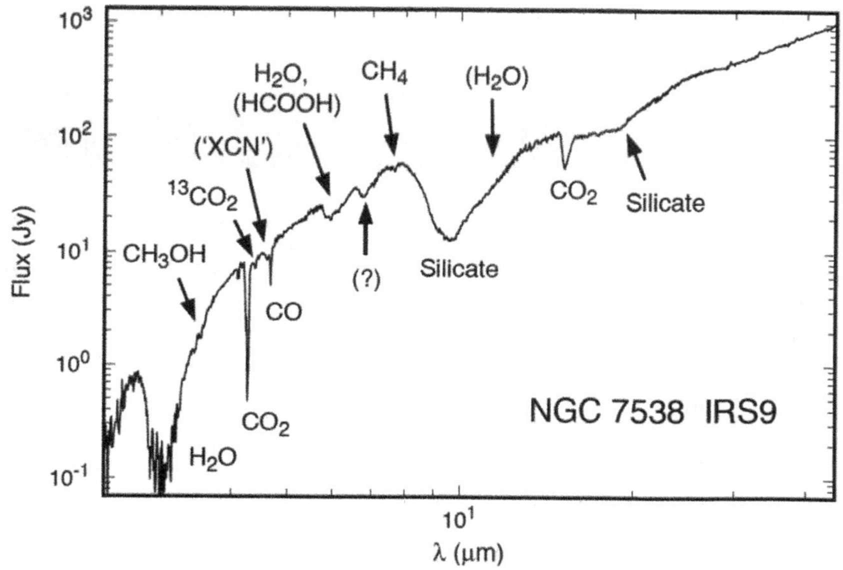

Figure 2.1. *Spectrum in the direction of a massive young star buried in a molecular cloud, obtained with the ISO satellite. The bands of silicates are due to the core of dust, the others to ice deposited on their surface. The band marked (?) is probably due to methyl alcohol (source: according to (Whittet* et al. *1996))*

Dust grains in dense star-forming regions are bigger than in the diffuse interstellar medium. This is because they are very cold (generally less than 10 K) and covered with different types of ice, as shown in Figure 2.1. However, this ice sublimates during the formation of the star. The abundances of elements in the most primitive meteorites, which are given in Table 1.5, reflect the abundances in the grains at the formation of the Solar System, and these abundances must be considered: hydrogen, helium, half of the oxygen, 90% of the carbon, most of the nitrogen, rare gases and probably some other non-refractory elements are found in gaseous, atomic (rare gases) or molecular form (H_2, CO, CH_4, N_2, NH_3 and H_2O in particular). One would have expected sulfur to be largely in the gaseous phase, but this does not seem to be the case. The mass in these grains represented about 0.5% of the total mass of interstellar matter. This fraction would reach 1 to 2% if the ice mantle covering the grains is taken into account.

2.3.3. *Physical properties*

Interstellar gas exists in a wide variety of physical conditions: density, temperature, turbulence, degree of ionization. Table 2.2 summarizes them.

Medium		Density (cm^{-3})	Temperature (K)	Total mass (M_{Sun})
Atomic (HI)	Cold	≈ 25	≈ 100	$4\ 10^9$
	Warm	≈ 0.25	≈ 8,000	$4\ 10^9$
Molecular (H_2)		≥ 1,000	≤ 100	> $3\ 10^9$
Ionized	HII Regions	$1 - 10^4$	≈ 10,000	$5\ 10^7$
	Diffuse	≈ 0.03	≈ 8,000	10^9?
	Hot	≈ $6\ 10^{-3}$	≈ $5\ 10^5$	10^8?

Table 2.2. *The different phases of the interstellar medium in the Galaxy. Note that the warm "atomic" medium is partially ionized. The mass of the molecular gas is poorly known because part of it could be in the form of undetectable solid hydrogen*

The cold atomic medium is bathed by ultraviolet radiation from hot stars in the Galaxy, at wavelengths longer than that of the Lyman discontinuity of atomic hydrogen at 91.1 nm, which corresponds to an energy of 13.6 eV (photons from the Lyman continuum are absorbed by the ubiquitous hydrogen atoms).

As a result, elements with an ionization potential lower than 13.6 eV are partially ionized: in fact, all elements except helium and other rare gases, nitrogen and oxygen are partially ionized. If two ionization states of the same element can be observed, for example, Ca° and Ca^+, the ionization equilibrium equation gives the electron density in the medium. In molecular clouds, the extinction by dust is high and ultraviolet radiation does not penetrate. However, cosmic particles produce a low degree of ionization in these clouds, which makes the medium conductive.

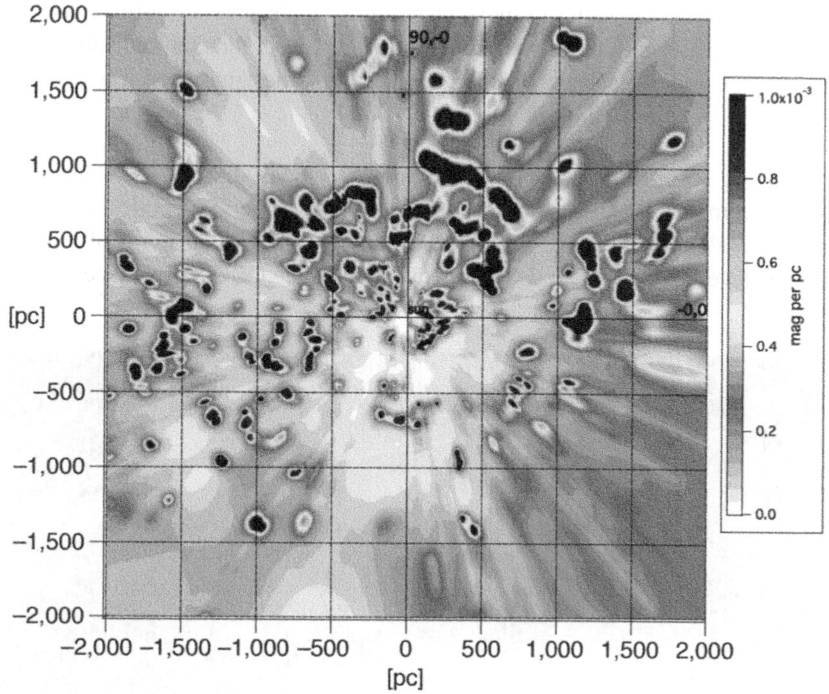

Figure 2.2. *Interstellar dust distribution, section in the plane of symmetry of the Galaxy. For a color version of this figure, see www.iste.co.uk/encrenaz/solar1.zip*

COMMENT ON FIGURE 2.2. – *The Sun is in the center and the galactic center is to the right. The units represent the excess color in magnitudes per parsec along the line of sight (the outer pink radial areas are artifacts of the method). One can deduce the average density of hydrogen atoms by multiplying the scale on the right by 1,880, but this does not make much sense in clouds that have a complex structure. On the other hand, the density of the warm medium between clouds, which is more uniform, is of the order of 0.2 atoms/cm*^{3} *(source: from (Lallement* et al. *2018)).*

For example, the neighborhood of the Sun contains cold atomic clouds and molecular clouds within a warm atomic medium, whose heating is due to X-rays emitted by the distant hot medium and by some stars. A 3D cartography of the dust has just been obtained with data from the Gaia satellite, supplemented by other data, based on the measurement of the extinction of a large number of stars of known distance (see Figure 2.2). As the dust is well mixed with the gas, this map is also a map of the interstellar gas distribution.

The distribution of matter in molecular clouds where stars are formed can also be obtained by measuring the thermal emission of dust in millimeter and sub-millimeter waves, or the radio emission of certain molecular lines, more or less sensitive to density.

There are significant uncertainties in both methods, but the geometry is well defined: the matter is not distributed in more or less spherical entities, but in dense clouds, 50–70% of which have the form of filaments, whose diameter is of the order of 0.1 pc whatever the density, and whose length very often exceeds 1 pc (see Figures 2.3 and 2.5).

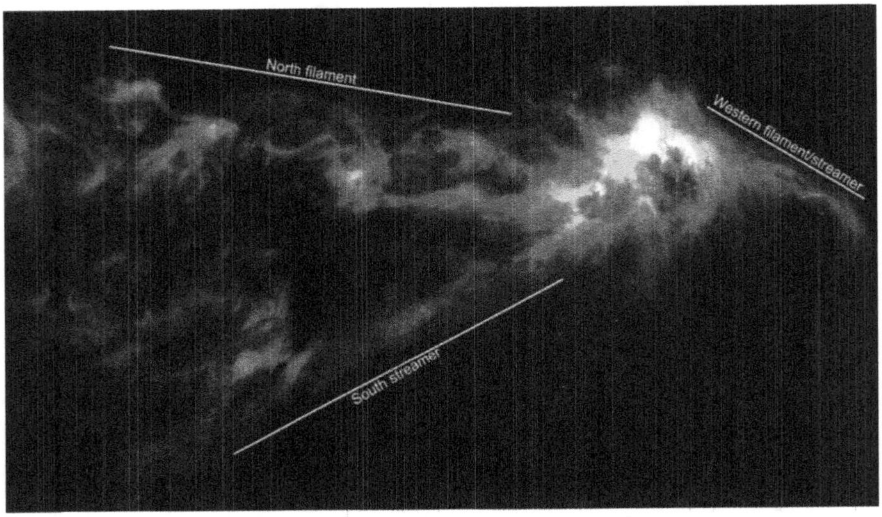

Figure 2.3. *Molecular cloud of Corona Australis. The image shows the emission of interstellar dust observed with the Herschel satellite at 250 μm (red) and 160 μm (yellow-orange), and with the Spitzer satellite at 24 μm (violet-blue) (source: (Bresnahan* et al. *2018)). For a color version of this figure, see www.iste.co.uk/encrenaz/solar1.zip*

Such filaments frequently appear in numerical simulations of turbulence in molecular clouds: they form at the junction between converging flows in regions of supersonic turbulence. The effects of gravity and the magnetic field are still somewhat uncertain, but it appears that filaments can grow by accreting the surrounding matter. The main filaments are generally perpendicular to the magnetic field and the turbulence is subsonic, while the secondary filaments, where the turbulence is supersonic, are generally parallel to the magnetic field (see Figure 2.4).

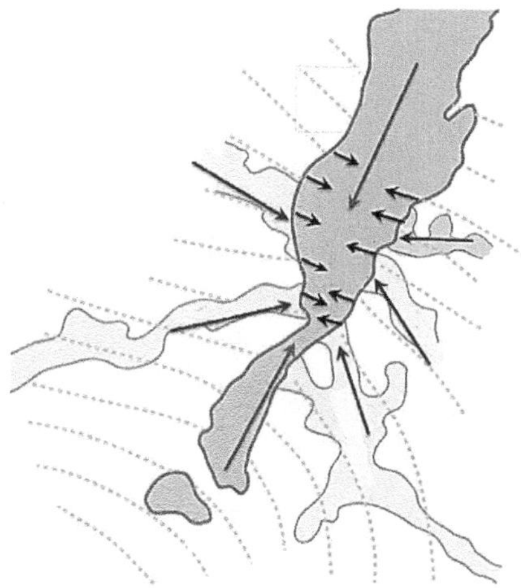

Figure 2.4. *Diagram of filaments and velocity of matter in the Serpens South molecular cloud. For a color version of this figure, see www.iste.co.uk/encrenaz/solar1.zip*

COMMENT ON FIGURE 2.4. – *The direction of the magnetic field, obtained from polarization measurements, is indicated by the green dotted lines. The blue arrows indicate the fall of matter along the magnetic field in the secondary filaments, the red arrows indicate the fall of matter in the main filament and the black arrows indicate the gravitational contraction of this filament, which has already started to form a cluster of very young stars (source: from (André 2002), adapted from (Sugitani* et al. *2011)).*

2.4. The formation of stars with masses close to that of the Sun

Two general reviews are complementary to one another: (André 2002; McKee 2007). Most isolated stars form in filaments by gravitational collapse when the mass per unit length of the filament exceeds about 16 $M_\odot$/pc, which corresponds to a critical density: $n(H_2) = 2\ 10^4$ mol.cm^{-3} for a filament of 0.1 pc in diameter. The filament fragments into condensations that are about 0.3 pc apart. Massive stars and star clusters are instead formed by "spherical" fragmentation in filament convergence zones (see Figure 2.4). This convergence is possibly aided or provoked by an external overpressure due to a supernova explosion or the stellar wind of a nearby massive star. Figure 2.5 shows two examples, where one can observe concentrations of gravitationally-unbound matter along filaments, which may or may not become pre-stellar, and then stars at different stages of development.

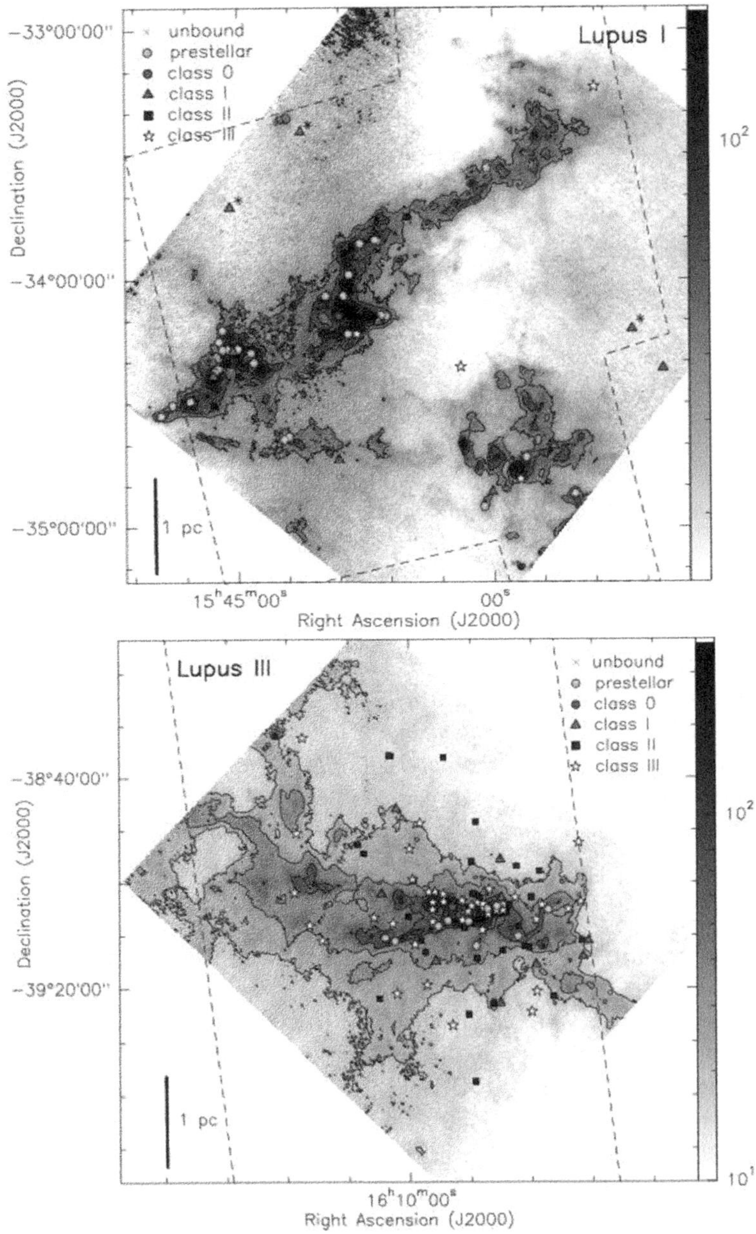

Figure 2.5. *Star formation in two molecular clouds in the constellation Lupus, observed with the Herschel satellite. For a color version of this figure, see www.iste.co.uk/encrenaz/solar1.zip*

COMMENT ON FIGURE 2.5. – *The column density of H_2, in units of 10^{20} mol.cm^{-2}, deduced from the observation of the dust, is in gray; note the filamentary structure. Stars in formation are represented by different symbols, from the protostellar condensation (green circles) to the star having dissipated its protostellar disk, the planets having been entirely formed (class III, yellow stars). The intermediate stages are indicated by red circles (class 0), purple triangles (class I) and blue squares (class II). See Figure 2.9 for the definition of these classes. Star formation is much more advanced in Lupus III than in Lupus I (source: based on (Rygl* et al. *2013)).*

Once the conditions of gravitational instability have been met, the condensation collapses on itself and it can be shown that the magnetic field cannot oppose it. The typical collapse time, or freefall time, is of the order of $(3\pi/32G\rho)^{1/2}$, or 4.3 $10^7/n_0^{1/2}$ years, where ρ is the density and n_0 is the initial number of hydrogen atoms per cm^3. This is only an order of magnitude, because in order to obtain it, the pressure and turbulence of the medium have been neglected. More realistic analytical calculations, of which an old example is given in Figure 2.6, show that a dense nucleus is formed, on which the surrounding matter falls practically at freefall speed. Recent simulations taking turbulence into account, such as (Murray and Chang 2015), broadly confirm this pattern, although there are no more self-similar solutions like the one in Figure 2.6.

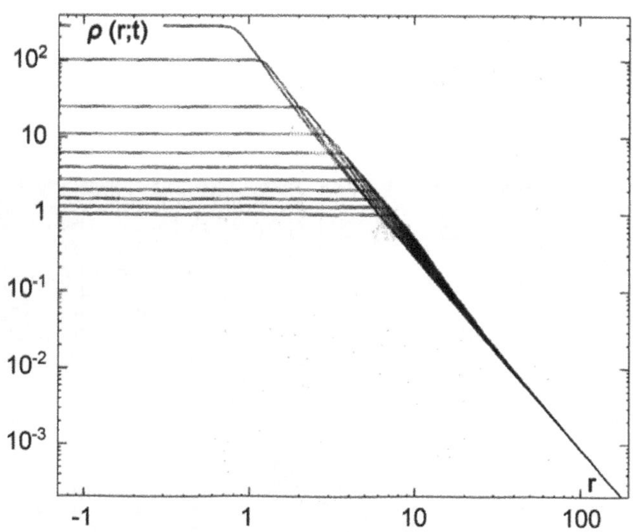

Figure 2.6. *Analytical solution for the collapse of a spherical cloud without a magnetic field or turbulence. The density is plotted against the radius r for different, evenly spaced periods of time. The units are arbitrary because the solution is self-similar (source: from (Blottiau* et al. *1988))*

The material heats up during the collapse. The resulting heat is first radiated by the millimeter-submillimeter emission of various molecules such as CO and by the thermal emission of dust, which is itself heated by the gas. However, the medium becomes increasingly opaque, even in the far-infrared, and the contraction becomes adiabatic. The thermal pressure stops the collapse of the core, leading to the first stage of a protostar, whose mass is of the order of 1% of the final mass. It grows progressively by accretion of the matter that continues to fall.

The rotation of the object, which seems to always be present because of the motions in the initial cloud and especially the turbulence, deeply affects the collapse. If the rotation is very strong, the object may split in two, finally forming a binary star by a mechanism that is still rather poorly understood. If it is smaller, a rotating disk forms around the core. The matter falls on this disk, and the accretion on the protostar itself is mainly done from inside this disk, which is therefore an accretion disk. The radial motion of the matter in the disk is due to the turbulent viscosity. In binary protostars, there may be a disk around each component or a common circumbinary disk.

Accretion from the disk requires a loss of angular momentum. Observation shows that this angular momentum is mainly carried away with part of the matter by a bipolar jet along the axis of rotation (see Figure 2.7). Such jets are so frequently observed that they seem to be the rule. The magnetic field seems necessary to generate the jet (see Figure 2.8). The modeling of the phenomenon, which needs magneto-hydrodynamic calculations, requires the consideration of both the disk and the outflow: for a discussion, see (Ferreira *et al.* 2006). The bipolar jet carries part of the residual molecular cloud with it, which makes it particularly spectacular (Bachiller 1996).

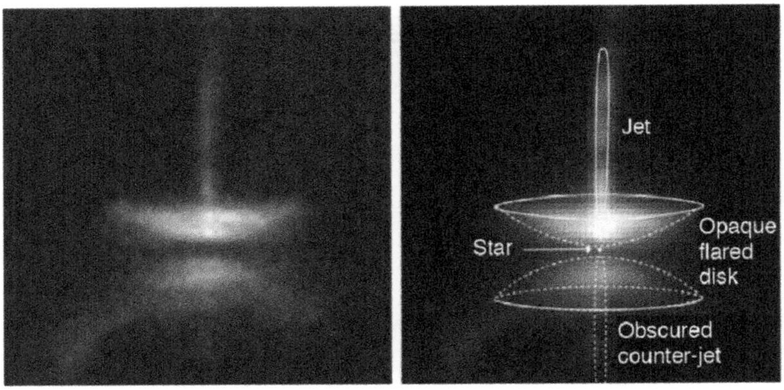

Figure 2.7. *Accretion disk and bipolar jet observed on the HH 30 object, in the Taurus constellation (source: NASA). For a color version of this figure, see www.iste.co.uk/encrenaz/solar1.zip*

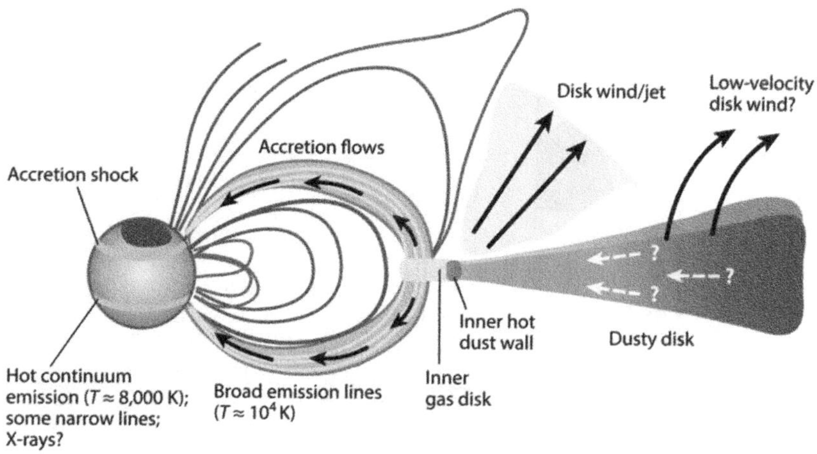

Figure 2.8. *Diagram of the accretion of matter by a protostar and formation of the outflow. The magnetic field of the matter of the protostar forms a magnetosphere, which conditions the accretion and the formation of the jet (source: from (Hartman* et al. *2016)). For a color version of this figure, see www.iste.co.uk/encrenaz/solar1.zip*

The observations make it possible to describe the evolution of the protostar towards the final star in the form of four classes (André, Ward-Thompson, Barsony, see (Bachiller 1996; McKee 2007) and Figure 2.9):

– class 0: objects with a very weak central protostar in the visible range and near- and mid-infrared, but detected in sub-millimeter waves by the emission of dust from the collapsing cloud. The protostar acquires most of its mass during this stage when it is hidden by this cloud, and a bipolar outflow is formed. The accretion seems to be very variable over time. The episodes where it is very intense correspond to the temporarily very luminous objects called FU Orionis, named after their prototype;

– class I: much brighter objects emitting in all the types of infrared. The cloud has almost completely disappeared, revealing the protostar and the accretion disk. The accretion on the protostar has greatly diminished;

– class II: the star is completely formed; nuclear reactions have begun. It is surrounded by a large disk, but the jets have disappeared. This stage corresponds to the pre-main-sequence stars, the classical T Tauri, named after their prototype; the emission lines produced by the gas of the disk are important;

– class III: the star has now completely expelled the gas from the central parts of the disk, and only a disk of debris remains where the planets are forming. This stage is that of T Tauri stars with low emission lines.

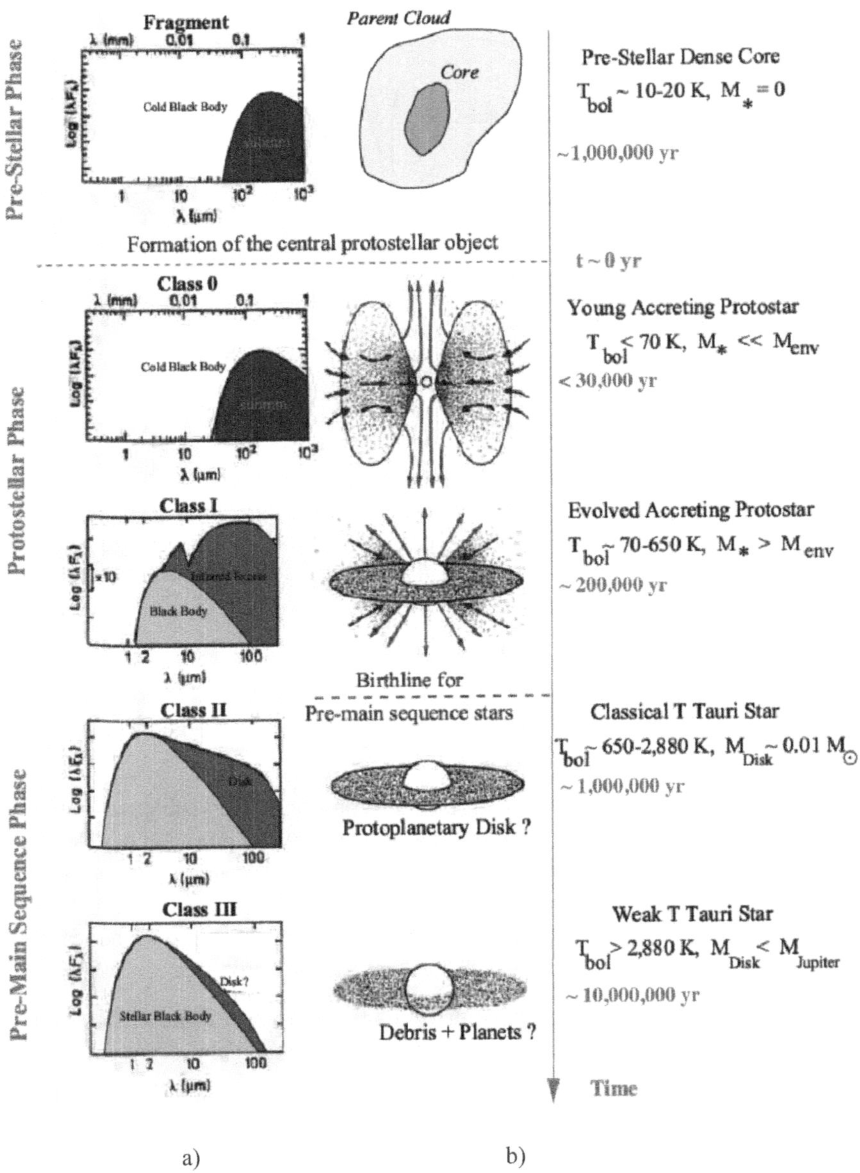

Figure 2.9. *Sequence of evolution of a solar mass star in the process of formation: (a) the aspect of the infrared and sub-millimeter spectrum; (b) diagram of the object (source: from (André 2002)). For a color version of this figure, see www.iste.co.uk/encrenaz/solar1.zip*

However, there is still some ambiguity in this classification: for example, a class II source seen from the side may appear to be class I when the protostar is hidden by the disk. However, this ambiguity is largely removed by radio wave or sub-millimeter observations, for which the clouds are little opaque or not opaque at all. The lifetime of these different classes can be estimated from their frequency (see Figure 2.9).

Let's take a more detailed look at what happens in the protostar itself (for a complete and pedagogical study, see (Maeder 2009)). In the phase corresponding to class I, it has an isothermal and initially cold core (about 10 K), on which the accreted matter falls (see Figure 2.6). A shock develops on the surface of the core, whose mass increases due to accretion. The gravitational energy of the accreted matter is released at this point: it is $(GM/R)\ dM/dt$, where M is the mass of the core and R its radius, dM/dt is the accretion rate and G is the gravitational constant. Half is radiated outwards and the other half heats the core. The hydrogen molecules dissociate when its temperature reaches 2,000 K, then the resulting hydrogen atoms ionize around 3,000 K. The opacity in the core becomes very high due to the Thomson scattering of light on the free electrons: the energy can only propagate there by convection and it therefore becomes totally convective.

The contraction of the core is now adiabatic. Its luminosity increases up to about 100 $L_\odot$ (solar luminosity) in the case of the proto-Sun, at the maximum of accretion, after 10^5 years: we enter class I. Then, the decrease of the radius is such that the luminosity decreases, despite an increase in effective temperature. The accretion decreases a lot due to lack of material or other causes and the star becomes visible after 10^6 years, as a classical T Tauri, named after the prototype: we then enter class II and have reached what is called "the birthline". The radius is then about 10 $R_\odot$, the temperature 4,000 K and the luminosity about 15 $L_\odot$, with a certain dependency on the residual rate of accretion.

The contraction will then continue at an essentially constant effective temperature close to 4,300 K, so the luminosity of the star decreases: this is the so-called Hayashi phase. The core temperature increases, and the star begins to burn deuterium at 10^6 K, then lithium at 2.5 10^6 K, which slows down the contraction. A brief phase of central combustion of carbon into nitrogen occurs, and, finally, the combustion of hydrogen by the pp chain: the star then enters the main sequence, after approximately 40 million years for 1 $M_\odot$. It will remain there in this state for approximately 11 10^9 years, while its luminosity slowly increases. Then, when the hydrogen in the core is exhausted, it will become a red giant and will end up as a white dwarf 13.3 10^9 years after its formation. Figure 2.10 describes this entire evolution.

T Tauri stars have emission lines in the visible and ultraviolet range, which are produced by gas accretion of the circumstellar disk (see Figure 2.11). They are weaker for class III because the gas has practically disappeared inside the disk. The T Tauri also produce an X-ray emission, most probably due to the magnetic activity on their surface, and an infrared emission at millimeter waves that comes from the disk itself, as we will see.

Since its formation, the Sun has continued to be magnetically active, producing strong bursts of charged particles that reach the Earth after a few days. It also emits a permanent wind of charged particles that, at Earth level, has a speed of 400 km/s. This wind interacts with the planets that have a magnetosphere, the Earth and the giant planets, but also with the other planets.

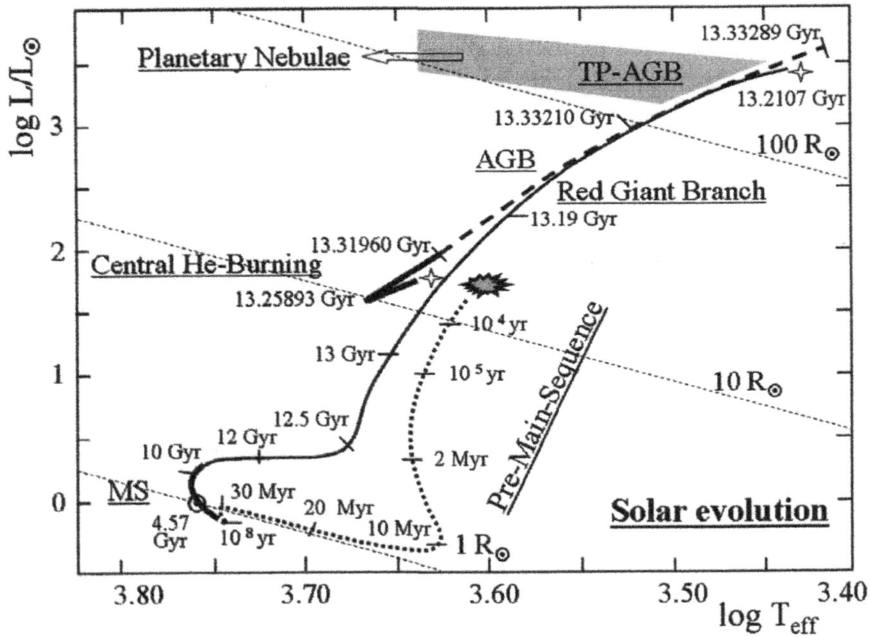

Figure 2.10. *The evolution of the Sun*

COMMENT ON FIGURE 2.10. – *The evolutionary trajectory of a star of the Sun's mass is plotted in the luminosity/effective temperature diagram, starting from the birthline. The time is indicated at the different stages. The two four-branch stars correspond to the helium flash and the very fast rearrangement of the star structure (source: from (Maeder 2009), data of C. Charbonnel).*

2.5. Circumstellar disks

Circumstellar disks were discovered around 1985 by the infrared excess of some stars, observed with the IRAS satellite. Soon after, millimeter wave observations showed the presence of dust grains around very young stars, revealing sufficient matter to form planetary systems. Interferometry has also made it possible to observe millimeter molecular lines and to measure the rotation of disks; the first visible light observations of disks in the Orion region with the Hubble Space Telescope in 1994 showed that these disks were flared towards the edges (see Figure 2.7).

The observation of circumstellar disks requires high angular resolution in millimeter and sub-millimeter waves, and has recently made great progress thanks to interferometers, notably Alma, while the Herschel satellite resolved some of them. The thermal emission of the dust is observed with these instruments, from which one can deduce the dimensions of the disk as well as its temperature and mass, assuming a dust/gas mass ratio similar to that of the interstellar medium (1/100 to 1/200). The mass of the disks is between about 0.02 and 0.1 $M_{\odot}$, for both class 0 and class I objects, which shows that almost all this mass is acquired during the first 100,000 years and that the transfer to the protostar is fast, in a very sporadic way (the mass of the protosolar disk itself is estimated by independent considerations between 0.01 $M_{\odot}$, in other words, 10 masses of Jupiter, and 0.07 $M_{\odot}$). The surface density of the disks varies approximately as $r^{-3/2}$. The radius is difficult to determine because the outer parts of the disks are cold and radiate very little; moreover, the disks generally have no sharp edges. For the disks observed in Orion in visible light with the HST, radii between 50 and 200 au are found, with some disks being two or three times larger. These values are confirmed by observations in millimeter waves, but with sometimes smaller radii, down to 15 au. The mass of the disk varies statistically as the power 1.6 ± 0.3 of the radius. There does not seem to be a clear relationship between the properties of the disk and the mass of the central star.

Figure 2.11 summarizes the structure of a circumstellar disk and its properties. The disks have a very strong temperature gradient between their central regions and the outer regions, which are much colder. The interferometric study of the CO lines of different excitations shows that the temperature varies with the radius r as r^{-q}, with q being between 0.4 and 0.7. There is also a temperature gradient in the thickness; the regions near the plane of symmetry can be very cold. This is mainly due to the heating by the radiation of the central star, and possibly by the X-rays emitted nearby. The heat released by the viscosity in the disk contributes somewhat to the heating.

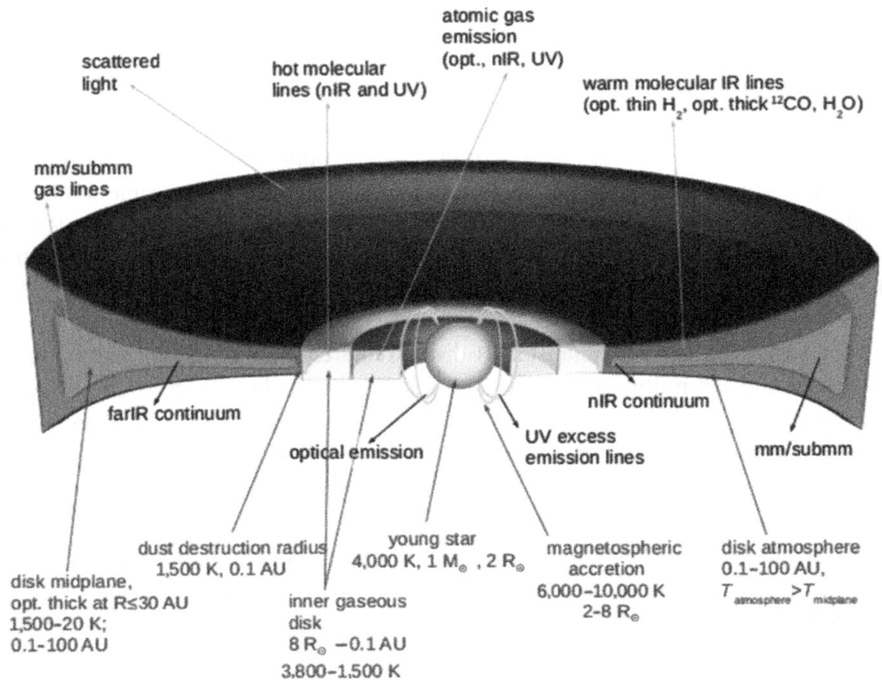

Figure 2.11. *Schematic section of a protoplanetary disk with its different emissions, during the classical T Tauri phase (class II). The dimensions are given as an indication, as they vary from one disk to another (source: from (Sicilia-Aguilar* et al. *2016)). For a color version of this figure, see www.iste.co.uk/encrenaz/solar1.zip*

Near the star, the temperature is high enough that all the non-refractory molecules of the disk, and even the refractory molecules very close to the star, are in a gaseous state. However, far from the star, the temperature of the dust becomes low enough for the less volatile molecules to condense: first, water vapor when the dust is between 100 and 130 K (the exact temperature depends on the ambient pressure), then ammonia, methane and carbon monoxide at lower and lower temperatures. The transition zone between the gas phase and the condensed phase is the ice line.

The H_2O ice line is generally too close to the star to have been directly identified so far by observation (it would be about 3 au from the Sun at the present time). Moreover, it can only be observed from space because of the very strong absorption by atmospheric water vapor, even in the driest sites. Although it did detect water vapor in protoplanetary disks, the Herschel satellite did not have, by far, sufficient angular resolution to map it.

On the other hand, the CO ice line, which corresponds to a dust temperature of about 20 K, was observed with the angular resolution of Alma in the disk surrounding the young star HD 163296 (see Figure 2.12), both directly by observing the distribution of the CO molecule, or rather the rare isotopomer $C^{18}O$ because the lines of the main molecule $C^{16}O$ are very optically thick, and indirectly by observing another molecule, N_2H^+. Indeed, N_2H^+ is destroyed in the presence of gaseous CO by the reaction $N_2H^+ + CO \rightarrow N_2 + HCO^+$, and its observation indicates the absence of CO. Figure 2.12 clearly shows that N_2H^+ and gaseous CO do not coexist. The molecule HCO^+ and its isotopomer DCO^+ coexist with CO, which is not surprising since they are formed from CO. The observations in Figure 2.12 show that CO is gaseous up to a distance of 90 au from the star, and solid beyond that. This is therefore the radius of the CO ice line.

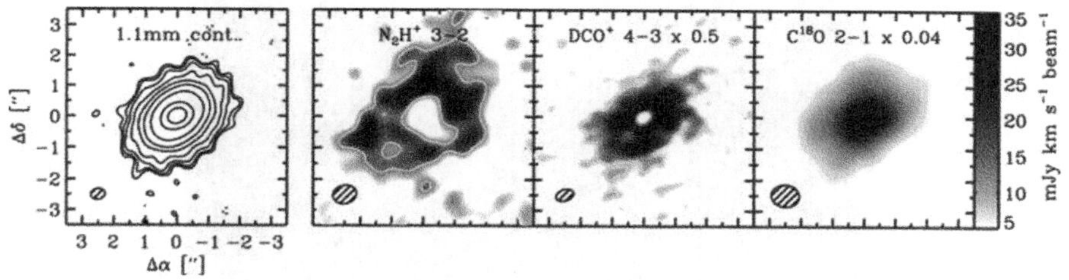

Figure 2.12. *Images of the disk surrounding the very young A1V star HD 163296, 119 pc away, obtained with Alma. For a color version of this figure, see www.iste.co.uk/encrenaz/solar1.zip*

COMMENT ON FIGURE 2.12. – *The ellipses indicate the angular resolution. From left to right, the continuum at 1.1 mm wavelength traces dust; the N_2H^+ molecule disappears in the central regions where CO is found in gaseous form; the DCO^+ molecule, an isotopomer of HCO^+, has a distribution that is close to that of CO but is deficient in the central regions because deuterated molecules are scarce at relatively high temperatures; the distribution of CO is given approximately by the distribution of its non-optically thick isotopomer $C^{18}O$ (source: according to (Qi* et al. *2015)).*

However, we were able to indirectly determine the position of the H_2O ice line in the disk surrounding the very young star V883 Orionis with Alma (see Figure 2.13), thanks to a particular circumstance: the star is currently undergoing a very important luminosity surge, reaching 400 times that of the Sun, which is due to an important fall of circumstellar matter on the star. The dust is heated much farther from the star than usual, so much so that the ice line is very distant, further than Neptune's orbit in the Solar System, and becomes observable with Alma despite the great distance of the star. As many young stars go through such a super-luminous stage, this observation brings an additional complication to the problem of the formation of giant planets.

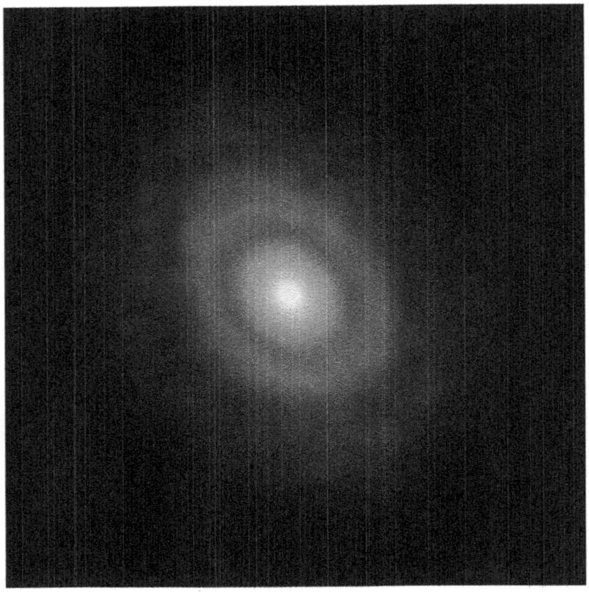

Figure 2.13. *Thermal radiation from the dust of the protoplanetary disk around the star V883 Orionis, observed with Alma at the frequency of 225 GHz. For a color version of this figure, see www.iste.co.uk/encrenaz/solar1.zip*

COMMENT ON FIGURE 2.13. – *The angular resolution is 0.03 seconds of a degree (12 au at the distance of the star). The H_2O ice line coincides with the inner dark ring, whose semi-major axis measures about 40 au, and where the dust temperature is about 10^5 K (source: Alma (ESO/NAOJ/NRAO)/L. Cieza).*

Figure 2.14 illustrates the evolution of a typical disk. The gas eventually disappears due to photoevaporation by the far ultraviolet radiation of the star, and only refractory solids remain inside the ice line and the ices beyond it. We are dealing with a disk of debris, where the grains have agglomerated to form larger solids that are difficult to detect. The intermediate stage corresponds to the so-called transition disks. However, there is a small amount of gas in older debris disks, where planets have already formed (Hugues *et al.* 2018). This gas is probably secondary, outgassed from the debris, as well as the fine dust sometimes found in the debris disks.

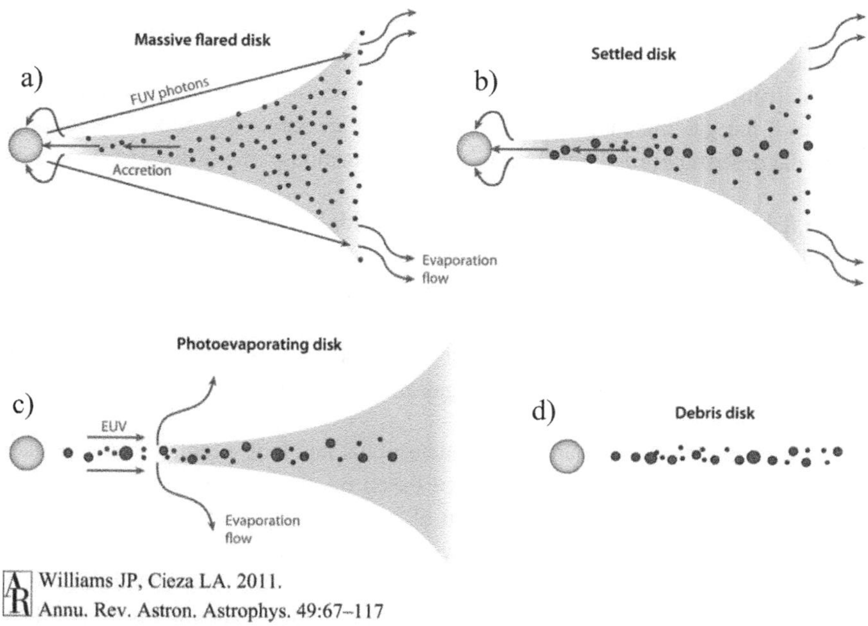

Figure 2.14. *Evolution of a circumstellar disk. For a color version of this figure, see www.iste.co.uk/encrenaz/solar1.zip*

COMMENT ON FIGURE 2.14. – *The gas is in blue and the dust in red. (a) The formed disk loses mass by accretion on the star and photoevaporation in its outer regions. (b) The size of the grains increases and they gather near the plane of symmetry. (c) The large amount of photoevaporation eliminates the gas. (d) The small grains have been driven away by the radiation pressure and carried away by the gas, and only large debris remains, the planetesimals (source: from (Williams and Cieza 2011)).*

We will not discuss the dynamics of disks here, which is a complex and still uncertain matter (Armitage 2011).

2.6. Formation of planetesimals and planetoids

The observation shows that grains grow in the planetary disks while there is still a lot of gas. Low-speed mutual collisions between grains first form aggregates, then larger and larger solid bodies, the planetesimals, by colliding mechanisms that have not yet been completely clarified (Blum and Wurm 2008).

An important element for this growth is the fact that gas and grains do not orbit at the same speed around the star. Indeed, the gas generally orbits at a speed slightly different from the Keplerian speed, because in the disk, the density and thus the pressure of the gas increases inwardly, producing an outwardly directed force which adds to the centrifugal force that is caused by the rotation of the disk.

The dust grains tend to orbit at Keplerian speed, but the smaller ones are strongly coupled with the gas by viscosity, and this has important consequences: their friction with the gas slows them down, causing them to migrate towards the interior of the disk and thus towards the star. For the planetesimals to form before the dissipation of the disk, the grains need to stick together fast enough to form larger bodies, of decimetric size, which will be less coupled with the gas and will not fall on the star.

The problem of the formation of planetesimals has long seemed insurmountable, but a solution has recently been proposed based on an idea put forward in 1972 by Fred Whipple: if there is a concentration of gas and dust in the disk, for example a dense ring as is often observed, the dust will not be able to come out since the pressure of the gas decreases inwardly, while the dust that drifts from the outside of the disk will accumulate there. In addition, large dust grains tend to drag the gas in its rotation, pushing it from the inside towards the dense zone and further stabilizing the system: this is called "feedback". A *dust trap* is thus formed. This can even form spontaneously, as shown in Figure 2.15. The dust or aggregates, which are now in a region of high density, will then have time to agglomerate to form the planetesimals.

Some planetesimals could also be formed from solids that are condensed at high temperature from the nebula gas during cooling of the disk, before the gas is dispersed (Dorn *et al.* 2019); similarly, pre-existing dust could be heated by stellar radiation until they are reduced to the less volatile compounds such as CaO, TiO and Al_2O_3, and then form planetesimals. The chemical composition of the planetesimals and the planets they would form would then depend on the condensation temperature. The diversity of the composition of carbonaceous chondrite meteorites could testify to this, but the radial migrations and mixing of planetesimals during the evolution of the early Solar System prevent any firm conclusion on this subject.

When the planetesimals reach a kilometer in size, they become what are called "planetoids": the forces of mutual gravitational attraction now dominate the electromagnetic forces and forces driven by gas, so that the planetoids continue to agglomerate by mutual collisions. This mechanism tends to accelerate, because large planetoids attract others more easily than small ones. However, this agglomeration seems much too slow to explain the formation in less than a million years of the nuclei of giant planets, which reach about ten solar masses. This problem has been solved by showing that a 500–1,000 km planetoid can grow by accreting solids – pebbles formed from grains – that are a few centimeters in diameter, and not by capturing other planetoids (Lambrechts and Johansen 2012): radio observations of protoplanetary disks show that these pebbles are numerous even at great distances from the star. The planetoid orbits around the star with a Keplerian speed, while the gas rotates faster or slower if it has a pressure gradient, as we have seen above. The pebbles are slightly carried along by the gas and thus pass close to the planetoid with a low relative velocity; they are then deflected by its field of gravity and can be captured. The small grains, on the other hand, are driven at too great a speed to be accreted, while the large stones are not driven by the gas and are in strict Keplerian revolution: they are therefore almost fixed with respect to the planetoid. Thus, only intermediate sized pebbles are captured. Numerical simulations show that the growth of a planetoid by this process is much faster than if it had been formed by the encounter and fusion of several large solids, all of which would have a Keplerian rotation.

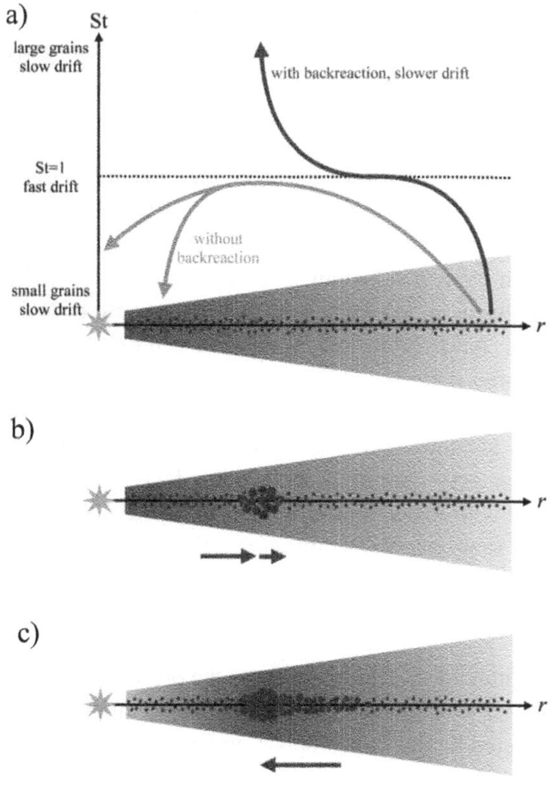

Figure 2.15. *Spontaneous formation of a dust trap, where dust can grow into planetesimals. For a color version of this figure, see www.iste.co.uk/encrenaz/solar1.zip*

COMMENT ON FIGURE 2.15. – *This is a sectioned view of the disk; the gas is blue and the dust is brown. In (1), the radial movement of the large grains is slowed down by their feedback on the gas. In (2), the large grains concentrate accordingly, while the gas is pushed outwards by the feedback (blue arrows). In (3), the trap is formed and captures the external gas and dust; it has the shape of a dense ring centered on the star (source: from (Gonzalez* et al. *2017)).*

The following events, in other words, the formation of the planets themselves, will be discussed in section 4.10 of Volume 2.

2.7. The environment of the Solar System at its birth

It is not possible to directly know what the environment of the Solar System was like at its birth, especially since its migration within the Galaxy means that it probably formed closer to the galactic center than today. We can, however, have some ideas about this environment based on some properties of the Solar System (Adams 2010). In particular, one can wonder whether it was born in isolation or as part of a star cluster.

If the Solar System was born in a cluster, this cluster might have affected the formation of the Sun and the planets through dynamic processes, such as disk truncation and perturbations in planetary orbits due to the close passage of stars. The truncation of the Solar System at a radius of 30 au could be due to the dynamical effect of nearby stars in a cluster, or to photoevaporation as a result of UV radiation from the cluster. The planets have orbits with small eccentricity, which eliminates the possibility of very large perturbations (we will see that this is not the case in many other planetary systems). However, some objects in the Kuiper belt show such perturbations, which are attributed to a hypothetical and very distant ninth planet, whose orbit would be very eccentric (see section 1.1): it could have been captured during an encounter with a neighboring planetary system. The comets of the Oort cloud are sometimes sent towards the interior of the Solar System under the effect of gravitational perturbations, perhaps due to the close passage of a star, but more likely as a result of density waves that travel through the Galaxy.

An interesting feature of the Solar System is the presence of short-lived radioactive isotopes, at the time of its formation. Of course, these isotopes have disappeared, but their presence is attested by their decomposition products found in meteorites, formed in the early stages of the Solar System's formation. They will be described in Chapter 3 of Volume 2, so we will only point out here that their existence indicates that a very short time – of the order of one million years – elapsed between their production and their incorporation into the matter that formed the Solar System. The most interesting is ^{26}Al, whose half-life is only 0.72 million years; its decomposition product is ^{26}Mg. ^{26}Al could have been produced by a supernova explosion or by the wind of a Wolf–Rayet star (a massive star at the end of its life) close to the Sun at its formation. These radioactive isotopes offer the most convincing evidence that the Solar System could have been born in a star cluster that has since dispersed.

2.8. Detection and properties of exoplanets and their systems

2.8.1. *First attempts*

Since we imagine that planets could exist around stars other than the Sun, we dream of discovering them. This is very difficult in practice, because these planets would be very close in angle to the star and millions of times less bright. However, if a planet and a star rotate around the center of gravity of the system, this can cause a small periodic displacement of the star in the sky. The first attempt to detect this motion involved the double star 70 Ophiuchi. In 1855, the two components of 70 Oph had almost completed one complete revolution around each other since their discovery by Herschel in 1779 (a period of 93 years), but it seemed difficult to represent the orbit by an ellipse. Was there a third body? This is what an English astronomer who ran the Madras Observatory in India, Captain William S. Jacob (1813–1862), claimed. He believed that the agreement between an elliptical orbit and the observed orbit could be improved by adding a body to the system, revolving in 26 years around the less massive of the two stars. Although his deduction turned out to be false, Jacob has the merit of being the first astronomer to have tried to discover an exoplanet. For nearly a hundred years, 70 Ophiuchi was observed by a whole series of astronomers, with contradictory results. Other stars are also followed, including Barnard's star, which is the closest to us after α Centauri and its neighbor Proxima Centauri. From 1938 to 1975, no less than 4,079 photographs of the field of this star were obtained by a Dutch astronomer, Peter van de Kamp (1901–1995), who worked at the Sproul Observatory in Pennsylvania. He deduced the presence of two planets orbiting around the star, with a mass slightly less than that of Jupiter. This would have been of the greatest interest if later observations had not invalidated this result. It is the same with all the results obtained on other stars, in particular those of the Danish astronomer Kaj Aa. Strand (1907–2000), who believed he had discovered four planets.

In fact, it was then practically impossible to detect exoplanets by classical astrometry: their effect is too weak. Other methods had to be tried. The first to have really understood this was an American astronomer of Russian origin, Otto Struve (1897–1963). In 1952, he wrote the following in *The Observatory*:

> One of the hottest questions in astronomy concerns the frequency of planetary objects in the Galaxy around stars other than the Sun.... There could be many such objects in the Galaxy.
>
> But how will we be able to detect them? Strand's direct photography method is, of course, excellent for close binary systems, but it is very limited[2]. It seems that there is, at present, no possibility to discover objects having the mass and size of Jupiter; and even then, there is not much hope to discover objects ten times more massive than Jupiter, if they are at distances of one or more astronomical units from the central star. However, there seems to be no reason why hypothetical planets cannot, in some cases, be much closer to their star than in the Solar System. It would be interesting to see if such objects exist. We know that stellar companions can be found at very small distances [from the star]. It is not unreasonable to think that a planet can exist at a distance of 1/50 astronomical unit, or 3 million km. Its period around a solar mass star would then be about 1 day.... Our hypothetical planet would have a velocity of about 200 km/sec. If its mass were equal to that of Jupiter, it would cause an oscillation of the radial velocity of the central star of ± 0.2 km/s, a quantity just about detectable with the most powerful Coudé spectrographs in existence. A planet ten times more massive than Jupiter would be very easy to detect, because it would cause oscillations of the radial velocity of the star of ± 2 km/s....
>
> There would also be eclipses [by the planet passing in front of the star]. Assuming that the density of the planet is five times that of the star (which could be optimistic for such a large planet), the eclipsed area would be about 1/50 of that of the star, and the decrease in light expressed in magnitudes would be 0.02. This too would be measurable with modern photometric methods, although the spectrographic test is probably more accurate.

Struve thus perfectly established the two main techniques that are used today to detect extrasolar planets. However, it does not seem that his visionary proposals had

2 The fact that all detections using astrometry were invalidated, especially those of Strand, was not yet known at the time.

much resonance in his time. It was probably thought that the eventual planetary systems had to resemble the Solar System, in which case detection would be impossible, even for a planet like Jupiter.

2.8.2. *The unexpected discovery of planets around a pulsar*

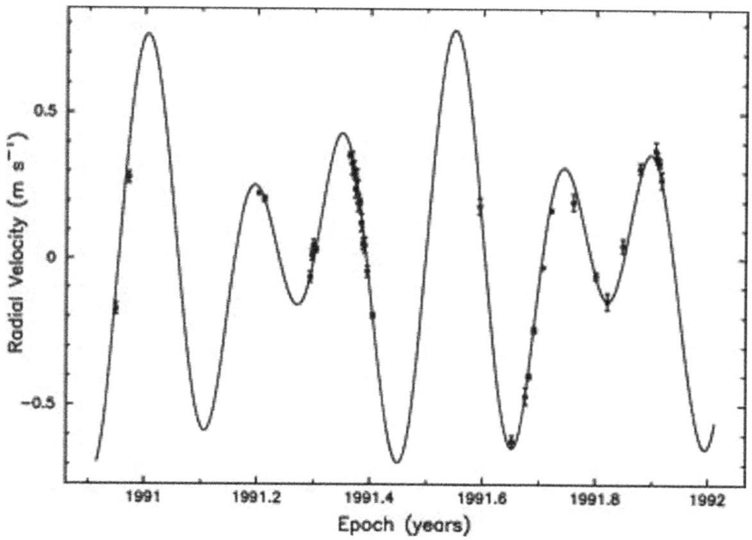

Figure 2.16. *Periodic variations of the radial velocity of the PSR B1257+12 pulsar over one year*

COMMENT ON FIGURE 2.16. – *The curve corresponds to the perturbations induced by the presence of two planets; the measurement points are indicated by bars. The fine-tuning of the measurement of the period variations of the pulsar not only allowed to detect a third planet, but also the very small effect of mutual perturbations of the first two, which somewhat modify their motion. It was thus possible to determine the period of the three planets (25.3, 66.5 and 98.2 days), the semi-major axis of their orbits (0.19, 0.36 and 0.56 au), their masses (0.020, 4.3 and 3.9 Earth masses, the last two are "super-Earths") and the inclination of the orbits on the sky plane (approximately 50°) (source: from (Wolszczan 2012)).*

Pulsars are rapidly rotating neutron stars that emit radio waves like a rotating lighthouse. Their period of rotation is extremely stable, so that a possible periodic variation of the apparent period of the received radio impulses indicates the presence of another celestial object orbiting around the pulsar: when it is moving towards or away from us, the apparent period changes due to the Doppler-Fizeau effect. Thus, in 1992, using the Arecibo radio telescope, the Polish astronomer Alexander Wolszczan discovered two exoplanets around the "millisecond" pulsar PSR B1257+12, whose rotation period is 0.0062 seconds (see Figure 2.16). A third planet, smaller and closer to the star, was discovered a little later. The formation of these planets, which seem quite rare, is poorly understood.

2.8.3. *Exoplanet detection methods*

To discover planets around stars that are more normal than pulsars, other methods are necessary. Astronomers first turned to the spectroscopic measurement of variations in the radial velocity of the star, called *velocimetry*. As in the case of astrometry, the principle is to detect the effect of the presence of a planet on the motion of the star: a star accompanied by a planet sees its motion modified periodically (see Figure 2.17). While astrometry tries to measure the lateral motion with respect to neighboring stars, velocimetry measures the radial velocity variations of the star, using the Doppler-Fizeau effect.

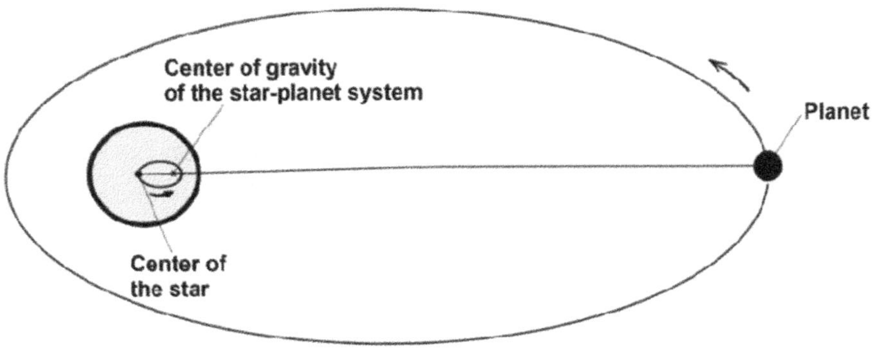

Figure 2.17. *Motion of a star and its planet around the center of gravity of the system. The center of the star defines an ellipse, which is the mirror image of the planetary orbit, reduced by a factor equal to the mass ratio. The center of gravity, which is fixed, is the common focus of these ellipses. For a color version of this figure, see www.iste.co.uk/encrenaz/solar1.zip*

As early as 1990, velocimetry made it possible to measure the radial velocity of a star with an accuracy of 10 m/s. However, in the case of the Solar System, Jupiter's motion induces a variation of the Sun's speed of 12.5 m/s with a period of 11.86 years, so that, with a lot of patience, it would have become possible to find a Jupiter around a nearby star. On the other hand, detecting an Earth 1 au from its star, which would only give a 9 cm/s variation in speed, was impossible at the time. It was therefore thought that the method would require long-term programs, with a long observation time and a perfectly stable spectrometer: in the case of Jupiter, it would be necessary to observe for at least a dozen years! It was discouraging: the method was only used to detect brown dwarfs around other stars, with increasingly powerful spectrographs. Today, these specialized spectrographs correlate the observed spectrum with a standard spectrum to measure the spectral shift, taking advantage of the gain in precision brought by the simultaneous observation of many lines. This works all the better the colder the star is, because its spectrum then contains a larger number of lines. The wavelength standard is carried out in Europe by the simultaneous measurement of the thorium spectrum. In the United States, it is preferred to superimpose, to the observed spectrum, the spectrum of a gas that produces many lines, iodine or hydrogen fluoride, in order to have a wavelength standard in the spectrum itself. These spectrographs, which must have a high stability over years, are fixed, placed on the ground in thermostatically controlled enclosures, possibly in a vacuum, and the light reaches them from the telescope through an optical fiber.

But luck would smile on the astronomers. While they imagined giant exoplanets far from their star, like Jupiter and Saturn, many planetary systems are very different: there are many giant exoplanets located in the immediate vicinity of their star. It is this totally unexpected fact, except for Otto Struve whose prediction had been forgotten, that would allow the Geneva astronomers Michel Mayor and Didier Queloz to discover the first exoplanet around a solar-type star in 1995. This planet, 51 Peg b, orbits the star 51 Pegasi (or 51 Peg a) at a distance of 0.052 au with a period of 4.23 days. Its mass is greater than or equal to 0.46 M_J, the mass of Jupiter[3].

The implications of this discovery are very important: not only is the Solar System not unique, but it cannot be taken as a model for other planetary systems. On the other hand, the planet 51 Peg b must be very hot and cannot have formed at the distance from the star where it is currently located. Therefore, it must have migrated inward from outside the ice line.

3 The masses $M_\odot$ of the Sun, M_J of Jupiter and M_T of the Earth are such that $M_\odot$ = 1,989 10^{30} kg = 1 047 M_J = 330 000 M_T.

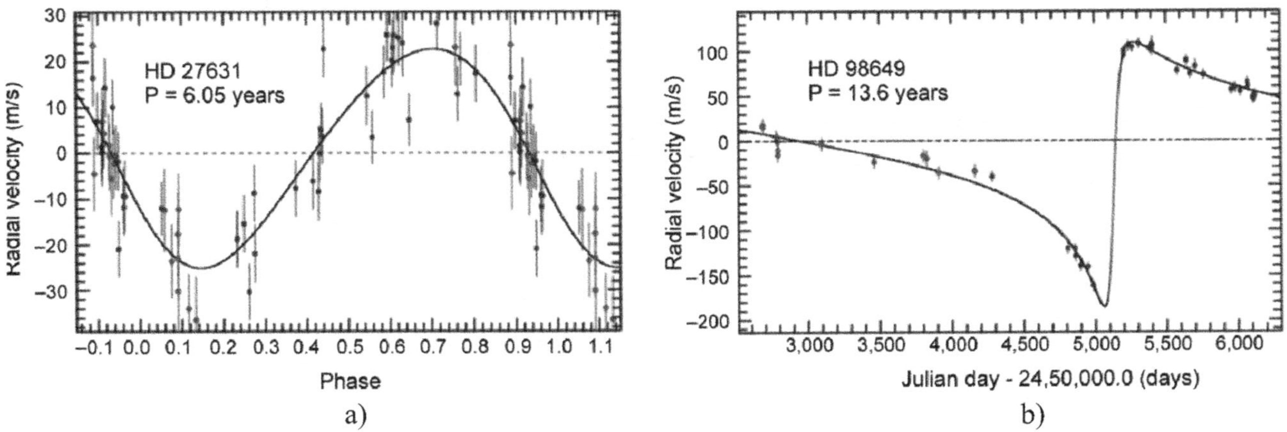

Figure 2.18. *Two examples of radial velocity curves for stars with planets of long period P. The radial velocity is in ordinate; (a) the phase in the revolution of the planet is in abscissa; (b) the revolution could not be totally observed, and the date of observation is in the abscissa. The orbit of HD 27631 b is slightly eccentric (e = 0.12) and the velocity curve is almost sinusoidal. For HD 98649 b, the orbit is very eccentric (e = 0.85) and the velocity curve is very distorted. For these planets, m sini is respectively 1.45 and 6.8* M_J *(source: based on (Marmier* et al. *2011)). For a color version of this figure, see www.iste.co.uk/encrenaz/solar1.zip*

What information does velocimetry provide us? The amplitude of the variation of the radial velocity of the star in the case of a circular orbit, inclined by the angle i on the line of sight, is written, by calling the mass of the star M and that of the planet m, supposedly very small in front of that of the star, T being the period of revolution and G the constant of gravitation:

$$\Delta v = (2\pi G)^{1/3}\, T^{-1/3}\, M^{-2/3}\, m \sin i \qquad [2.1]$$

or numerically:

$$\Delta v = 28.4 \text{ m/s } (T/\text{yr})^{-1/3}\, (M/\text{M}_{\odot})^{-2/3}\, (m \sin i)/\text{M}_J \qquad [2.2]$$

Knowing the period T, one can obtain the quantity m sin$i$ if one can determine the mass of the star, which is generally possible from the characteristics of its spectral type. Since i is not known, we only obtain a lower limit of the mass of the planet. There is, however, an exception: if one observes passages of the planet in front of the star, the observer is practically in the orbital plane and i is very close to 90°: the mass of the planet is then known. The third law of Kepler makes it possible to obtain the semi-major axis a of the orbit:

$$(a/\text{au})^3 = (M/\text{M}_{\odot})\, (T/\text{yr})^2 \qquad [2.3]$$

In the more general case of an elliptical orbit of eccentricity e, m sin$i$ is to be replaced by $m \sin i\, (1-e^2)^{-1/2}$. The eccentricity[4] can be deduced from the shape of the curve, which gives the radial velocity versus time (see Figure 2.18).

From the year 2000, another technique is proving its worth: the *transit method*, which allows the detection of an exoplanet passing in front of its star by the small decrease in light of the star, which can be measured at each passage. This relative decrease is obviously equal to the ratio of the square of the radius of the planet to the square of the radius of the star. If an observer outside the Solar System were to observe a transit of Jupiter in front of the Sun, the diameter of Jupiter being about one tenth of the solar diameter, the solar flux would be decreased by 1%. In the case of the Earth, whose diameter is 10 times smaller, the decrease would only be 0.01%. Techniques of stellar photometry make it possible to obtain an accuracy of the order of 0.1%, or slightly better, from terrestrial observatories, which makes it possible to detect giant exoplanets by transit from the Earth. Many research programs are currently devoted to these detections, often with small telescopes because the limitation is due more to the atmosphere than to the number of photons. However, the

4 The eccentricity e of an elliptical orbit is such that the distance from the center to one of the foci of the ellipse is ae, where a is the semi-major axis of the orbit.

detection of terrestrial type exoplanets – the *exo-Earths* – is only possible with terrestrial telescopes if the star is much smaller than the Sun. Only observation from space makes it possible to obtain the stability of measurements necessary for the photometric precision that is required for small exoplanets orbiting solar-type stars. The French space mission CoRoT, and then NASA's Kepler satellite, were implemented with this objective. Whatever the telescope, it is necessary to observe several transits because the variability of the star can lead to false detections, and moreover, one obtains the period of revolution of the exoplanet. It is also recommended, if possible, to observe the system by velocimetry, which brings a confirmation of the detection and very important additional information.

The observation of the transits thus gives the period of revolution of the exoplanet, as well as the semi-major axis of its orbit. Eventual velocimetry allows, as we have seen, the determination of the mass of the planet, this time without the uncertainty linked to the inclination i of the orbit, which is now close to 90°. One also obtains the radius of the planet if the radius of the star is determined, hence its density, which gives access to its physical characteristics (gaseous, icy or rocky) with some ambiguities.

When the system includes several planets, the study of the mutual interactions which modify the orbital parameters, thus the period of the transits, makes it possible to obtain their masses, at least to a certain extent. Several exoplanets have, moreover, been discovered by the perturbations they bring to the orbit of an already known planet: this method is called "TTV" (for Transit Timing Variations).

A particular class of transit of a planet in front of a star is the gravitational transit. This time, it is not in front of the star with which the planet is associated, but in front of a more distant star. In this case, general relativity predicts the curvature of light rays when they pass close to a massive object that is interposed between the source and us. The gravity field of the object can be compared to a lens that curves the path of light rays passing nearby, hence the name *gravitational lens* given to this phenomenon. If a star passes directly in front of a more distant star, the light of the latter is amplified by the time of the passage, the star which plays the role of the lens making the light rays coming from the distant star (Figure 2.19) converge towards the observer. Furthermore, if the star which plays the role of the lens is endowed with an exoplanet, the system behaves like a double lens and, in addition to the amplification due to the closer star, the light curve of the distant object presents a narrow peak characteristic of the presence of the exoplanet associated with it.

Gravitational transits allow the discovery of planets of all masses, including those close to the mass of the Earth, and at all distances from the central star. They can detect planets located at very large distances, several thousand parsecs. However, once the star and the planet have passed in front of the background source, the phenomenon cannot be re-observed; this is the main limitation of the method. The method has a great future, however, thanks to projects of deep and repeated mapping of portions of the sky from the ground, and in space.

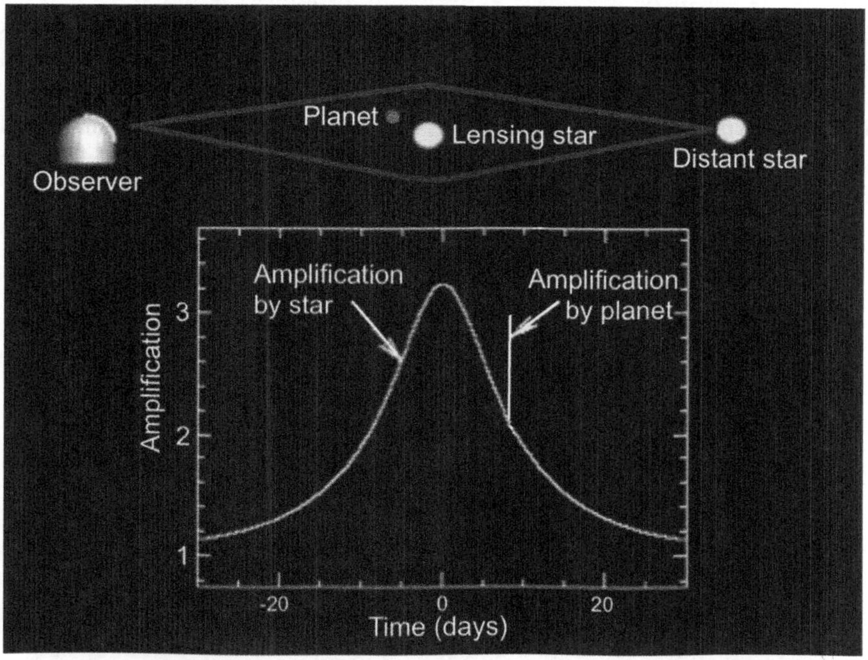

Figure 2.19. *Schematic diagram of the detection of an exoplanet by gravitational transit. For a color version of this figure, see www.iste.co.uk/encrenaz/solar1.zip*

COMMENT ON FIGURE 2.19. – *At the top, the deflection of light rays when the lens star is exactly aligned with the source star is shown (deflection by the planet is not shown). At the bottom, a light curve of the lens star can be seen, showing the peak associated with the presence of an exoplanet, when the star-planet assembly is moving downwards with respect to the source star.*

Lastly, a small number of exoplanets could be observed directly. This direct observation is very difficult: the star around which the planet orbits is enormously brighter, by a factor of about 10^9 in visible light, and the angular distance between the star and the planet is very small. Current techniques fortunately make it possible to partly overcome these obstacles. First of all, it is necessary to correct the image spread produced by atmospheric turbulence by adaptive optics. This is done by analyzing the image of a reference source (a natural star or an artificial star produced by laser illumination of a point in the Earth's upper atmosphere) close to the object to be observed, and by correcting the distortions of the wave plane in real time thanks to a deformable mirror placed on the optical path in the telescope: we thus obtain near-infrared images limited by the diffraction of the primary mirror of the telescope, which can be better than those of the Hubble Space Telescope that is placed outside the atmosphere. It is also necessary to occult the image of the star as well as possible, without occulting that of the planet: the technique used, derived from the *coronagraph* of Bernard Lyot (1897–1952), makes it possible to attenuate the infrared image of the star by a factor reaching 1.5 10^8. Figure 2.20 shows a recent example of direct observation of an exoplanet and its orbital displacement.

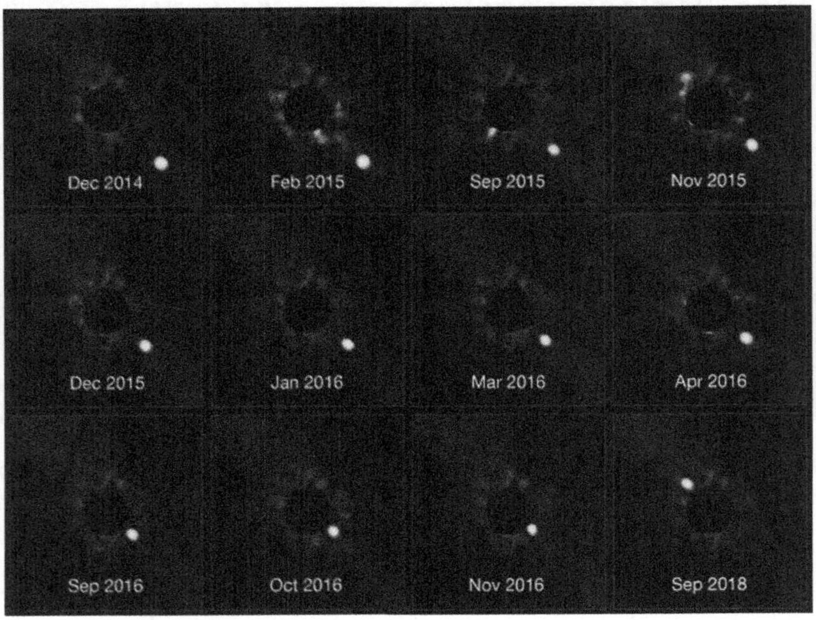

Figure 2.20. *The orbital displacement of the planet β Pictoris b, observed in near-infrared with ESO's Sphere instrument. For a color version of this figure, see www.iste.co.uk/encrenaz/solar1.zip*

COMMENT ON FIGURE 2.20. – *The side of each square measures 1 second of a degree. The star itself, in the center, is obscured. From November 2016 to October 2018, the planet was too close to the star to be separated from it. The planet has a mass of about 12.7 M_J and orbits in 20.3 years at 8.9 au from the star (source: ESO/Lagrange/SPHERE consortium).*

At the time of writing, 912 exoplanets have been observed by velocimetry, 3,407 by transit, 140 by gravitational transit and 23 by the TTV method; 145 have been seen directly. Astrometry, which once seemed so promising, has not yet provided any reliable detection. However, the results of the European astrometric satellite Gaia are eagerly awaited, which should detect a very large number of exoplanets by astrometry (Perryman *et al.* 2014), and also by transit because it has photometric capabilities, when enough observations will have been accumulated. Figure 2.21 compares the possibilities of detection by velocimetry and astrometry.

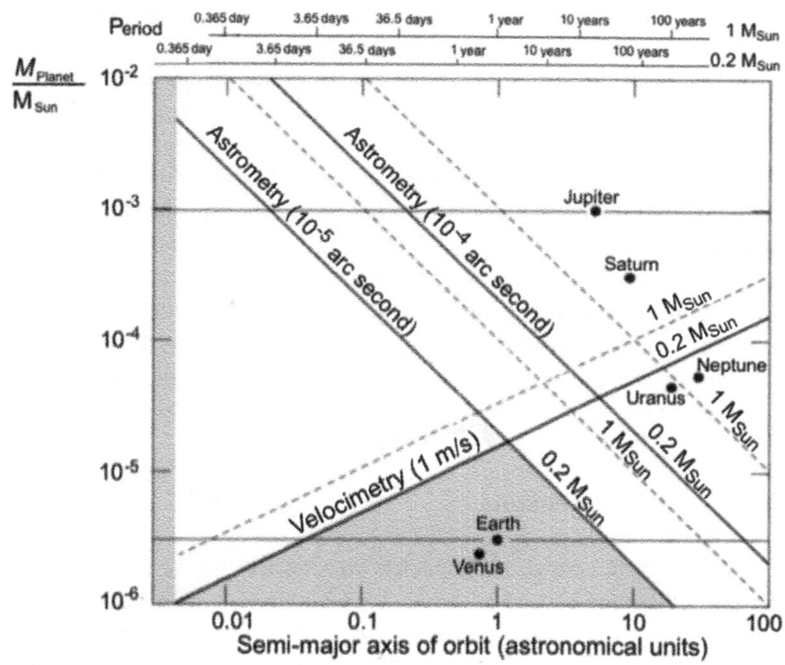

Figure 2.21. *The possibilities of exoplanet detection by dynamical methods*

COMMENT ON FIGURE 2.21. – *The mass of the planet is given as a function of the semi-major axis of its orbit, in logarithmic scales. The scale of the upper part gives the period of revolution as a function of the semi-major axis, around a star of 1 and 0.2* $M_\odot$*. The lines going up to the right, which correspond to the velocimetric method, are drawn for a half-amplitude of the radial velocity of the star of 1 m/s, for stars of 1 and 0.2* $M_\odot$*. In this case, the distance does not intervene. The descending lines, for the astrometric method, indicate a half-amplitude of the position of respectively* 10^{-4} *and* 10^{-5} *seconds of a degree, for a distance from the star of 10 parsecs and star masses of 1 and 0.2* $M_\odot$*. The first value could be reached from the ground, the second is reached* *with the Gaia satellite. An Earth with an orbit of 1 au radius around one of the nearest stars would be marginally detectable (source: according to the TOPS report).*

For astrometry, the measured quantity is the angular projection α of the semi-major axis a_p of the planetary orbit (assumed to be circular for simplicity):

$$\alpha = (m/M)\,(a_p/1\text{ ua})\,(d/1\text{ pc})^{-1}\text{ arc second} \qquad [2.4]$$

where m and M are respectively the masses of the planet and the star, and d the distance of the system. The distance d being well-known for these nearby stars, the measurement gives $(m/M)a_p$; if we manage to determine the period of revolution of the planet and if the mass M of the star is deduced from its spectral type, Kepler's third law (see equation [2.3]) gives a_p and, lastly, the mass m of the planet.

Each method of detecting exoplanets has its biases. For all of them, the detection of planets of terrestrial or lower mass is currently difficult. It is also difficult to detect planets with a long period of revolution. Velocimetry works all the better as the period is smaller, while the opposite is true for astrometry, but there are limits because observation with Gaia will not be extended beyond the lifetime of the satellite. Velocimetry does not depend on the distance of the star (however, the further away the star is, the weaker the signal is); on the other hand, astrometry directly depends on it since it is an angle that is measured. The same applies to direct observation.

2.8.4. *Some statistical results*

The most complete census of exoplanets was carried out by the Kepler satellite, which observed the same portion of the sky for several years in order to detect transits. The distribution of these planets according to their volume (the least biased of the quantities that can lead to statistics is the radius of the planet), brings major surprises: roughly 50% are super-Earths with volumes between 3 and 30 times that of the Earth. Larger planets make up just 30% of the total, of which only 12% have a volume larger than our planet Jupiter, whereas the first discoveries suggested that these large planets were the most frequent. The proportion of telluric planets with a volume less than three times that of the Earth (30%) is certainly very underestimated, as they are the most difficult to detect. Nevertheless, it should be kept in mind that all these planets have periods of revolution of less than a few years, because at least three transits must be observed to be sure of the result.

Kepler's census also showed that there is a 50% probability that a star has a planet in our Galaxy. This does not take into account the planets of terrestrial or lower mass, which are difficult to detect, and it seems certain that there are at least as many planets as there are stars in the Galaxy, thus more than 10^{11}. It is also necessary to add isolated planets, which were probably ejected from planetary systems, about which little is known, but which seem quite numerous both in star clusters and in complete isolation. These objects seem rather massive, but this is a selection effect and their separation from the brown dwarfs is quite blurred, because the smallest masses determined are of the order of 10 M_J. Some are surrounded by a disk.

These results are confirmed by the velocimetric study of a complete sample of 822 dwarf stars from M0 to F0 (Mayor *et al.* 2011): 155 planets belonging to 102 planetary systems have been discovered, in other words, in 12% of the stars. However, the detection is certainly very incomplete, especially for low mass planets. After correcting the biases, it is found that more than 50% of these stars have at least one planet with a mass less than 30 Earth masses and a period of less than 100 days. This fraction is certainly underestimated, because only very small numbers of planets have been detected with $m \leq M_T$. Moreover, seven giant planets of mass $m \geq M_J$ and long period (6 to 15 years) were later discovered in this sample (Marmier *et al.* 2011): Figure 2.18 shows the light curve for two of them.

2.8.5. *The diversity of exoplanets*

Figure 2.22 shows exoplanets whose mass could be determined in a mass/period diagram. This is not a complete sample, but a qualitative examination reveals the full diversity of exoplanets. The giant planets, of mass $m \geq M_J$, are divided into two groups. Those with a period longer than a hundred days are cold and resemble Jupiter, but many have masses up to 10 times greater than Jupiter. Those with a smaller period are close to the star and are therefore very hot: these are the *hot Jupiters*.

There are many planets whose mass is of the order of 10 M_T: these are the *super-Earths*, of which there is no equivalent in the Solar System (with the possible exception of a hypothetical 9th planet). Earths are certainly much more numerous than the figure suggests. Between the super-Earths and Jupiters, there are planets whose mass is reminiscent of Uranus and Neptune. They are often called *Neptunes*. Many, like the super-Earths, have periods of less than 100 days.

The Solar System had accustomed us to almost coplanar and near-circular orbits around the Sun. This is not the case for neighboring stars: exoplanets are often located in very elliptical orbits, whose eccentricity can be greater than 0.9 (see Figure 2.23). Planets of all masses have the full range of eccentricities; this fact has profound implications in the evolution of planetary systems.

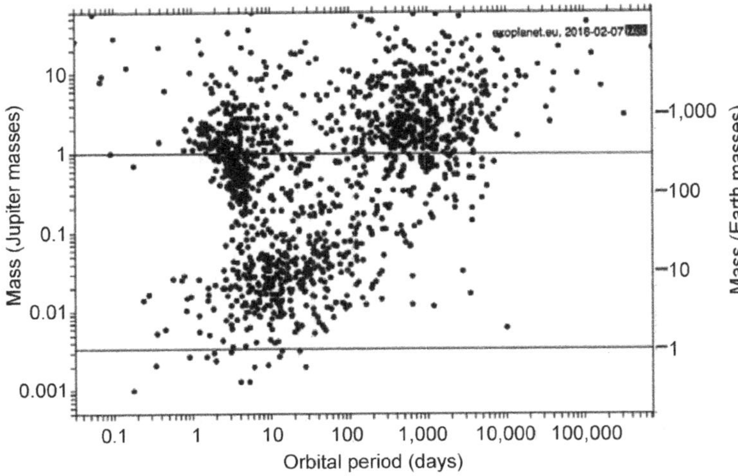

Figure 2.22. *Mass of exoplanets where this quantity could be determined, according to their orbital period; this figure shows very few low-mass planets due to selection effects*[5]

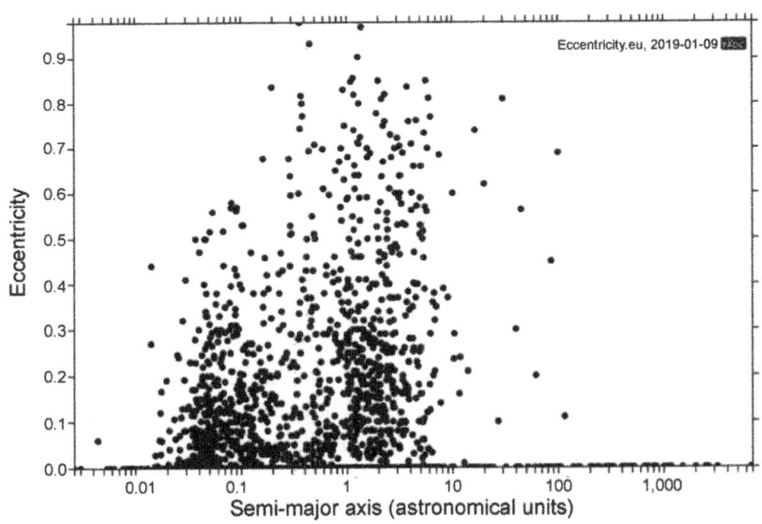

Figure 2.23. *Eccentricity of the orbits of exoplanets according to their semi-major axis*[6]

5 Source: from exoplanets.eu.catalog.

6 Source: from exoplanets.eu.catalog.

Figure 2.24 shows the density of exoplanets whose radius and mass could be determined, as a function of their mass.

The correlation observed between the density and mass of the giant planets is remarkable. It corresponds to an approximately constant radius (to about 30%), whatever the mass, which is predicted by theory for planets consisting essentially of hydrogen and helium, like Jupiter and Saturn. This shows that all hot and cold Jupiters are made up this way.

However, hot Jupiters have a larger radius than cold ones, because in this case, we can observe the radius of the gaseous envelope resulting from the evaporation of the planet.

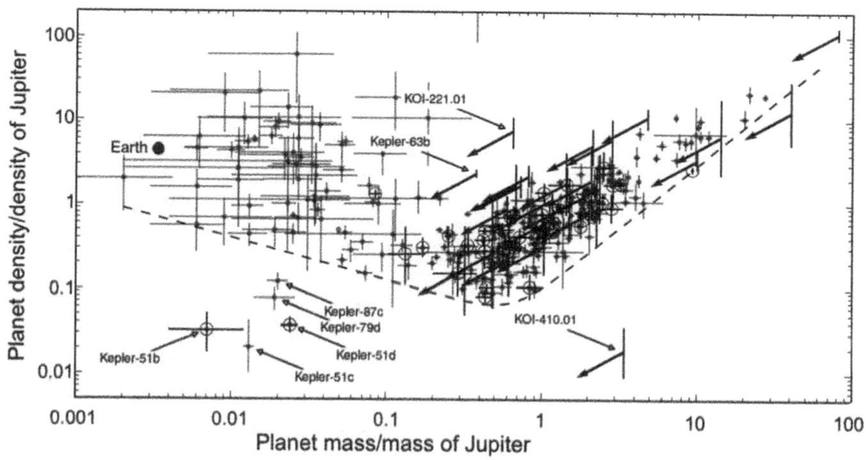

Figure 2.24. *Mean density of the planets according to their mass*

COMMENT ON FIGURE 2.24. – *The density is relative to that of Jupiter (1.33 g/cm^3), and the mass is in units of Jupiter mass (logarithmic scales). The arrows indicate planets for which only an upper limit of mass could be obtained. The circles are relative to a complete sample of giant planets, and the points to the other exoplanets, with error bars. The broken line is an empirical envelope of density for planets of different masses. Planets of very low density may be of a special nature, but their mass has been determined by the TTV method, the results of which are often uncertain (source: from (Santerne* et al. *2016)).*

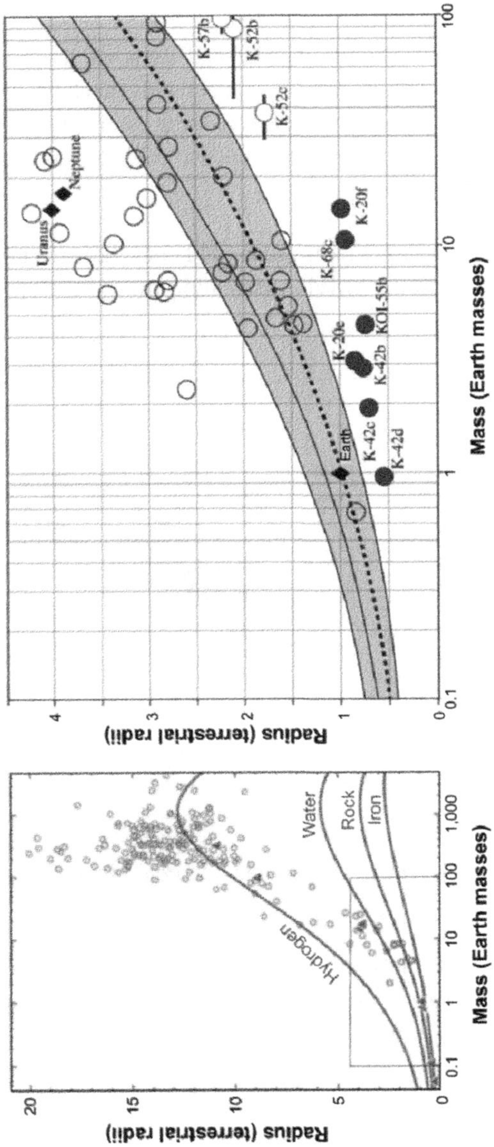

Figure 2.25. *Radius of exoplanets as a function of their mass. For a color version of this figure, see www.iste.co.uk/encrenaz/solar1.zip*

COMMENT ON FIGURE 2.25. – *(a) Mass-radius diagram of exoplanets observed by transit; the curves in blue show different internal composition patterns predicted for these planets. The planets of the Solar System are represented by green triangles (source: based on (Howard* et al. *2013)). (b) Magnification of the rectangle of the left diagram, with more data. From top to bottom, the curves correspond to the pure water planets, the ocean planets with 50% water and 50% silicates, the rocky planets of terrestrial composition (broken line) and the pure iron planets. The planets above the curves contain hydrogen gas. The data for the planets represented by the solid circles are unclear. It can be seen that existing models do not adequately account for the observed diversity of the planets (source: from (Mocquet* et al. *2014)).*

The planets of the Solar System fall into two quite distinct categories: the giants ($m > 10$ M_T) and the telluric ones. This dichotomy is simply explained with the nucleation formation scenario (see section 1.3.2.1), which predicts that an object with a mass greater than a dozen Earth masses has a sufficient gravity field to accrete the surrounding protoplanetary gas, composed essentially of hydrogen and helium; the object then becomes a giant planet. However, the mass distribution of all the exoplanets detected (see Figure 2.22) does not reflect this dichotomy: a continuous range of masses from M_T to 100 M_T is observed. This is not at all the simple scenario one might have expected. We can also observe small exoplanets, whose density reaches that of silicates and even that of iron. Located very close to their star, they can be hot enough (more than 2,000 K) that their surface facing the star is in the form of magma. This is particularly the case for CoRoT-7 b, one of the first small planets detected by transit. Other objects are even more astonishing, because their density exceeds that of iron. This is the case of Kepler-52b, Kepler-52c and Kepler-57c. They could be the residues of former giant planets, initially formed far away from their star, which would then have migrated towards their star by gradually getting rid of their gaseous envelope, only to be left with their initial core, which had been subjected to a high compression rate in the first phase of formation of these objects.

Other planets could have a peculiar chemical composition, if they were formed from planetesimals condensed at high temperature from nebula gas before this gas was dispersed (Dorn *et al.* 2019), or from interstellar dust deprived of their most volatile compounds by the heating due to the protostar. If the temperature is lower than 1,200 K, an Earth-like planet, rich in Mg, Si and Fe would be found; if it is much higher, the planet would be rich in Ca and Al, the first solids to condense as CaO and Al_2O_3, and it would be deficient in Fe. Some super-Earths very close to the star, whose specific mass seems abnormally low, could belong to this class. One could jokingly say that these planets are covered with sapphire and ruby. However, is there enough material available to form these super-Earths?

2.8.6. *Exoplanet atmospheres*

The planetary transits observed by spectroscopy provide valuable information on the planet's possible atmosphere. By subtracting the spectrum of the star outside the transit from the spectrum during the transit, we obtain the absorption spectrum of the planet's atmosphere at the terminator, in other words, the night-day transition.

One can also make the difference between the spectrum of the star + planet sum when the planet is going to pass or has passed behind the star, and the spectrum of the star when the planet is hidden behind it (secondary transit), acquiring the emission spectrum of the illuminated side of the planet. In the case of a primary transit, the signal is stronger if the planet is larger, less dense and closer to its star; in the case of a secondary transit, it is mainly the temperature of the planet that plays a role, so the signal is strong for very hot planets.

In all cases, small stars provide the largest signals: thus, cold stars (K and M types) are preferred. However, current measurements are still rather rudimentary given the weakness of the signal.

The first spectroscopic measurements in primary transit were performed in the early 2000s, in the ultraviolet and visible ranges with the spectrometers of the Hubble Space Telescope, as well as from the ground in the visible range. This is how sodium and potassium atoms, as well as oxygen, carbon, and hydrogen were detected, escaping from the exosphere of hot Jupiters such as HD209458 b and HD189733 b, the two most easily observable exoplanets because of their relatively close proximity to us. These measurements also revealed the presence of clouds and mists, attributed, given the high temperature of these planets, to the oxides of titanium TiO and vanadium VO in solid form.

A little later, the infrared range, which is better adapted to the detection of neutral molecules, became accessible to transit observations thanks to two very powerful instruments: the NICMOS-WFC3 spectrometers of the Hubble Space Telescope, in the near-infrared, and IRAC-IRS-MIPS of the Spitzer Space Telescope, up to the mid-infrared (10 μm). Water, methane, ammonia, carbon dioxide and/or carbon monoxide molecules have been detected in about 20 exoplanets, mainly hot Jupiters. An example is shown in Figure 2.26.

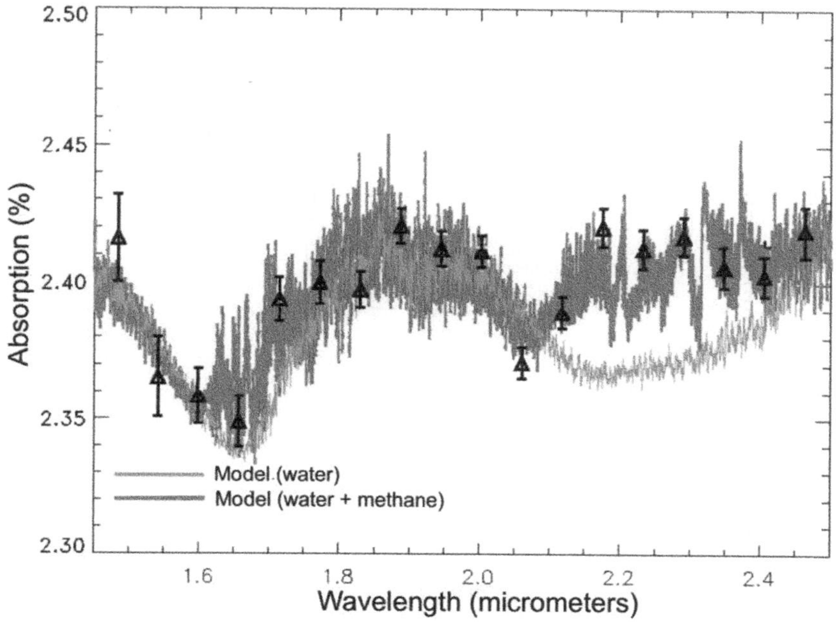

Figure 2.26. *Detection of water vapor and methane in the atmosphere of hot Jupiter HD 189733 b by transmission spectroscopy with the HST. The black triangles are the measurement points. The curves represent models of the atmosphere containing H_2O (blue) and H_2O + CH_4 (orange) (source: from (Swain* et al. *2008)). For a color version of this figure, see www.iste.co.uk/encrenaz/solar1.zip*

Some observations by primary transit have been made on objects smaller than hot Jupiters; these are possible when the host star also has a small radius. This is particularly the case of GJ 1214 b, a super-Earth in close orbit around a M star; its equilibrium temperature, of the order of 600 K, suggests that it could contain water vapor. However, the observed spectrum is surprisingly devoid of structure, probably due to mists.

An example of an emission spectrum obtained in secondary transit is shown in Figure 2.27. The interpretation of these spectra is more complicated, but also richer in information than that of absorption spectra, because it gives information on the temperature of the planet and its atmosphere. The spectrum depends on the temperature of the emitting region; if the exoplanet has a stratosphere with temperature inversion (which becomes higher above a certain altitude), molecular bands may appear in emission, possibly superimposed on bands formed in absorption, deeper in the stratosphere.

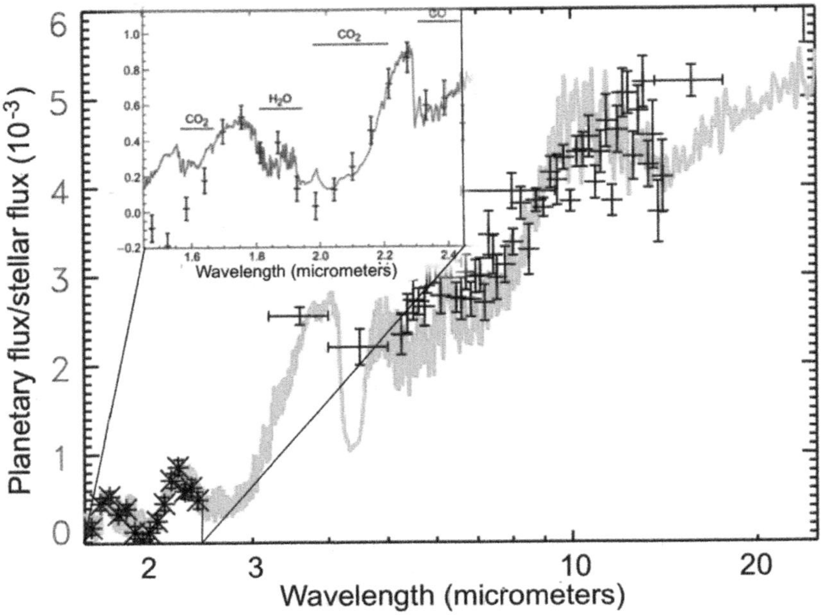

Figure 2.27. *Example of an emission spectrum obtained on the exoplanet HD 189733 b. For a color version of this figure, see www.iste.co.uk/encrenaz/solar1.zip*

COMMENT ON FIGURE 2.27. – *The green and red curves correspond to the predictions of an atmosphere model, and the points with error bars to the observations. The enlarged part, in linear wavelength scale, is relative to the data obtained in the near-infrared with the Hubble Space Telescope. The mid-infrared data come from the Spitzer satellite (source: from (Tinetti* et al. *2013)).*

An original method has allowed the detection of the CO molecule in the hot Jupiter HD 209458 b (Snellen *et al.* 2010), from the ground in the near-infrared. It consists of measuring the Doppler-Fizeau effect of the CO lines of the planet, due to its revolution around the star, with a very large telescope and a very high spectral resolving power. The method also provides an estimate of the atmospheric wind speed between the day and night sides of the planet. The same method has allowed the detection of CO and H_2O on other objects, including non-transiting exoplanets such as τ Boo b and 51 Peg b. Finally, using instruments such as Sphere at ESO, it is now possible to obtain images of planets that are located a few dozen or a few hundred astronomical units away from their star, and then perform low-resolution direct spectrophotometry.

2.8.7. *Habitable planets*

We do not know how life appeared on Earth, but we have been able to identify some factors favorable to its emergence (see Chapter 5 of Volume 2): the presence of a source of energy and the presence of liquid water on a solid surface, to allow the development of a complex chemistry within the liquid medium. In the absence of other information, astronomers in search for extraterrestrial life will prefer planets where the conditions for the appearance of life are present. These are necessarily Earths or super-Earths. Indeed, beyond a mass of about 10 M_T, the gravity field of the object at its formation would be sufficient to capture gas from the surrounding protostellar nebula, and it would become a giant planet not very conducive to life. The presence of liquid water implies that the surface temperature of the planet is between 0°C and about 100°C (273 and 373 K), with the upper limit depending on pressure. Planets that meet these criteria are said to be habitable, and are located in a zone of distance from the star such that their temperature is adequate: this is the *habitable zone*. In the case of the Solar System, it is centered around 1 au; it is closer to the star if the star is less bright (which is the case of stars K and M, which represent the great majority of the stellar population) and further away in the opposite case, for massive and brighter stars, which are rare (see Figure 2.28). Since the Sun's luminosity has increased by about 30% since its formation, the habitable zone has moved outward since the birth of the planets.

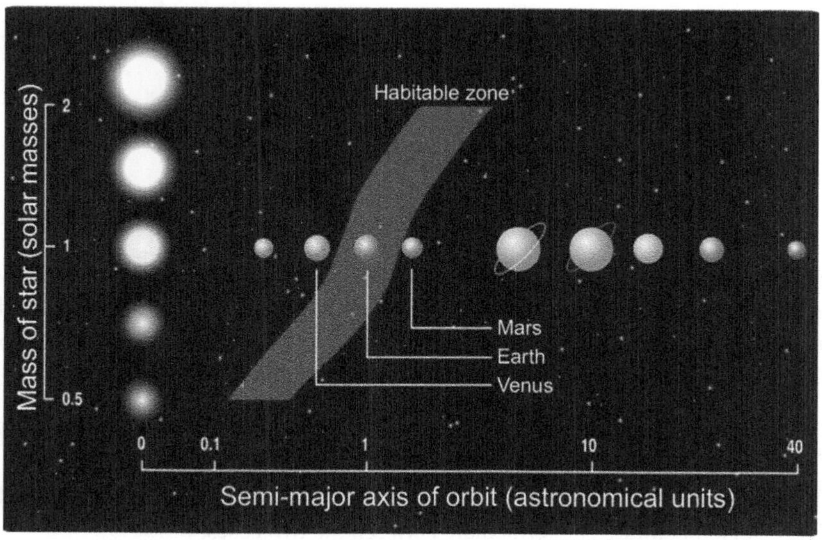

Figure 2.28. *Habitable zone of a star according to its mass. Its extension may vary according to the age of the star and according to the criteria chosen to define its limits (source: from Wikimedia Commons). For a color version of this figure, see www.iste.co.uk/encrenaz/solar1.zip*

It is obvious that the habitable zone is only a criterion indicating favorable conditions for life, and in no way implies that life will appear there. On the other hand, one cannot exclude the presence of life in environments such as oceans of possibly salt water, which are probably present inside certain satellites (Europa, Enceladus and perhaps Ganymede).

The equilibrium temperature of an exoplanet can be estimated as follows, in order to determine whether or not it is in the habitable zone (see also section 1.3.1). R_* and T_* being the radius and the effective temperature of the star, the flux received by the planet at distance a per unit area, perpendicular to the direction of the star, is $\sigma T_*^4 R_*^2/a^2$, where σ is the Stefan-Boltzmann constant. The planet only captures the fraction 1-A of the radiation of the star, A being its albedo, on a surface πR_p^2, R_p being its radius; at thermal equilibrium, if it rotates fast, it emits an energy σT_p^4 per unit area equally over its entire surface, that is to say, a total of $4\pi R_p^2 \sigma T_p^4$. We thus have the following:

$$(1\text{-}A)\pi R_p^2 \sigma T_*^4 R_*^2/a^2 = 4\pi R_p^2 \sigma T_p^4 \qquad [2.5]$$

or by simplifying:

$$(1\text{-}A) T_*^4 R_*^2/a^2 = 4 T_p^4 \qquad [2.6]$$

and thus:

$$T_p = (1\text{-}A)^{1/4} (R_*/2a)^{1/2} T_* \qquad [2.7]$$

and numerically, if R_* is in solar units and a in astronomical units:

$$T_p = 0{,}0482\ (1\text{-}A)^{1/4} (R_*/a)^{1/2} T_* \qquad [2.8]$$

If the planet is in synchronous rotation with its revolution, which is generally the case if its period of revolution is less than several dozen days (the value of this limit depends on the nature of the planet), it emits only, in principle, through half of its surface and the temperature of this illuminated half is numerically:

$$T_p = 0{,}0573\ (1\text{-}A)^{1/4} (R_*/a)^{1/2} T_* \qquad [2.9]$$

However, there is a partial redistribution of temperature by the winds, and the actual mean temperature is intermediate between the two cases.

T_p is not very sensitive to the exact value of the albedo A. However, A can be very different from one planet to another, and is difficult to assess. For example, the bolometric albedos (in other words, the fraction of reflected solar energy) of Mercury,

Venus, Earth, Mars, and Jupiter or Saturn are 0.12, 0.75, 0.31, 0.25 and 0.34 respectively. If our planets were illuminated by the main infrared radiation of an M star, their albedos would be different: for example, a rocky planet illuminated by such a cold star must have an albedo close to 0. Theoretically speaking, one could determine the temperature and the albedo of exoplanets by observing them near the secondary transit, but for this, the measurement of their flux would not only have to be made in the infrared, but also in the visible range, which is not currently possible.

On the other hand, the temperature can be increased by the greenhouse effect: for example, the greenhouse effect of H_2O and incidentally CO_2 and CH_4, raises the temperature of the Earth from 254 K to 288 K, while CO_2 raises the temperature of the surface of Venus from 328 K to 730 K.

Let us now compare the estimated temperature of the exoplanet with the condensation temperature of the water vapor at low pressure, which is of the order of 200 K. If it is lower, the planet is an icy object. If it is higher, we are dealing with a rocky planet, which could be habitable if it is not too massive and has an atmosphere and water. If it is massive, it is a hot Neptune or Jupiter, neither of which are habitable.

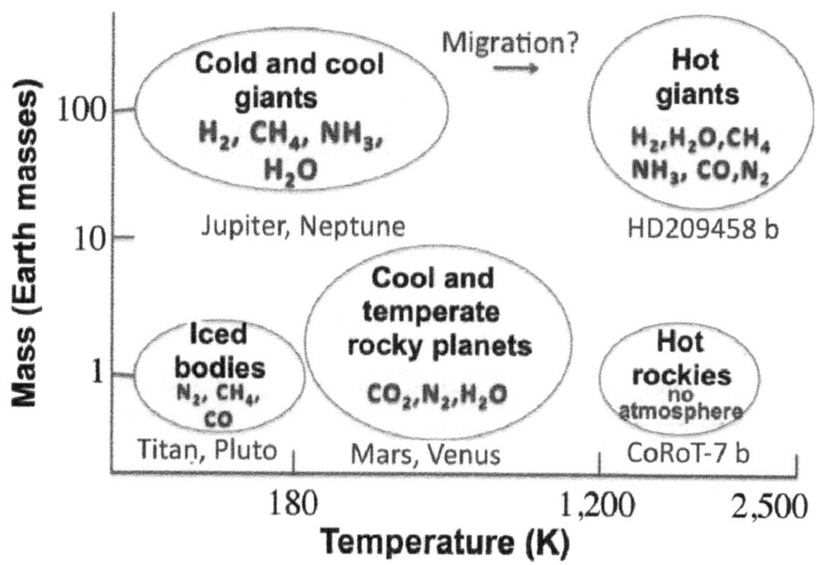

Figure 2.29. *Classification of exoplanets according to their mass and temperature. Young planets in the process of cooling are missing. Giant planets can become hot as they approach the star, through a migration phenomenon that will be discussed in Chapter 4 of Volume 2*

If the temperature is high and the pressure is low, the chemical equilibrium predicts that the atmosphere is essentially composed of CO, N_2, H_2O and CO_2 formed by CO + $H_2O \rightarrow CO_2 + H_2$. If the temperature is low and the pressure is high, one would mainly be able to find CH_4 and NH_3 and possibly H_2O, in addition to H_2 and He. Figure 2.29 illustrates the classification of exoplanets according to their temperature and mass.

2.8.8. *Some extrasolar planetary systems*

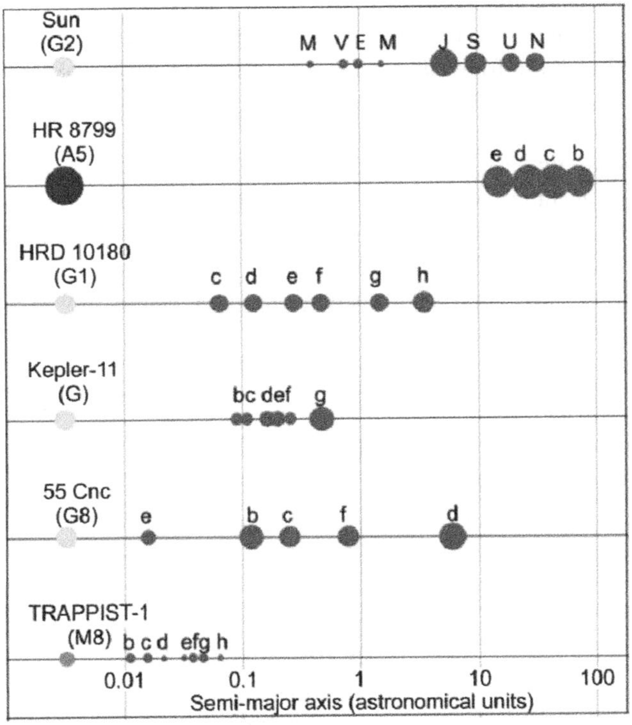

Figure 2.30. *Some exoplanetary systems, compared to the Solar System (top). Stars are shown on the left, with their spectral type. The planets are placed along the semi-major axis of their orbit (logarithmic scale), the size of the circle qualitatively indicating their mass. For a color version of this figure, see www.iste.co.uk/encrenaz/solar1.zip*

Exoplanet systems are so varied that it is difficult to make a selection, and even more to explain their origin. Figure 2.30 shows some of them as examples. Note, in particular, the very small system of seven Earths around the red dwarf star

TRAPPIST-1, two members of which are in the habitable zone. For more details, see the Exoplanet site[7] and (Lequeux *et al.* 2017).

2.9. References

Adams, F.C. (2010). The birth environment of the Solar System. *Annual Review of Astronomy and Astrophysics*, 48, 47–85.

André, P. (2002). The initial conditions for protostellar collapse: Observational constraints. In *Star Formation and the Physics of Young Stars – Vol. 3*, Bouvier, J., Zahn, J.-P. (eds). EDP Sciences, Les Ulis.

Armitage, P.J. (2011). Dynamics of protoplanetary disks. *Annual Review of Astronomy and Astrophysics*, 49, 195–236.

Bachiller, R. (1996). Bipolar molecular outflows from young stars and protostars. *Annual Review of Astronomy and Astrophysics*, 34, 111–154.

Bally, J. (2016). Protostellar outflows. *Annual Review of Astronomy and Astrophysics*, 54, 491–528.

Blottiau, P., Bouquet, S., Chieze, J.P. (1988). An asymptotic self-similar solution for the gravitational collapse. *Astronomy & Astrophysics*, 207, 24–36.

Blum, J. and Wurm, G. (2008). The growth mechanisms of macroscopic bodies in protoplanetary disks. *Annual Review of Astronomy and Astrophysics*, 46, 21–56.

Bresnahan, D., Kirk, J.M., Ward-Thompson, D., Pattle, K., Eyres, S., White, G.J., Könyves, V., Menshchikov, A., André, P., Schneider, N. *et al.* (2018). The dense cores and filamentary structure of the molecular cloud in Corona Australis: Herschel SPIRE and PACS observations from the Herschel Gould Belt survey. *Astronomy & Astrophysics*, 615(A125).

Dorn, C., Bonsor, A., Hands, T.O., Harrison, J.H.D. (2019). A new class of Super-Earths formed from high-temperature condensates: HD219134 b, 55 Cnc e, WASP-47 e. *Monthly Notices of the Royal Astronomical Society*, 484, 712–727.

Ferreira, J., Cabrit, S., Dougados, C. (2006). Which jet launching mechanism(s) in T Tauri stars? *Astronomy & Astrophysics*, 453, 785–796.

Gonzalez, J.-F., Laibe, G., Maddison, S.T. (2017). Self-induced dust traps: Over-coming planet formation barriers. *Monthly Notices of the Royal Astronomical Society*, 472, 1162.

Hartman, L., Calvet, N., Herczeg, G. (2016). Accretion onto pre-main-sequence stars. *Annual Review of Astronomy and Astrophysics*, 54, 135–180.

7 exoplanet.eu/catalog/.

Howard, A.W., Sanchis-Ojeda, R., Marcy, G.W., Johnson, J.A., Winn, J.N., Isaacson, H., Fischer, D.A., Fulton, B.J., Sinukoff, E., Fortney, J.J. (2013). A rocky composition for an Earth-sized exoplanet. *Nature*, 503, 381–384.

Hugues, A.M., Duchêne, G., Matthews, B.C. (2018). Debris disks: Structure, composition, and variability. *Annual Review of Astronomy and Astrophysics*, 56, 541–591.

Lallement, R., Capitanio, L., Ruiz-dern, L., Danielski, C., Babusiaux, C., Vergely, L., Elyajouri, M., Arenou, F., Leclerc, N. (2018). Three-dimensional maps of interstellar dust in the Local Arm: Using Gaia, 2MASS and APOGEE-DR14. *Astronomy & Astrophysics*, 616(A132).

Lambrechts, M. and Johansen, A. (2012). Rapid growth of gas-giant cores by pebble accretion. *Astronomy & Astrophysics*, 544(A32).

Lequeux, J. (2005). *The Interstellar Medium*. Springer, Berlin.

Lequeux, J. (2013). *Birth, Evolution and Death of Stars*, EDP Sciences, Les Ulis and World Scientific, Singapore.

Lequeux, J., Encrenaz, T., Casoli, F. (2020). *The Exoplanet Revolution*. EDP Sciences, Les Ulis.

Maeder, A. (2009). *Physics, Formation and Evolution of Rotating Stars*. Springer, Berlin.

Marmier, M., Ségransan, D., Udry, S., Mayor, M., Pepe, F., Queloz, D., Lovis, C., Naef, D., Santos, N.C., Alonso, R. *et al.* (2011). The CORALIE survey for southern extra-solar planets XVII. New and updated long period and massive planets. *Astronomy & Astrophysics*, 551(A90).

Mayor, M., Marmier, M., Lovis, C., Udry, S., Ségransan, D., Pepe, F., Benz, W., Bertaux, J.-L., Bouchy, F., Dumusque, X. *et al.* (2011). The HARPS search for southern extra-solar planets XXXIV. Occurrence, mass distribution and orbital properties of super-Earths and Neptune-mass planets. Available at: https://arxiv.org/ abs/1109.2497.

McKee, C.E. and Ostriker, E.C. (2007). Theory of star formation. *Annual Review of Astronomy and Astrophysics*, 45, 565–687.

Mocquet, A., Grasset, O., Sotin, C. (2014). Very high-density planets: A possible remnant of gas giants. *Philosophical transactions of the Royal Astronomical Society A*, 372(2013-064).

Murray, N. and Chang, P. (2015). Star formation in self-gravitating turbulent fluids. *Astrophysical Journal*, 804(1).

Perryman, N., Hartman, J., Bakos, G.A., Lindegren, L. (2014). Astrometric exoplanet detection with GAIA. *Astrophysical Journal*, 798(14).

Qi, C., Öberg, K.I., Andrews, S.M., Wilner, D.J., Bergin, E.A., Hughes, A.M., Hogherheijde, M., D'Alessio, P. (2015). Chemical imaging of the CO snow line in the HD 163296 disk. *Astrophysical Journal*, 813(2).

Rygl, K.L.J., Benedettini, M., Schisano, E., Elia, D., Molinari, S., Pezzuto, S., André, P., Bernard, J.P., White, G.J., Polychroni, D. *et al.* (2013). Recent star formation in Lupus as seen by Herschel. *Astronomy & Astrophysics*, 549(L1).

Santerne, A., Moutou, C., Tsantaki, M., Bouchy, F., Hébrard, G., Adibekyan, V., Almenara, J.-M., Amard, L., Barros, S.C.C., Boisse, I. *et al.* (2016). SOPHIE velocimetry of Kepler transit candidates. XVII. The physical properties of giant exoplanets within 400 days of period. *Astronomy & Astrophysics*, 587(A64).

Schaller, G., Schaerer, D., Meynet, G., Maeder, A. (1992). New grids of stellar models from 0.8 to 120 solar masses at Z = 0.020 and Z = 0.001. *Astronomy & Astrophysics Supplement Series*, 96, 269.

Sicilia-Aguilar, A., Banzatti, A., Carmona, A., Stolker, T., Kama, M., Mendigutía, I., Garufi, A., Flaherty, K., van der Marel, N., Greaves, J. (2016). A "Rosetta Stone" for protoplanetary disks: The synergy of multi-wavelength observations. *Publications of the Astronomical Society of the Pacific*, 33, 59.

Snellen, I.A.G., De Kok, R.J., De Mooij, E.J.W., Simon, A. (2010). The orbital motion, absolute mass and high-altitude winds of exoplanet HD 209458 b. *Nature*, 465, 1049–1051.

Sugitani, K., Nakamura, F., Watanabe, M., Tamura, M., Nishiyama, S., Nagayama, T., Kandori, R., Nagata, T., Sato, S., Gutermuth, R.A. *et al.* (2011). Near-infrared imaging polarimetry toward Serpens South: Revealing the importance of the magnetic field. *Astrophysical Journal*, 734(1).

Swain, M.R., Vasusht, G., Tinetti, G. (2008). The presence of methane in the atmosphere of an extrasolar planet. *Nature*, 452, 329–331.

Tinetti, G., Encrenaz, T., Coustenis, A. (2013). Spectroscopy of planetary atmospheres in our Galaxy. *Astronomy & Astrophysics Reviews*, 21(63).

Whittet, D.C.B., Schutte, W.A., Tielens, A.G.G.M., Boogert, A.C.A., de Graauw, T., Ehrenfreund, P., Gerakines, P.A., Helmich, F.P., Prusti, T., van Dishoeck, E.F. (1996). An ISO SWS view of inter-stellar ices: First results. *Astronomy & Astrophysics*, 315(L357).

Williams, J.P. and Cieza, L.A. (2011). Protoplanetary disks and their evolution. *Annual Review of Astronomy and Astrophysics*, 49, 67–117.

Wolszczan, A. (2012). Discovery of pulsar planets. *New Astronomy Reviews*, 56(1), 2–8.

3

The Interaction of Solar System Bodies with the Interplanetary Medium[1]

Laurent LAMY

LESIA, Paris Observatory, PSL University, Paris, France

The different objects of the Solar System interact directly with the interplanetary medium. This chapter describes the physical processes that govern the main types of interaction and the associated exchanges of energy, matter and momentum between the different media.

The interplanetary medium includes four components: dust from the Solar System or the interstellar medium, cosmic radiation (flux of very high energy atomic nuclei) of solar, galactic or extragalactic origin, neutral gas of interstellar origin and, lastly, the solar wind. The latter is a plasma, mainly composed of H^+, He^{++} ions and electrons, produced and accelerated in the solar corona from where it extends into the interplanetary medium. The region of space where the solar wind flows forms the heliosphere, a plasma cavity embedded in the interstellar medium that extends well beyond the Kuiper belt.

After recalling the properties of the solar wind and the heliosphere (section 3.1), this chapter will describe the different types of interactions between the solar wind and Solar System bodies, which vary according to the nature of their gas envelope

1 This chapter has been adapted and updated from the one written in French by P. Zarka and M. Blanc in *Le Système solaire* (Encrenaz *et al.* 2003, Chapter 4).

The Solar System 1,
coordinated by Thérèse ENCRENAZ and James LEQUEUX.

and the presence of an intrinsic magnetic field (section 3.2). When the intrinsic magnetic field is strong enough, a magnetosphere is shaped around the object, a large-scale magnetic cavity dug in the solar wind. Non-magnetized conductive bodies possess an induced magnetosphere instead. The activity of planetary magnetospheres can be studied *in situ* with space exploration probes and remotely via their auroral electromagnetic emissions (section 3.3).

3.1. Interplanetary plasma: origin and properties of the solar wind

3.1.1. *Coronal expansion*

Beyond a few solar radii, the plasma of the solar corona, the outermost part of the solar atmosphere, reaches considerable temperatures, from 10^5 to 10^6 K, much higher than that of the photosphere at 6,000 K. The resulting pressure of the plasma is such that it can no longer be contained by the coronal magnetic field. Part of the solar magnetic tubes opens onto interplanetary space, forming a nozzle along which the corona plasma can escape as a supersonic flow: this is the solar wind.

The coronal structure can be studied from images from space in extreme UV, which allows the temperature to be probed, or in soft X-rays, mainly produced by energetic electrons, as illustrated in Figure 3.1a. The bright areas map closed magnetic arches, populated by a dense thermal plasma, observed in equatorial and mid-latitude regions. At higher latitudes, the dark regions correspond to magnetic tubes that are open to the interplanetary medium and filled with a plasma about three times less dense than that of the magnetic arches: these are the coronal holes. The corona cannot be observed as easily in the visible range, due to an overly-weak contrast with the photosphere. When the solar disk is occulted during eclipses, or with the help of coronagraphs, one can nevertheless directly observe the external part of the corona which continuously flows into interplanetary space, as shown in Figure 3.1b.

This expansion of the coronal plasma forms the solar wind, a completely ionized gas whose chemical composition is close to that of the photosphere, although depleted in heavy ions. This is undoubtedly a consequence of the escape mechanism, which is thermal at its origin, and favors light ions. We thus find more than 99% of H^+ and He^{++} ions (protons and alpha particles), about 3.5% to 4.5% of which are helium. The solar wind flow was initially calculated by E. Parker with a hydrodynamic approach, considering the plasma as a fluid. For a radial magnetic field, the solutions of the equation of motion along the magnetic field, coupled to the continuity equation, are represented by the network of curves in Figure 3.2. There are two types of solution that connect the solar surface to infinity:

– a family of subsonic solutions, for which the escape velocity is zero at the origin, increases to a limit value lower than the sound velocity (of protons) and then decreases to zero at infinity: this is the solar breeze;

– a singular solution, for which the escape velocity increases monotonically, reaches the sound velocity (Mach number $M = 1$) at a critical distance r_C and remains supersonic beyond. This solution describes the solar wind that is actually observed. In the case of the solar corona, the critical point is at an altitude between three and ten solar radii. The velocity of protons there is 30 km s^{-1}. It continues to increase considerably thereafter, reaching the approximate velocity observed at the Earth's orbit, about 400 km s^{-1}, and which remains more or less constant in the outer Solar System.

a) b)

Figure 3.1. *(a) X-ray image of the solar corona obtained with the Yokkoh satellite (source: ISAS). (b) Solar eclipse in 2017 observed in the visible range, revealing the expansion of the corona in the interplanetary medium (source: N. Lefaudeux). For a color version of this figure, see www.iste.co.uk/encrenaz/solar1.zip*

The choice between these two types of solution depends on the boundary conditions of the flux tube: for the flow to remain subsonic, the pressure at infinity must remain above a given critical pressure. The low pressure prevailing in the interstellar medium, on the contrary, imposes supersonic expansion.

However, Parker's fluid approach has some limitations. The solar wind becomes collisionless beyond the exobase, at about six solar radii of altitude, when the mean free path of the charged particles becomes larger than the mid-scale height. The terminal speed of the solar wind is, moreover, solely determined by the thermal energy of the corona. However, a coronal temperature of 10^6 K can account for a terminal flow at 400 km s^{-1}, but not for the faster solar wind associated with the coronal holes, revealed by the interplanetary Ulysses probe at high latitudes.

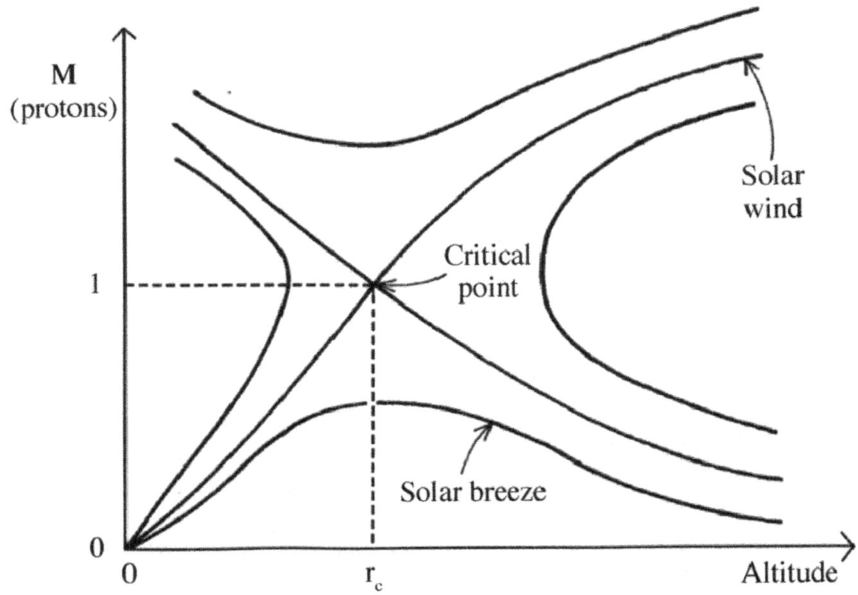

Figure 3.2. *Vertical profile of coronal expansion velocity, expressed in Mach number of protons, for the different solutions of the hydrodynamic problem examined by E. Parker. Only the "solar wind" solution, which crosses the sound barrier at the critical point, is compatible with the very low pressure imposed at great distance by the interstellar medium (source: J.-C. Cerisier)*

Models of expansion based on an exospheric kinetic approach, dealing with the ballistic motion of a collection of individual particles, have attempted to overcome these limitations. Considering the particle thermal velocity $V_{th} = (2kT/m)^{1/2}$ for a balanced medium (Maxwell's velocity distribution) at temperature T, k being Boltzmann's constant, V_{th} should be $m_p/m_e \sim 43$ times higher for electrons than for protons. The electrons would thus escape much more efficiently from the Sun, which would be positively charged without the appearance of a restoring electrostatic field E_{es}, necessary to maintain the quasi-neutrality of the medium. This field slows down the electrons, accelerates and extracts the protons and positive ions from the solar atmosphere. At the exobase, it can be shown that the available electrostatic energy, proportional to eE_{es}, is much higher than both the thermal energy of the electrons and the gravitational energy of the protons, and thus pushes the solar wind. With this electrostatic energy replacing the heat flux of the fluid expansion, the energy outcome of the solar wind shows that a 10^6 K corona can reach a terminal velocity of about 250 km s^{-1}, with Maxwellian distributions of coronal electrons (at thermodynamic equilibrium). However, measurements reveal that the observed distributions deviate significantly from a Maxwellian distribution (with copious suprathermal tails), which the theory explains by a mean free path of the charged particles increasing in proportion to the square of their energy E, thus much larger for the more energetic particles. Under these conditions, an expansion to 700 km s^{-1} becomes possible for an ionized wind. This would be impossible for a neutral wind, where the mean free path is independent of energy.

In reality, the mean free path of the particles is close to the mid-scale height in the critical region of the acceleration, so that the solution of the expansion is probably in an intermediate “middle path” between the fluid and kinetic approaches, for example a kinetic approach with collisions.

A key question remains: how does one explain a coronal temperature of the order of 10^6 K with a photosphere at only 6,000 K? Here, again, the E^2 (or v^4) dependency of the mean free path of charged particles of velocity v, associated with non-Maxwellian distributions, may provide an explanation: only the fastest particles reach higher altitudes, implying a “velocity filter”. Thus, the temperature being only the second-order moment of the particle distribution function, it increases with altitude.

Definitively elucidating coronal heating and solar wind acceleration is the scientific objective of two space missions that will work in synergy: Parker Solar Probe (NASA) and Solar Orbiter (ESA). The first one, launched on August 12, 2018, has completed four perihelion surveys to date (it will get as close as possible to nine solar radii to probe the critical region); the second one, launched on February 10, 2020, is beginning its operation phase.

3.1.2. *The structure of the heliosphere in the ecliptic plane*

The volume occupied in space by the solar wind, or in an equivalent way by the solar magnetic field lines it carries, forms the heliosphere, whose structure we will describe on a large scale.

If one is far beyond the altitude of the critical point, the magnetic forces acting on the coronal plasma are no longer sufficient to carry it with the solar rotation. Instead, the coronal plasma, a magnetohydrodynamic (MHD) fluid in which the frozen-in condition of the magnetic field in the flow is realized, has sufficient inertia to carry the coronal magnetic field with it along its radial expansion. The footprint of the magnetic field lines, however, remains anchored in the photosphere, and these continue to be driven into the solar rotation. The combination of these rotational and radial expansion motions results in the interplanetary magnetic field wrapping around the Sun. In the ecliptic plane, close to the solar equator, this wrapping forms an Archimedean spiral whose pitch is equal to the radial distance travelled by a fluid element of the solar wind in one solar rotation, in other words, about 10^9 km or six astronomical units (au), as shown in Figure 3.3.

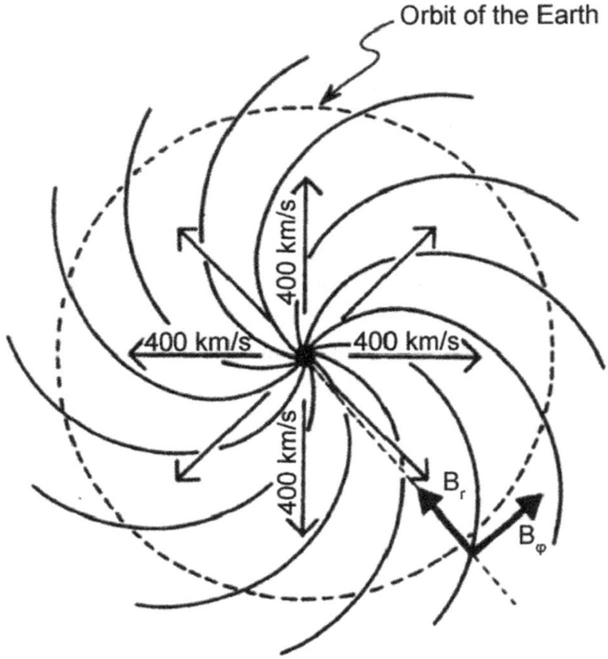

Figure 3.3. *The radial expansion of the corona, combined with the solar rotation, wraps the interplanetary magnetic field lines around the Sun to form an Archimedean spiral. At Earth's orbit, the B_r radial and B_φ azimuthal components of the interplanetary magnetic field are approximately equal (source: (Encrenaz* et al. *2003))*

From these simple remarks, we can deduce the radial dependence of the main parameters of the plasma and magnetic field in the ecliptic plane according to the distance R to the Sun, and reference them to their value at the Earth's orbit, which is well-known from numerous *in situ* observations. For a constant radial velocity, the conservation of the flux of matter leads to a variation in R^{-2} of the ionic concentration of the solar wind n, with $n_0 \sim 5$ cm^{-3} at 1 au:

$$n\ (\text{cm}^{-3}) \sim n_0 / R^2\ (\text{au}) \quad [3.1]$$

The conservation of the magnetic flux imposes the same law of radial dependence for the B_r radial component of the interplanetary magnetic field B:

$$B_r = B_{r0} / R^2 \text{ (au)} \tag{3.2}$$

The structure of the Archimedean spiral then imposes the radial dependence of the other main component of the magnetic field, the B_φ azimuthal component:

$$B_\varphi = B_{\varphi 0} / R \text{ (au)} \tag{3.3}$$

Measurements from various interplanetary missions have shown that these laws are, on average, well verified (see Figure 3.4). At Earth's orbit, we have $B_{r0} \sim B_{\varphi 0} \sim$ 3 nT and the spiral thus makes an angle of 45° with the Sun–Earth direction.

As for the other plasma parameters, it was observed that the velocity V and the temperature T vary little with distance, respectively, in proportion to $R^{-2/7}$ and $R^{-1/7}$, with typical values at 1 au of 400 km s^{-1} and 10^5 K.

Characteristic solar wind velocities include the Alfvén velocity V_A and the acoustic velocity V_S. The Alfvén velocity, defined by:

$$V_A = B / (\mu_0 n m_{eq})^{1/2} \tag{3.4}$$

is directly related to B and the ionic concentration n of the plasma, μ_0 being the vacuum permeability and the equivalent mass $m_{eq} \sim 1.1\ m_{proton}$. For its radial variation, we find:

$$V_A \text{ (km s}^{-1}) \sim V_{A0}\, [1/2 + 1/2R^2(\text{au})]^{1/2} \tag{3.5}$$

with $V_{A0} \sim$ 50 km s^{-1} at 1 au. In the outer Solar System, the Alfvén velocity tends towards a uniform average value.

The acoustic velocity is written as:

$$V_S = (\gamma P / n m_{eq})^{1/2} = (\gamma k T / m_{eq})^{1/2} \tag{3.6}$$

γ being the polytropic index, P the pressure, k the Boltzmann constant and T the temperature. For its radial variation, we find:

$$V_S \sim V_{S0} / R^{-1/7} \tag{3.7}$$

with $V_{S0} \sim$ 60 km s^{-1} at 1 au, which indicates a slight decrease with distance.

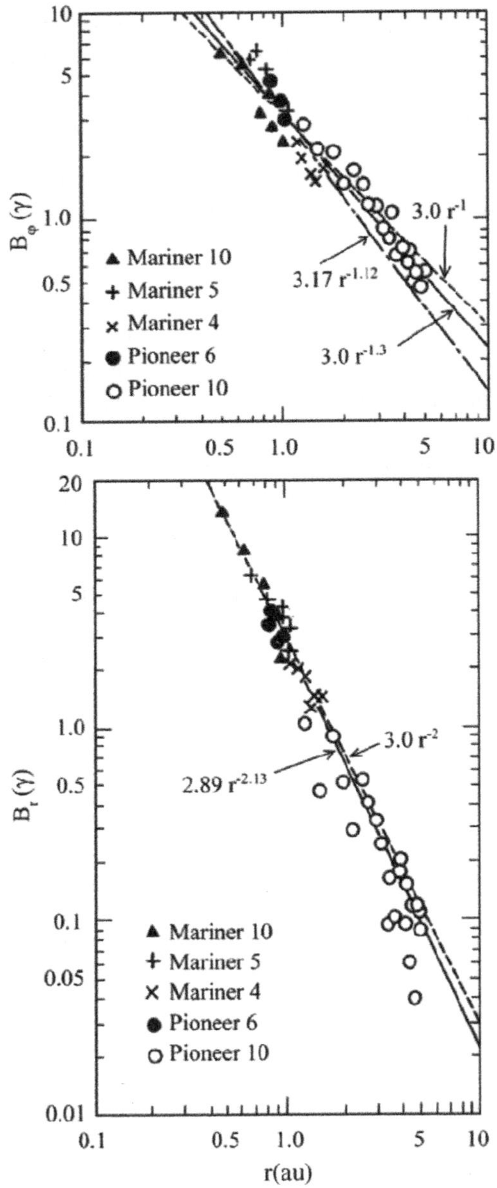

Figure 3.4. *Various interplanetary probes have made it possible to explore the radial variation of the planetary magnetic field from 0.4 to several dozen au*

COMMENT ON FIGURE 3.4. – *The Mariner and Pioneer measurements show that the variations in R^{-2} of B_r and in R^{-1} of B_φ, predicted by the simple Parker spiral model, are in fairly good agreement in average with the observations. Both components are of the order of 3 nT at the Earth orbit (source: from (Behannon 1978)).*

3.1.3. *The three-dimensional structure of the heliosphere*

When the interplanetary magnetic field is measured in the ecliptic plane, it is structured in sectors of opposite magnetic polarities, the magnetic field lines being oriented either towards the Sun or in the opposite direction. By leaving the ecliptic plane for the first time after its flyby of Jupiter, the interplanetary probe Pioneer 11 made it possible to explore the heliosphere up to an unprecedented solar latitude of 16° and discover that this sector structure hides a much simpler three-dimensional reality, illustrated by Figure 3.5.

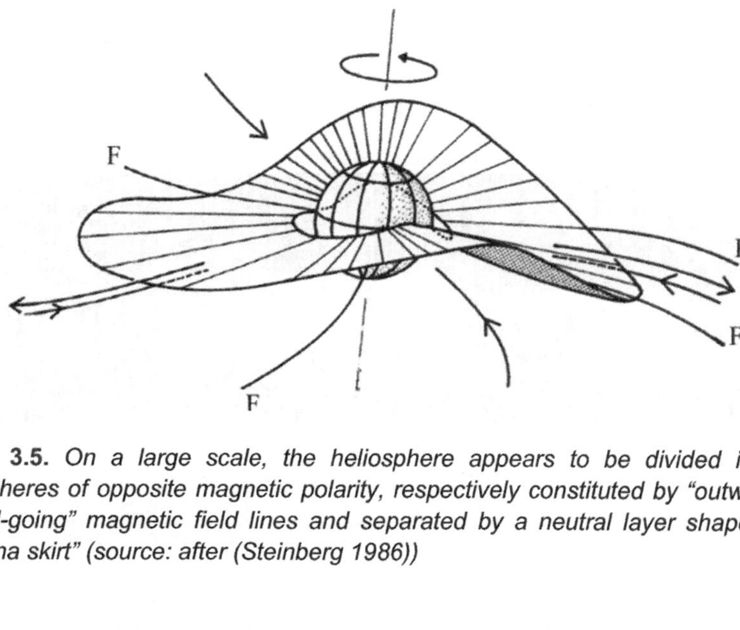

Figure 3.5. *On a large scale, the heliosphere appears to be divided into two hemispheres of opposite magnetic polarity, respectively constituted by "outward-" or "inward-going" magnetic field lines and separated by a neutral layer shaped as a "ballerina skirt" (source: after (Steinberg 1986))*

The heliosphere is, in fact, composed of two magnetic hemispheres of opposite polarities, in each of which the field lines are connected, probably for the most part, to coronal holes of north or south polarities. These two hemispheres are separated by a neutral layer shaped as a "ballerina skirt", which rotates with the Sun by undulating on either side of the solar equatorial plane according to the complex geometry of the photospheric and coronal fields. The boundaries of the observed sectors are thus only the intersection of the ecliptic plane with this neutral layer.

The Ulysses probe has explored the very high latitude regions of the solar wind, successively flying over the two solar poles in 1994–1995, 2000–2001 and 2007–2008. This mission significantly advanced our understanding of the relations between coronal holes and solar wind.

Two distinct states of the solar wind have been characterized:

– a "slow" wind confined near the ecliptic and the neutral layer, with an average speed of ~ 400 km s^{-1}, resulting in reality from the superposition of jets of different speeds;

– a "fast", more homogeneous wind at high latitude, reaching speeds of 700–750 km s^{-1}.

A sharp boundary separates these two regimes in the interplanetary medium and extends through the corona and chromosphere. Furthermore, the high-latitude expansion is strongly non-radial, fanning out toward the heliographic equator, with the solar wind measured at a few au being on average connected to heliographic latitudes that are higher than those of the measurements. A global north-south symmetry exists on a large scale (see Figure 3.6).

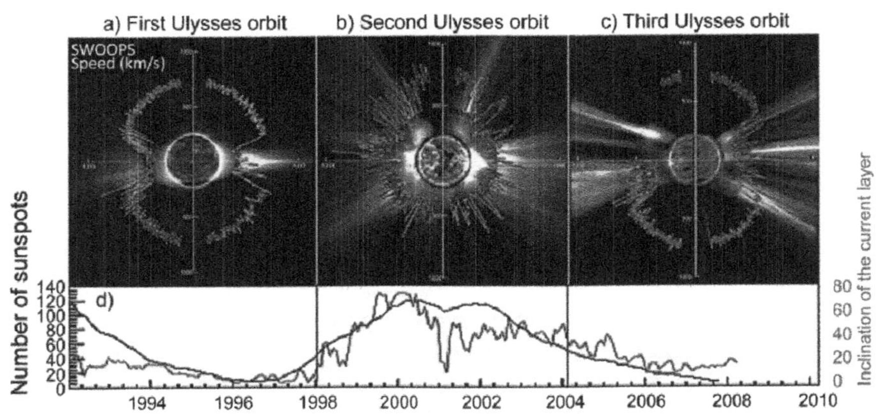

Figure 3.6. *Plots of the solar wind speed as a function of latitude for the three orbits of the interplanetary Ulysses probe around the sun. For a color version of this figure, see www.iste.co.uk/encrenaz/solar1.zip*

COMMENT ON FIGURE 3.6. – *The blue and red lines indicate magnetic field lines oriented in the antisolar and solar directions, respectively. The first and third orbits, corresponding to periods of solar minimum, reveal a "slow" solar wind near the equator and a "fast" wind at higher latitudes, with a polarity inversion between the two. The second orbit, corresponding to a solar maximum, shows a mixture of the two types of wind at all latitudes (source: adapted from (McComas* et al. *2008)).*

3.1.4. *Transient structures of the solar wind*

Large-scale transient variations of the solar wind superimpose on its mean structure described above. These transient phenomena result from inhomogeneous solar wind velocities at its coronal source. They can be classified in two main categories.

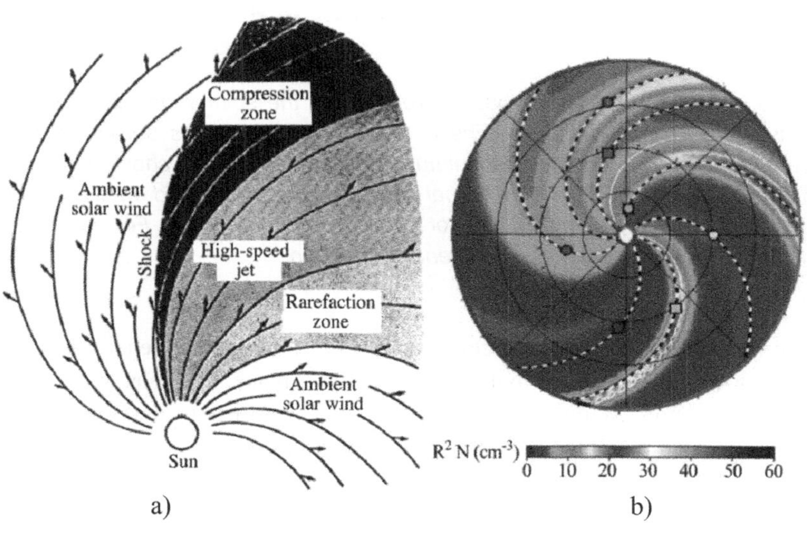

Figure 3.7. *Interaction between the fast solar wind and the slow solar wind. For a color version of this figure, see www.iste.co.uk/encrenaz/solar1.zip*

COMMENT ON FIGURE 3.7. – *(a) The interaction between a high-speed solar wind jet with the slower ambient solar wind creates a corotating interaction region, consisting of a downstream plasma compression region and an upstream rarefaction region. These may be preceded and followed, respectively, by a shock region. The whole structure rotates rigidly with the Sun, as long as the source of the fast solar wind remains stationary. Source: after (Hundhausen 1972). (b) MHD simulation of the solar wind density in the ecliptic plane on September 7, 2011 at 00:00 UT, showing several corotating interaction regions. Symbols indicate the position of planets and space probes (source: CCMC).*

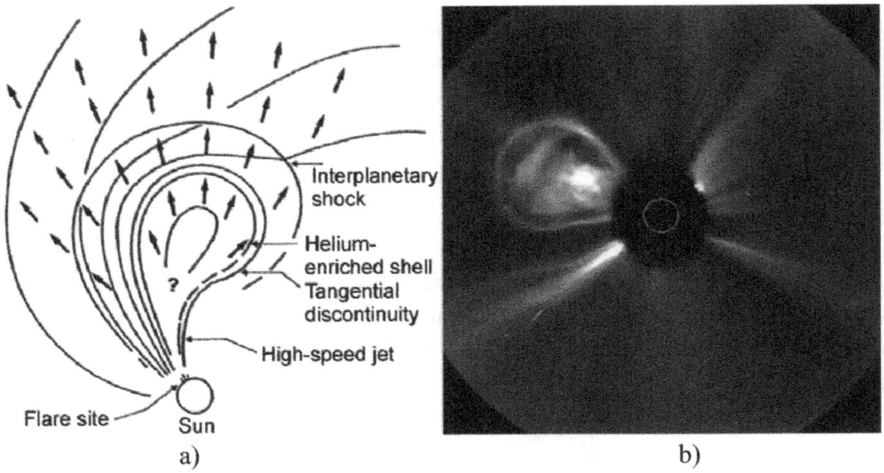

Figure 3.8. *(a) Transient perturbation produced in the solar wind by the ejection of coronal matter following a flare. The coronal mass ejection is separated from the ambient medium by a contact discontinuity and preceded by a shock wave (source: from (Hundhausen 1972)). (b) Coronal mass ejection observed with the Stereo-A coronagraph on September 23, 2011 at 23:24 UT (source: ESA). For a color version of this figure, see www.iste.co.uk/encrenaz/solar1.zip*

The first type of transient structures comes from the inhomogeneity of the radial escape velocities of the solar wind. In the ecliptic plane, the solar wind is indeed structured in sectors of longitude, where fast solar wind jets (~ 600 km s^{-1} or more) flow are separated by regions of ambient solar wind (~ 400 km s^{-1}).

It has been rather accurately shown that the high-speed sectors are connected by the interplanetary magnetic field lines to the equatorial coronal holes. Because of the solar rotation, the slow and fast jets follow each other on the same radial direction, on which the slow solar wind is compressed in front of the fast jet and becomes rarefied behind it (see Figure 3.7). This interaction region, which is in corotation with the Sun like the sources to which it is linked, can be accompanied by an interplanetary MHD shock wave upstream and downstream of the compression region, if its velocity exceeds the characteristic speeds of the medium.

The observations of Ulysses have shown that these structures in corotation, separated by shocks, are not confined to low latitudes and produce effects up to high latitudes (modulation of the incident cosmic ray flux, injection of low-energy accelerated particles in the solar polar regions).

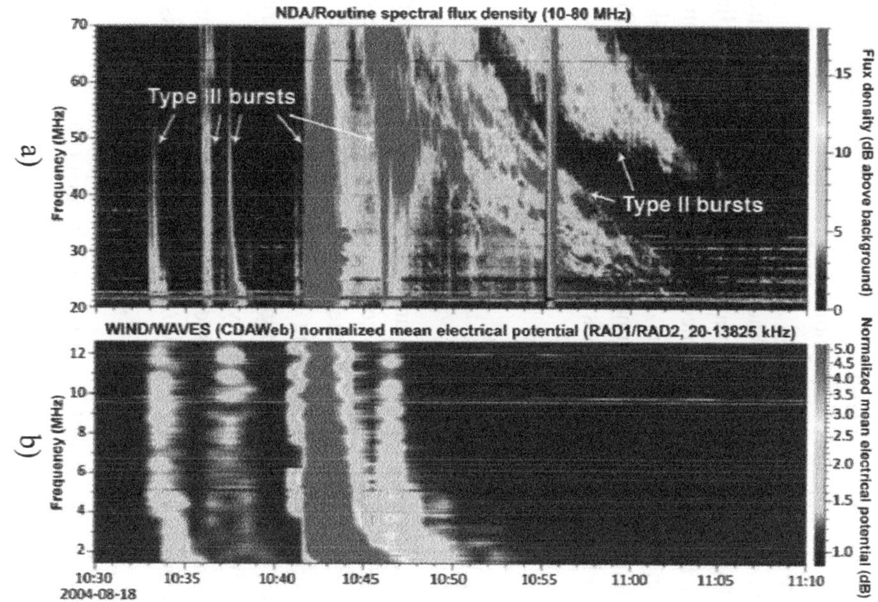

Figure 3.9. *Time-frequency plots of the flux density of solar radio observations obtained with: (a) the Nançay Decametric Array; (b) the Wind satellite. For a color version of this figure, see www.iste.co.uk/encrenaz/solar1.zip*

COMMENT ON FIGURE 3.9. – *Two types of emission are visible: type II and III solar radio bursts. The first are due to shock waves propagating at ~ 1,000 km s^{-1} and generating radio emission at the local electron plasma frequency f_{pe}, as well as its harmonic $2f_{pe}$ (producing two parallel bands on the figure). The second ones are due to relativistic electron beams (v ~ 0.1-0.3c) that also excite emissions at f_{pe} and $2f_{pe}$ (source: Paris Observatory/ Nançay Radio Observatory, ESA/NASA).*

The second type of transient structures is produced by coronal events linked to solar activity: a flare, or a sporadic coronal phenomenon, results in the localized and high-speed ejection of a large mass of coronal plasma into interplanetary space. Figure 3.8 provides an example of it and labels the different types of discontinuities. In this case, the perturbation has a limited duration and is confined to a limited sector of longitude in a fixed frame of reference. This type of perturbation, also associated with interplanetary shock waves, is at the origin of the well-known category of solar/interplanetary radio bursts, named "type II" (see Figure 3.9). The long-term remote monitoring of solar radio bursts thus forms a direct proxy of solar activity.

3.1.5. *The boundaries of the heliosphere*

At the edge of the heliosphere, the interplanetary medium interacts with the interstellar medium, which is formed of three components: the neutral interstellar gas, the cosmic rays and the magnetized interstellar plasma. The local interstellar medium could be remotely probed by studying the profile of absorption lines in stellar spectra (such as Ca II in the visible range), the resonant backscattering of solar lines in the extreme UV radiation (H Lyman-α, He I at 58.4 nm, O^+ at 83.4 nm), by direct measurements from Ulysses (neutral helium, hydrogen and deuterium, ions from interstellar atoms) and by *in situ* plasma measurements from the Voyager probes.

The local neutral interstellar gas has a density of about 0.1 particle cm^{-3} (~ 0.06 cm^{-3} for hydrogen and ~ 0.015 cm^{-3} for helium), a temperature of about 7,000 K, and a composition close to the solar one. Its relative speed with respect to the Sun, close to the ecliptic plane, is about 25 km s^{-1}. Being insensitive to electromagnetic forces, it can penetrate freely in the heliosphere at this average speed. Its interactions with the Solar System are reduced to the deviation imposed on the trajectory of each particle by gravity and solar radiation, and to the exchange of charges between neutral interstellar particles and solar wind ions. This last mechanism, which consists of replacing a fast ion of the solar wind by a slow ion from the interstellar medium, constitutes a loss of energy and momentum for the flow of the solar wind. The effect of this charge exchange has been demonstrated by the modulation it induces on the Lyman-α emission of interstellar hydrogen, and by the direct detection by Ulysses of H^+, He^{++}, $^3He^+$, $^4He^+$, C^+, N^+, O^+ and Ne^+ ions that are produced by the interaction of interstellar atoms with the solar wind (and also probably by photoionization).

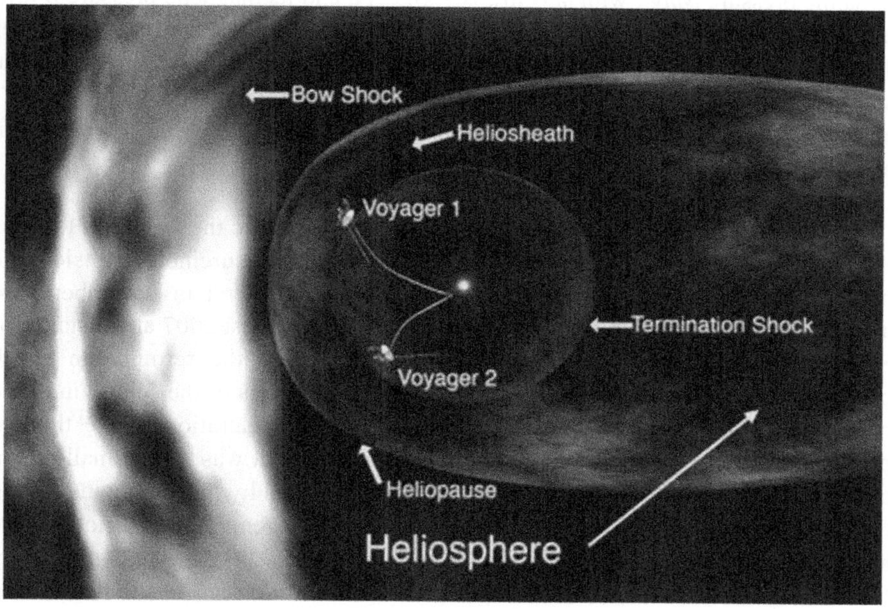

Figure 3.10. *Representation of the heliosphere and the main discontinuities formed by the interaction between the solar wind and the interstellar magnetized plasma. The Voyager probes are the only ones to date to have passed through the interstellar medium (source: NASA/JPL-Caltech). For a color version of this figure, see www.iste.co.uk/encrenaz/solar1.zip*

The heliosphere is separated from the interstellar medium by a discontinuity, the heliopause. The spatial extension of this boundary is the result of a pressure balance. The expansion of the solar wind is, indeed, stopped as soon as its total pressure is no longer higher than that of the interstellar plasma and magnetic field. A few au before reaching the heliopause, the solar wind goes through a "termination" shock, when it becomes subsonic again, before being deflected towards the region of the heliosphere located downstream from the flow of interstellar wind, forming a region called the heliosheath. Outside the heliosphere, the interstellar plasma comes to rest at 20–25 km s^{-1} on the heliopause, on the sides of which it flows. An illustration of this representation is given by Figure 3.10.

The characteristics of the interstellar thermal plasma are still poorly known, with a concentration of the order of some 10^{-2} particles cm^{-3}, a temperature and a velocity similar to that of the neutral gas and a magnetic field close to 0.35 nT. With these parameters, the thermal pressure of the interstellar plasma NkT is ~ 10^{-15} Pa, its dynamic pressure Nmv^2, with $m \sim m_p$, is ~ 10^{-14} Pa and its magnetic pressure $B^2/2\mu_o$ is ~ 5 10^{-14} Pa. It is thus the latter which dominates the total pressure, which reaches 10^{-13} Pa. An equal pressure is reached by the predominant dynamic pressure of the solar wind at a distance d such that: $(N/d^2)m_pV^2 = 10^{-13}$ Pa, with d in au, $N \sim 5$ cm^{-3} and $V \sim 400$ km s^{-1}. This simple calculation yields a heliocentric distance $d \sim 115$ au, about three times the distance from the Sun to Pluto.

The Voyager probes, which were launched in 1977, were the first ones to cross these boundaries and to provide the first *in situ* measurements outside the heliosphere. The termination shock was crossed by Voyager 1 in December 2004 at a distance of 94 au, and by Voyager 2 five times in August 2007 at a distance of 84 au. These values illustrate the dynamic nature of the termination shock position, which depends on the time-variable characteristics of the solar wind. The measurements revealed an unusual shock, with the deceleration of the thermal plasma being accompanied by a temperature increase that was much smaller than expected. These results suggest that the energy dissipated by the solar wind during the passage of the shock is mainly transferred to energetic particles. The heliopause was crossed by Voyager 1 on August 25, 2012 at 122 au and by Voyager 2 on November 5, 2018 at 119 au. This transition has been identified in particular by an increase in plasma density, magnetic field and galactic cosmic ray flux. As the two probes have symmetrical trajectories on both sides of the ecliptic plane, inclined at 45° with respect to the direction of the nose of the heliopause (see Figure 3.10), these results reveal a remarkably symmetrical heliosphere. However, the position of the heliopause must also vary in time and space due to the variability and inhomogeneities of the solar wind and the interstellar medium, particularly with the 11-year solar activity cycle. It could be lower than 100 au in a period of minimum activity and higher than 150–200 au in a period of maximum activity. Since their crossing of the heliopause, the Voyager probes have obtained the first *in situ* measurements of interstellar plasma, although it was in a boundary layer corresponding to a disturbed interstellar wind. The plasma density at the edge of the heliopause was thus ~ 0.04–0.06 cm^{-3} at ~ 120 au, although Voyager 1 observed this value continuously increasing to ~ 0.12 cm^{-3} at 146 au (Gurnett and Kurt 2019).

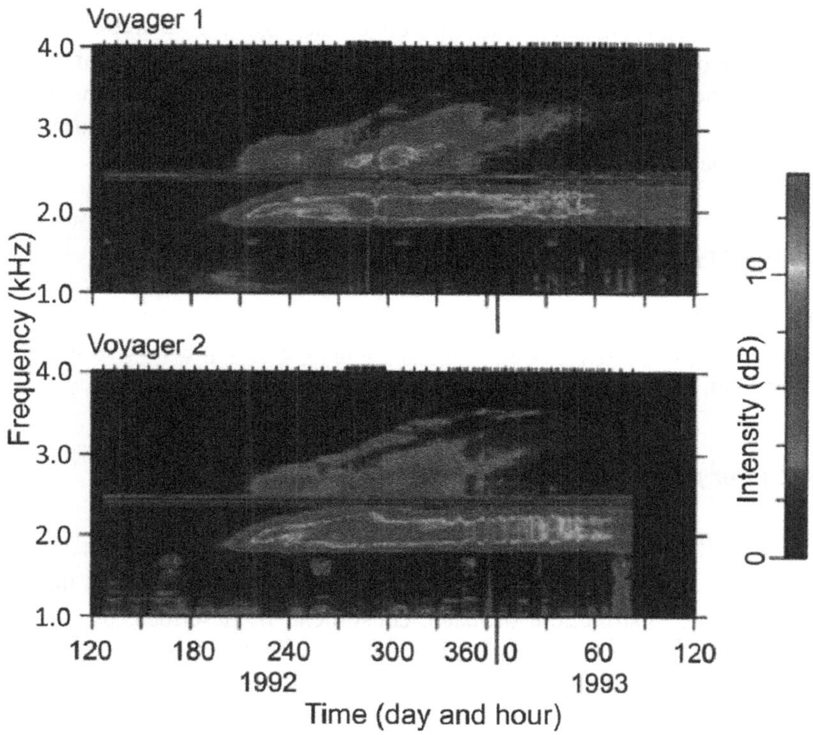

Figure 3.11. *Time-frequency flux spectrogram showing the heliospheric radio emission detected by Voyager 1 and 2 in 1992–1993 (source: from (Gurnett* et al. *1993)). For a color version of this figure, see www.iste.co.uk/encrenaz/solar1.zip*

Lastly, it is interesting to note that radio emissions at low frequencies, between 1.8 and 3.6 kHz, were measured by the Voyager probes (see Figure 3.11) when they passed a heliocentric distance of 12 au, when the density-proportional plasma frequency of the ambient solar wind fell below ~ 2 kHz. Two periods of intense solar activity, in 1982 and 1991, characterized by coronal mass ejections at 500–600 km s^{-1} and decays of the galactic cosmic ray flux detected in the inner Solar System, were followed by strong intensifications of this radio radiation, with a delay of ~ 400 days. This delay corresponds to a boundary located at 110–160 au. The radiation is attributed to plasma oscillations excited near the nose of the heliopause by incident interplanetary shocks – with a mechanism comparable to that of type II solar bursts (see Figure 3.9). In this framework, the emitted frequencies correspond to a local plasma density of 0.04–0.16 cm^{-3}, in excellent agreement with the measurements cited above, and well above the expected density of the solar wind at such distances (< 0.001 cm^{-3}). The total emitted power has been estimated at 10^{13} W, which is much higher than the radio powers emitted by other sources in the Solar System (Sun or magnetized planets). The time-frequency structure of the radio emission, particularly the drifts from ~ 2 to ~ 3.5 kHz in a few months, could be interpreted in terms of radial density gradients between the heliopause and its associated shocks.

3.2. Planetary envelopes

Interaction between the solar wind and the Solar System bodies depends on the conductive nature of the planetary envelope and/or on the presence of an intrinsic magnetic field, whose description is the subject of the following sections.

3.2.1. *Upper planetary atmospheres*

3.2.1.1. *The neutral upper atmosphere*

Planetary atmospheres are complex gaseous envelopes. In the upper atmosphere, the distribution of each constituent is described by its partial hydrostatic equilibrium, which is controlled by molecular diffusion, the turbulent mixing of the atmosphere, and chemical and photochemical reactions associated with incident solar radiation. The entire atmospheric gas obeys the law of global hydrostatic equilibrium, which describes the vertical variation of the pressure of the neutral atmosphere P_n by an exponential function decreasing in $-z/H_n$, z being the altitude and $H_n = kT_n(z)/Mg(z)$ being the characteristic scale height of the medium, T_n being the temperature of the neutral atmosphere and M the average molecular mass. The planetary atmospheric structure is described in detail in Chapters 1 and 4. For now, we will only remind readers of its vertical stratification in layers defined by the variation of T and M as a function of z, as illustrated in the Earth's case by Figure 3.12, in order to focus on its upper part.

The temperature variation defines several domains depending on the altitude of absorption of solar radiation. Most of the visible and infrared radiation is absorbed by the Earth's surface. Ultraviolet radiation (200–300 nm) is absorbed by the ozone layer in the stratosphere, resulting in a first rise in temperature. Above the mesopause (around 90 km altitude), the absorption of far ultraviolet radiation (100–170 nm) and extreme ultraviolet radiation (< 100 nm) by photodissociation and photoionization produces another rise in temperature, asymptotically reaching its maximum value of 900 K or more, named the exospheric temperature. This region is called the thermosphere.

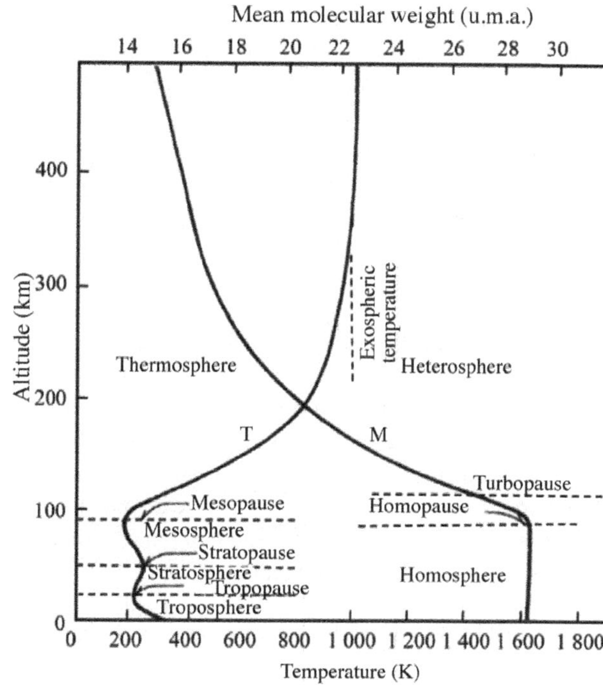

Figure 3.12. *Vertical distribution of temperature T and average molecular mass M in the Earth's atmosphere and associated atmospheric layers (source: (Encrenaz* et al. *2003))*

The variation of the molecular weight defines two main layers. From the ground to the homopause (around 90 km altitude), the chemical composition of the air remains constant ($M = 29$) due to the turbulent mixing of the air, the main cause of which is the preponderant heating of the atmosphere by the ground. The homosphere forms the convection zone of a planetary atmosphere. With the clear positive vertical temperature gradient, which is encountered from 90 km in the thermosphere, the atmosphere becomes very stable convectively, and the vertical mixing of constituents ceases rapidly, becoming negligible above the turbopause (around 110 km). This is the heterosphere, where a gravitational separation of the different chemical constituents takes place, and therefore a progressive decrease of M with altitude, the lighter constituents becoming progressively dominant. In the absence of chemical reactions, the concentration of each constituent decreases exponentially with altitude. In these areas of the thermosphere and heterosphere, most of the ionospheric plasma is formed, under the effect of photoionization by solar UV radiation.

The fluid description of the vertical distribution of atmospheric constituents (by hydrostatic equilibrium) is no longer valid in the highest layers of the atmosphere. Indeed, the fluid description assumes that the different species of velocity distribution functions do not deviate very much from a Maxwellian distribution. This is only true if the collisions between particles are sufficiently numerous. In a heterogeneous medium such as the atmosphere, this situation of local thermodynamic equilibrium (LTE) is achieved when:

$$L_{col} < L \qquad [3.8]$$

L_{col} (characteristic length of collision) being the mean free path, and L the characteristic length of the spatial variations of the macroscopic parameters of the medium. Let us take $L = H_n$, which is, by definition, the characteristic scale of vertical pressure variation. Above the altitude defined by $L_{col} = H_n$, called the exobase, we enter a layer of the atmosphere that is no longer maintained by collisions in the vicinity of the local thermodynamic equilibrium: this is the exosphere. In the terrestrial atmosphere, the exobase is located between 500 and 600 km of altitude. The exosphere is thus essentially made up of oxygen, helium and hydrogen.

To calculate the distribution of constituents in the exosphere, we can no longer use hydrostatic equilibrium. Instead, Liouville's theorem is used to compute the species distribution functions from "source" distribution functions at a reference level (usually the exobase, where Maxwellian distributions are most often assumed), assuming that all particles have ballistic trajectories, free of collisions. This kinetic method has important limitations. Liouville's theorem simply states that the numerical value of the distribution function $f_i(r,v)$ of each species i is conserved along each trajectory. Initially knowing f_i only at the exobase, we can only calculate it at the points in phase space (r,v) that are actually connected to the exobase by a trajectory.

As shown in Figure 3.13, there are three types of trajectories that differ in terms of their connection to the exobase:

– "ballistic" trajectories, with both starting and arrival points at the exobase. These are those for which the initial velocity v of the particle leaving the exobase is lower than the escape velocity v_l;

– the so-called "escape" trajectories, with their starting point only at the exobase. These are hyperbolas or parabolas in the field of a spherically symmetric central body such as $v > v_l$;

– the so-called "satellite" trajectories, which do not intercept the exobase.

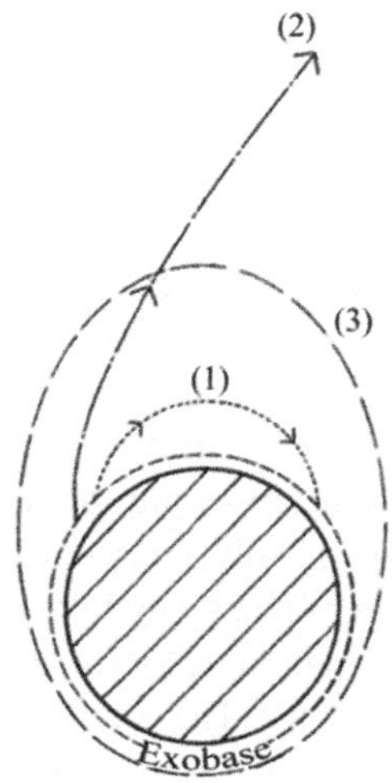

Figure 3.13. *Examples of trajectories of neutral particles in the exosphere: (1) ballistic trajectory from exobase to exobase; (2) escape trajectory from the exobase to the interplanetary medium; (3) satellite trajectory that does not intercept the exobase (source: (Encrenaz* et al. *2003))*

The method described above allows the calculation of $f_i(r,v)$ along the so-called "ballistic" and "escape" trajectories. In particular, the total flux of particles travelling on "escape" trajectories corresponds to a clear-cut escape flux for the planetary atmosphere. This escape flux is all the greater as the considered component is lighter (because its thermal velocity is inversely proportional to $m_i^{1/2}$) and as the attraction of the central body, which determines the escape velocity v_l, is weaker. This escape explains why the Earth and Mars permanently lose hydrogen, and why most of the lightest bodies of the Solar System (Moon, Mercury, small satellites of giant planets) cannot retain an atmosphere. On the other hand, the kinetic method does not allow $f_i(r,v)$ to be calculated along the "satellite" trajectories since, in the absence of collisions, they are not connected to the source that constitutes the exobase. These trajectories being unpopulated, one should simply make the assumption that $f_i = 0$. However, this conclusion is inaccurate, because collisions, even if they are very rare in the exosphere, allow particles that are initially placed on "ballistic" or "escape" trajectories to transfer to satellite trajectories. This effect, which is very weak but cumulative over long periods of time, tends to restore the thermodynamic balance, even in rarefied regions like these. However, a conceptually correct model of this situation, halfway between fluid and kinetic description, is very difficult to build, as in the case of solar wind expansion.

3.2.1.2. *The ionosphere*

Solar UV and X-ray electromagnetic radiation, the precipitation of energetic particles (especially auroral electrons) and meteorites are at the origin of the ionization of the outermost envelopes of planetary atmospheres. A plasma component is formed in a region called the ionosphere. The ions have a finite lifetime, controlled by the reactions of ionic chemistry and recombination with electrons that return it to the upper neutral atmosphere. Figure 3.14 gives an illustration of the vertical ion distribution of the Earth's ionosphere. The ionospheric plasma is a minor component of the upper atmosphere, immersed in the neutral gas. Its various constituents are dynamically coupled to both the neutral fluid of the thermosphere, which tends to drag them along by friction, and to the magnetic field, which tends to "trap" each charged particle by forcing it, through gyration, to remain in the vicinity of the same field line. The competition between these two effects totally governs the dynamic behavior of the ionospheric ions and electrons. A detailed description of the structure and dynamics of the ionospheric layers is provided in (Encrenaz *et al.* 2003).

These ionospheric plasma layers play a fundamental role in the dynamics of ionized envelopes, illustrated in Figure 3.15:

– they are conductive. The ionospheric shell, immersed in the upper atmosphere of each planet, thus ensures the circulation of intense electric currents, originating from the upper atmosphere and/or the magnetosphere;

– they are sources of plasma. By diffusing along the magnetic field lines, the ionospheric plasma, whose energies are typically between 0.1 and 1 eV, fills a very large part of the magnetospheric cavity, where it behaves like an MHD fluid "frozen" in the magnetic field, due to its low temperature. As its motion obeys the following equation:

$$\boldsymbol{E} + \boldsymbol{V} \times \boldsymbol{B} = 0 \quad [3.9]$$

It thus also materializes the "motion" of the magnetic field lines, in the MHD sense;

– the magnetic field lines become very conductive where the ionospheric plasma has diffused. The component parallel to $\boldsymbol{B}$ of equation [3.9] is simply written as:

$$\boldsymbol{E}.\boldsymbol{B} = 0 \quad [3.10]$$

because of the very high mobility of the ionospheric cold electrons along $\boldsymbol{B}$. These field lines are very efficient electrical connectors between the ionospheric conductor and the dynamos (electrical generators), distributed in the magnetospheric cavity or at its interface with the solar wind. The currents delivered by these dynamos therefore tend to close up along the field lines (we refer to "field-aligned currents" with $\boldsymbol{B}$), and then through the ionospheric conductive shell.

The presence of an electric field $\boldsymbol{E}$ applied to the ionospheric conductor, or of an electromotive force $\boldsymbol{V}_n \times \boldsymbol{B}$ caused by the motion of the conducting atmospheric fluid through the planetary magnetic field lines, induces an electric current in the plasma.

The conductive properties of the ionosphere allow us to express this current via Ohm's ionospheric law:

$$\boldsymbol{j} = \sigma_{//} (\boldsymbol{E}.\mathbf{b})\, \mathbf{b} + \sigma_P\, \boldsymbol{E}'_\perp + \sigma_H\, (\boldsymbol{E}'_\perp \times \mathbf{b}) \qquad [3.11]$$

$$\boldsymbol{E}' = \boldsymbol{E} + \boldsymbol{V}_n \times \boldsymbol{B} \qquad [3.12]$$

In the neutral gas frame of reference, $\sigma_{//}$ is the conductivity parallel to the field lines, σ_P and σ_H are the Pedersen and Hall conductivities that are perpendicular to the field lines and **b** is the unit vector that is collinear to $\boldsymbol{B}$. The conductivity $\sigma_{//}$, essentially carried by electrons, is always much higher than the perpendicular Pedersen and Hall conductivities, which are limited by the mobility of ions.

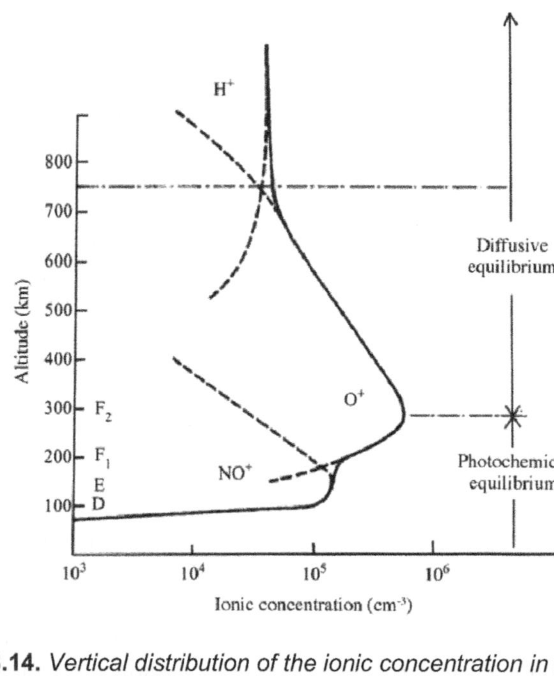

Figure 3.14. *Vertical distribution of the ionic concentration in the Earth's ionosphere, whose layers are labelled by the letters on the left-hand side*

COMMENT ON FIGURE 3.14. – *In the lower ionosphere, the characteristics of the* NO^+ *ion layer and that of the lower* O^+ *layer are controlled by photochemical equilibrium. The characteristics of the upper part of the* O^+ *layer and that of the higher proton layer are controlled by diffusive (hydrostatic) equilibrium (source: (Encrenaz* et al. *2003)).*

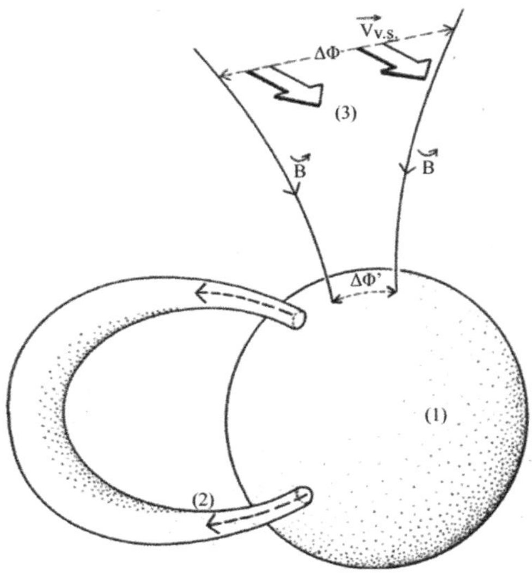

Figure 3.15. *Functions of the ionospheric plasma source in the electrodynamics of planetary ionized envelopes*

COMMENT ON FIGURE 3.15. – *(1) The ionospheric layers form a conductive shell, covering the neutral upper atmosphere; (2) for magnetized planets, the ionospheric plasma diffuses into the magnetic flux tubes; (3) the cold electrons that have diffused make the field lines highly conductive and provide electrical coupling of the ionospheric conductor with the magnetosphere. A fraction* $\Delta\Phi'$ *of the electric potential* $\Delta\Phi$ *induced by the solar wind between two magnetic field lines is thus applied to the ionosphere (source: (Encrenaz* et al. *2003)).*

In ionospheres without a strong magnetic field (Venus and Mars), the conductivity is naturally isotropic and equal to $\sigma_{//}$. The ionosphere is therefore a very good conductor, which surrounds the planet and opposes the penetration of the interplanetary magnetic field, as we will see later. In ionospheres of strongly magnetized planets, the electrodynamic properties of the ionosphere are equally important, as illustrated in Figure 3.15. Figure 3.16 shows a vertical profile of the ionospheric conductivities that is typical of the diurnal terrestrial ionosphere, which reveals that the layer of strong conductivities perpendicular to $\boldsymbol{B}$ forms a conductive shell that surrounds the Earth at the base of the ionosphere, a few neutral scale heights wide: this is the ionospheric dynamo layer. This highly conductive layer that intersects the magnetic field lines is electrically connected by these field lines to the regions farther away, in other words, to the magnetosphere and to the solar wind, due to the very high values of the parallel conductivity. The ionosphere is thus electrodynamically coupled with the magnetosphere and the solar wind. Through these different mechanisms, it will play an important role in the coupling of the planetary atmosphere with the interplanetary medium, which will be described in section 3.3.

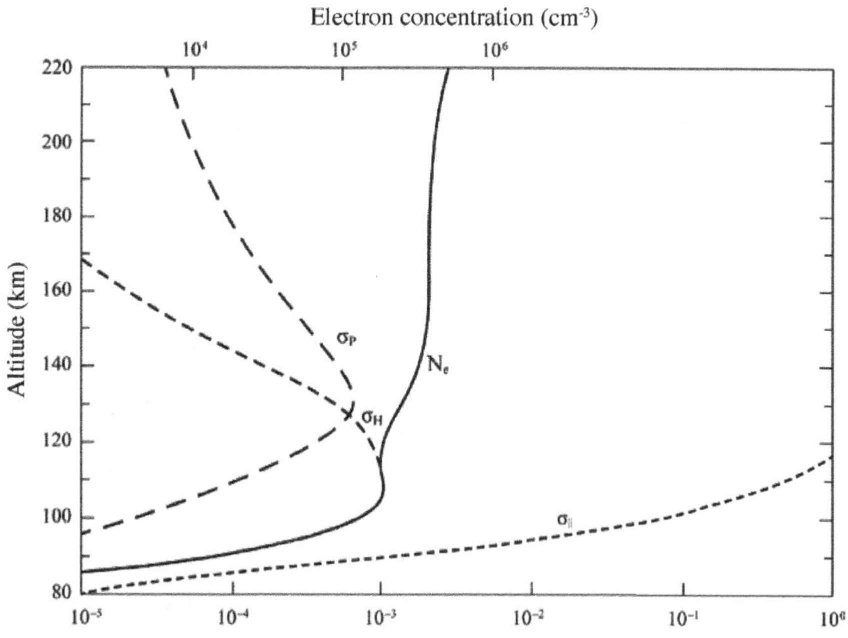

Figure 3.16. *Vertical profiles of parallel $\sigma_{//}$, Hall σ_H and Pedersen σ_P conductivities in the Earth's ionosphere for typical mid-latitude conditions on the day side*

COMMENT ON FIGURE 3.16. – *The electron concentration profile is superimposed with a different scale. $\sigma_{//}$ is several orders of magnitude larger than σ_H and σ_P at all ionospheric altitudes. The layer where there are high perpendicular conductivities extends from an altitude of 90 to 160 km. This is the dynamo layer of the Earth's atmosphere (source: after (Richmond 1983)).*

3.2.2. *Planetary magnetic fields*

The interior of Mercury, Earth and of the giant planets (see the following chapters) is the source of a large-scale magnetic field. The vector field $\boldsymbol{B}$ can be described as the gradient of a scalar potential ψ:

$$\boldsymbol{B} = -\nabla\psi \qquad [3.13]$$

This potential is expressed as the sum of the contributions of internal (noted as S_i^n) and external (S_e^n, including the magnetopause currents, the magnetospheric equatorial current disk, etc.) sources or currents:

$$\psi = R_P \Sigma_{n=1\to\infty} (r/R_P)^n S_e^n + (R_P/r)^{n+1} S_i^n \qquad [3.14]$$

r being the vector radius and R_P being the mean planetary radius. Each contribution is an expansion to degree n as a function of polar coordinates (θ, Φ):

$$S_i^n = \Sigma_{m=0\to n} P_n^m(\cos\theta)\,[g_n^m \cos m\Phi + h_n^m \sin m\Phi]$$

$$S_e^n = \Sigma_{m=0\to n} P_n^m(\cos\theta)\,[G_n^m \cos m\Phi + H_n^m \sin m\Phi] \qquad [3.15]$$

The $P_n^m(\cos\theta)$ are the orthogonal Legendre functions (with Schmidt normalization) and the g_n^m, h_n^m, G_n^m, H_n^m are the internal and external Schmidt coefficients. This description of the magnetic field is valid in the region between the internal and external sources, assumed to contain no currents. A simplification of this representation consists of describing the unavoidable currents (such as the magnetospheric equatorial disk) with an explicit model. We will therefore limit ourselves hereafter to the discussion of the internal potential coefficients, g_n^m and h_n^m. The degree n = 1 corresponds to the dipole, n = 2 to the quadrupole, n = 3 to the octupole, etc.

Modeling a planetary magnetic field consists of determining the g_n^m and h_n^m coefficients that provide the best fit to measurements made in the planetary environment (mainly by onboard magnetometers). As these measurements do not sample a closed surface encompassing all sources, and as they are inevitably accompanied by uncertainties, the solution is generally not unique. The models are thus improved iteratively as new observational constraints become available. The coefficients of the internal Earth field are known to a high degree (n = 13, in other words, 195 coefficients).

Table 3.1 gives the g_n^m and h_n^m coefficients of the most recent internal magnetic field models for Mercury, Earth and the four giant planets up to degree n = 3. A temporal variation of the internal magnetic field has been measured for the Earth and Jupiter. In the case of the Earth, this variation is measurable on a yearly scale and the field models are regularly updated.

Planet	Mercury	Earth	Jupiter	Saturn	Uranus	Neptune
R_P (km)	2,439	6,378	71,492	60,268	25,559	24,766
Model	Messenger	IGRF-12	JRM09	Cassini 11	AH5	O8
Order	4	13	10	11	4	8
g_1^0	- 0.00190	- 0.29442	+ 4.10245	+ 0.21140	+ 0.11278	+ 0.09732
g_1^1	0	- 0.01501	- 0.71498	0	+ 0.10928	+ 0.03220
h_1^1	0	+ 0.04797	+ 0.21331	0	- 0.16049	- 0.09889
g_2^0	- 0.00075	- 0.02445	+ 0.11670	+ 0.01581	- 0.09648	+ 0.07448
g_2^1	0	+ 0.03013	- 0.56836	0	- 0.12284	+ 0.00664
h_2^1	0	- 0.02846	- 0.42027	0	+ 0.06405	+ 0.11230
g_2^2	0	+ 0.01677	+ 0.48690	0	+ 0.01453	+ 0.04499
h_2^2	0	- 0.00642	+ 0.19353	0	+ 0.04220	- 0.00070
g_3^0	- 0.00022	+ 0.01351	+ 0.04019	+ 0.02260	- 0.01265	- 0.06592
g_3^1	0	- 0.02352	- 0.37791	0	+ 0.02778	+ 0.04098
h_3^1	0	- 0.00115	- 0.32957	0	- 0.01548	- 0.03669
g_3^2	0	+ 0.01226	+ 0.15926	0	- 0.04535	- 0.03581
h_3^2	0	+ 0.00245	+ 0.42085	0	- 0.02165	+ 0.01791
g_3^3	0	+ 0.00582	- 0.02711	0	- 0.06297	+ 0.00484
h_3^3	0	- 0.00583	- 0.27544	0	- 0.03036	- 0.00770
M^t dipole ($G.R_P^3$)	0.002	0.299	4.170	0.211	0.225	0.142
Inclination (B/Ω)	0.8°	+ 9.7°	- 10.3°	0°	- 59.8°	- 46.9°
Dipole center/planet center offset (R_P)	0.2	0.08	0.07	0.05	0.31	0.55
Reference	(Anderson *et al.* 2012)	(Thebault *et al.* 2015)	(Connerney *et al.* 2018)	(Dougherty *et al.* 2018)	(Herbert 2009)	(Connerney and Acuna 1991)

Table 3.1. *Internal Schmidt coefficients of the main internal planetary magnetic field models (in gauss), and parameters of the equivalent inclined and eccentric dipole model*

To obtain a complete magnetospheric field model, it is necessary to complete the considered planetary internal field model with the field produced by external magnetospheric current sources (equatorial current, magnetopause currents, magnetotail currents, etc.).

3.3. The solar wind's interaction with objects of the Solar System

3.3.1. *The different types of interaction*

The various objects of the Solar System form various physical obstacles to the flow of the solar wind, which bumps into them at supersonic and super-Alfvénic speed. This interaction can be understood in terms of pressure.

The physical nature of the obstacles encountered is very diverse, but relies on three main characteristics:

– the existence of an atmosphere around the planetary body, and the pressure prevailing in its external ionized component, the ionosphere, created by the ultraviolet and X-ray ionizing radiation of the Sun;

– the magnetic pressure created in the environment of the central body by its possible intrinsic magnetic field;

– the conductivity of the central body (liquid or solid). This last characteristic only really plays a role if the sum of the ionospheric pressure and the magnetic pressure is not sufficient to balance the total pressure (dynamic, thermal and magnetic) of the solar wind.

For supercritical flows such as the solar wind, it is in fact the dynamic pressure term that dominates, and it is therefore what we have to compare the different pressure terms of the obstacle to.

The geometry of the four main types of interaction is shown in Figure 3.17. In cases (a) and (c), the solar wind directly collides with the central body, without atmosphere or intrinsic magnetic field. In case (b), it collides with its atmosphere and, in case d), with its magnetized envelope.

In case (a), the central body is an insulator adsorbing the solar wind ions. The best-known example is the Moon. The lunar soil directly adsorbs the solar wind ions and the lunar material, as a whole, is not conductive enough to be penetrated through diffusion by the interplanetary magnetic field. The magnetic field lines or the solar wind plasma cannot therefore accumulate upstream of the obstacle, which remains free of any perturbation. On the contrary, downstream of the obstacle, a virtually empty plasma cavity is formed. It is progressively filled by the diffusion of ions and electrons as one moves downstream. As electrons diffuse more rapidly due to their low mass, a sheath of negative charges tends to form on the borders of this conical cavity.

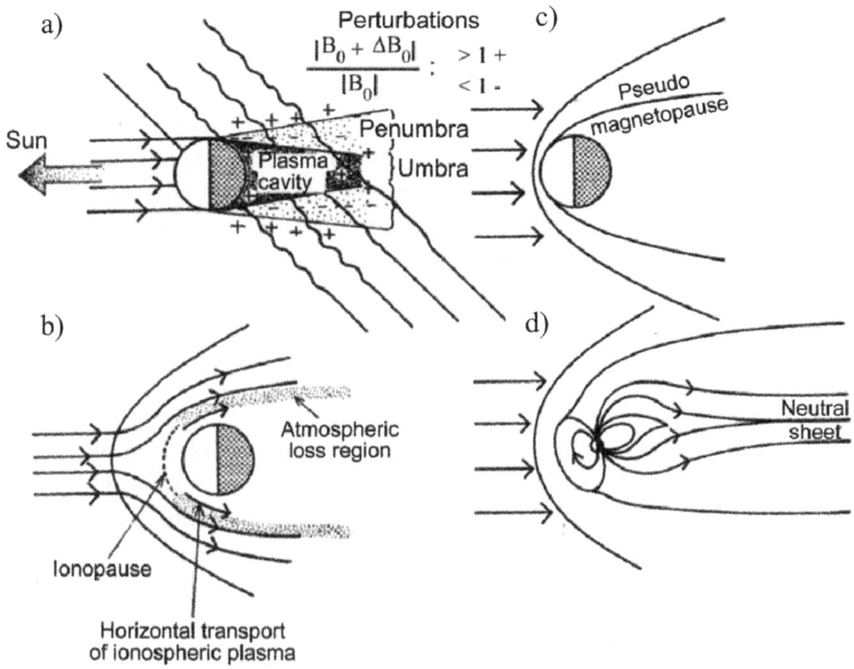

Figure 3.17. *Main cases of interaction between the solar wind and a planetary body (source: from (Lepping 1986))*

In all other cases, there is no simultaneous adsorption of the solar wind on the obstacle and diffusion of magnetic field lines through it. There is therefore an accumulation of either magnetic flux or plasma, or generally of both upstream of the obstacle. In a flow with a Mach number much higher than one, such as the solar wind, this leads to the formation of a stationary shock wave upstream of the obstacle, through which the plasma is brutally compressed and decelerated in order to pass into a subsonic and sub-Alfvénic flow regime, in which it can be deflected to flow around the obstacle. The plasma parameters on both sides of the shock wave are described by the Rankine-Hugoniot continuity relations. The dynamic pressure upstream of the shock is partially converted into thermal and magnetic pressures downstream: there is compression of the plasma and of the magnetic field, and thus adiabatic heating of the plasma and accumulation of interplanetary magnetic field lines in the region between the shock wave and the obstacle, called the magnetosheath.

The case in Figure 3.17c could theoretically occur, in the absence of an intrinsic atmosphere and field, if the central body was sufficiently conductive to prevent the diffusion of the interplanetary magnetic flux in its interior. The magnetic field lines would then bend and stretch around the central body in the magnetosheath. None of the planetary or cometary bodies of the Solar System seem to show this behavior. However, measurements by the Galileo, Cassini and HST space missions have revealed that many satellites (such as Europa, Ganymede and Callisto for Jupiter, or Enceladus for Saturn) have an ocean under their crust, which is probably conductive and could hinder the diffusion of the planetary magnetic field flux. Their interaction with the magnetosphere would be similar to the case shown in Figure 3.17c, except that the incident flow, being strongly magnetized, is, in this case, subsonic and sub-Alfvénic, and thus leads to a lesser stacking and stretching of the planetary magnetic field lines.

In the case of Figure 3.17b, the central body has a dense atmosphere but no magnetic field: the solar wind interacts directly with the upper atmosphere. This is the situation that prevails for comets, for Venus, and also for Mars. Saturn's satellite Titan, which also has a dense atmosphere, undergoes an interaction of this nature, sometimes directly with the solar wind, sometimes with the plasma and the magnetic field of Saturn's magnetosphere. Jupiter's Io satellite, whose atmosphere is of volcanic origin, also interacts in this way with the Jovian magnetosphere.

Lastly, in the case of Figure 3.17d, the central body has an intrinsic magnetic field that is strong enough to deflect the solar wind. The planetary magnetic field then digs a cavity in the wind, inside which it is mostly confined: this is the magnetosphere. Mercury, Earth, Jupiter, Saturn, Uranus and Neptune have a large-scale magnetosphere. By extension, the term *induced magnetosphere* defines the cavity formed by the non-magnetized conduction obstacles in Figure 3.17b.

We will now describe the cases of Figures 3.17b and 3.17d in more detail, which are the most common, but also the most interesting in terms of the physical mechanisms they involve.

3.3.2. *The case of non-magnetized gaseous envelopes*

3.3.2.1. *Comets*

Comets interact with the solar wind in a spectacular way (see Chapter 2), whose principle is given by the theoretical diagram of Figure 3.18. While approaching the Sun, the surface of the cometary nucleus, which is barely a few kilometers wide and composed of ice and various dusts, sublimates under the effect of solar heating. The produced gaseous envelope, the coma, enters supersonic expansion in all directions, then goes into subsonic flow after an internal shock.

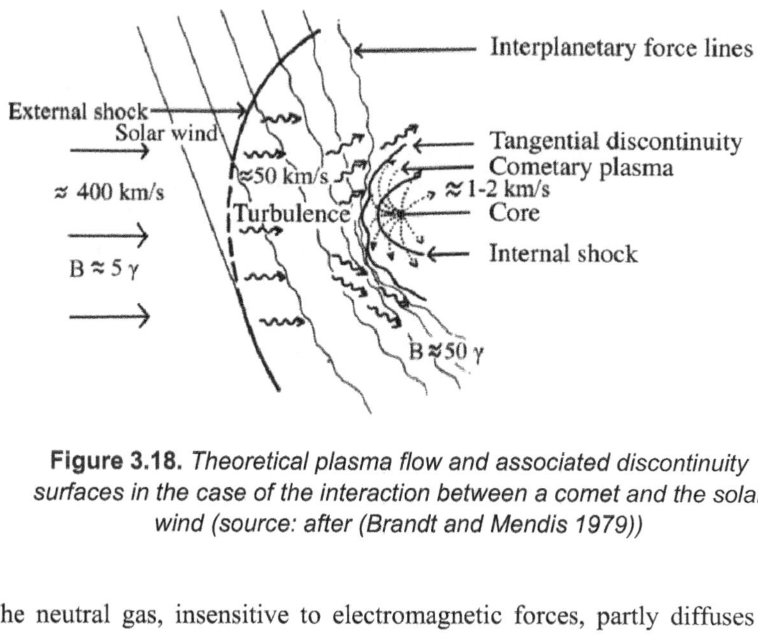

Figure 3.18. *Theoretical plasma flow and associated discontinuity surfaces in the case of the interaction between a comet and the solar wind (source: after (Brandt and Mendis 1979))*

The neutral gas, insensitive to electromagnetic forces, partly diffuses into the solar wind. Photometric observations have shown that a cloud thus formed could extend up to 10^6 km for the neutral atomic hydrogen produced by the photodissociation of the cometary water. The size of the cometary obstacle is thus much larger than that of the nucleus. The part of the neutral cometary gas that diffuses into the interplanetary medium is ionized by two main mechanisms: by charge exchange between the protons of the solar wind and the cometary atoms, and by "critical" ionization. This second process, studied by Alfvén in 1960, occurs when a magnetized plasma and a neutral gas interpenetrate at a sufficiently high speed, so that the kinetic energy of the neutral gases, with respect to the plasma, becomes greater than or equal to their ionization energy. The heavy cometary ions of low initial kinetic energy thus formed (as opposed to the light and high kinetic energy ions of the solar wind) are then accelerated by the convective electric field of the solar wind reaching its flow velocity. The impulse that is provided to them is, of course, taken from the solar wind flow. This latter is thus slowed down far upstream of the coma and weighed down in cometary ions. An external shock can appear if the flow becomes sub-Alfvénic, while the interplanetary magnetic field lines, slowed down by the effect of the cometary ions, bend around the obstacle.

In the coma itself, a fraction of the neutral gas is directly photoionized by solar EUV and X-rays, and then by various ionic chemistry reactions. This produces a highly conductive cometary ionosphere, therefore not very permeable to the external magnetic field, which is dragged along with the neutral gas in the coma expansion. In this explanation, the magnetized plasma of the solar wind (contaminated and weighed down with cometary ions) and the non-magnetized plasma of the cometary ionosphere are separated by a tangential discontinuity, the cometary ionopause, which is impermeable to the magnetic field. The draping of interplanetary magnetic field lines, which cannot penetrate the ionospheric conductor, around the obstacle is certainly at the origin of the formation of cometary plasma tails (called type I tails), whose anti-solar direction was interpreted since the 1950s as an indirect proof of the existence of solar wind. Figure 3.19 illustrates the formation of the radial structure of cometary tails by this bending mechanism, according to H. Alfvén. Observations accurately reveal that cometary tails are often structured in lines arranged radially with respect to the nucleus, downstream of the comet with respect to the solar wind. These lines are generally arranged in symmetrical pairs that bend towards the tail axis through time, as expected for interplanetary magnetic field lines bending around the obstacle.

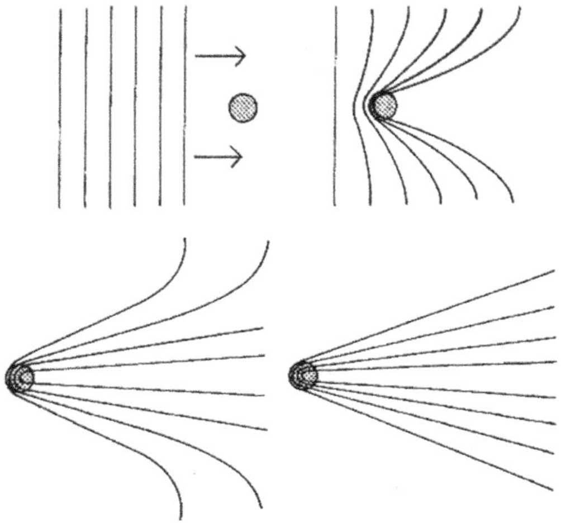

Figure 3.19. *The interaction of a comet with the interplanetary magnetic field explains the formation of a cometary plasma tail and the motion of cometary lines (source: after (Alfvén 1957))*

In 1986, five interplanetary probes flew over Halley's comet: the Vega 1 and 2 Soviet probes, the ESA's Giotto probe, and the two Japanese interplanetary probes, Sakigake and Suisei. In 1987, NASA's ICE (International Cometary Explorer) satellite performed the first *in situ* exploration of a cometary environment by crossing the tail of comet Giacobini–Zinner. Finally, between 2014 and 2016, with the help of its Philae lander, the ESA's Rosetta mission explored the environment and surface of comet 67P/Churyumov-Gerasimenko in detail, during its perihelion passage. These explorations have shown good consistency between the measurements and the theoretical predictions described above. However, they revealed a much more complex and dynamic cometary environment than expected. In the case of Halley, at least three successive discontinuities have been identified in the region of the external shock, where the interplanetary magnetic field lines progressively stack up on the cometary obstacle. The contact discontinuity, the ionopause, has been observed very clearly by Giotto in all the plasma parameters, as well as in the magnetic field intensity, becoming almost zero inside this boundary. The structure of this discontinuity seems to result from a complex dynamic and chemical interaction between the plasma, the neutral gas and the magnetic field. The continuous exploration of 67P/Churyumov-Gerasimenko for several years has made it possible to accurately map the solar wind interaction region and its different plasma boundaries, as well as their temporal evolution for different illumination and solar wind conditions (Glassmeier 2017). Plasma waves and turbulence play a major role in heating cometary ions. The latter are accelerated by the convective electric field of the solar wind: $\boldsymbol{E}_{conv} = -\boldsymbol{V} \times \boldsymbol{B}$, according to equation [3.9], in the direction perpendicular to the flow and the interplanetary magnetic field, carry an electric current that forms an electrodynamic obstacle to the solar wind, deflected in the direction antiparallel to $\boldsymbol{E}_{conv}$, producing an asymmetric structure. The same type of interaction as the one described here applies to the case of Pluto.

3.3.2.2. *Venus and Mars*

3.3.2.2.1. Venus

The Venus example stands as the archetype of the interaction between the solar wind and the atmospheres of non-magnetized planets, which has been studied by more than two dozen space probes, including the Pioneer Venus Orbiter probes (NASA) between 1978 and 1992, and the Venus Express (ESA) probe from 2006 to 2014. The geometry of this interaction, shown in Figure 3.20, is similar to that of comets: beyond the outer shock zone, where the solar wind slows down to a sub-Alfvénic speed, the interplanetary magnetic field lines and plasma accumulate on the day side of the planetary ionosphere. They remain separated by a boundary, the ionopause, which is not very permeable to the external magnetic field, and which delineates the induced magnetosphere.

The two interaction mechanisms mentioned above for comets also apply here. On one hand, the solar wind upstream of Venus is weighed down by the diffusion of the neutral components of its upper atmosphere (mainly composed of atomic oxygen), although the scale of this effect is diminished by the gravity which binds the atmosphere to the planet, and makes its concentration decrease very quickly with altitude. On the other hand, the planetary ionosphere interacts directly with the compressed solar wind of the magnetosheath through the ionopause, along which the magnetic pressure of the magnetosheath is balanced by the pressure of the ionospheric plasma (Futaana *et al.* 2017).

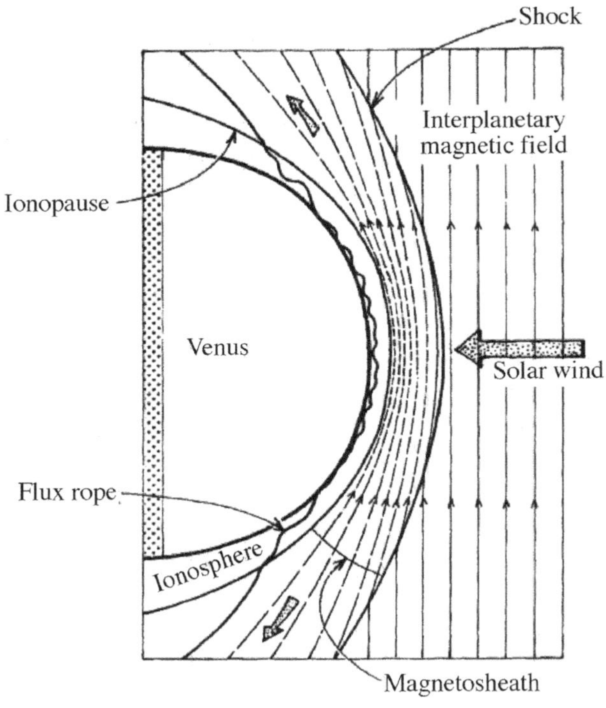

Figure 3.20. *Simplified diagram of the Venus/solar wind interaction (source: after (Saunders* et al. *1986))*

Figure 3.21 shows three vertical profiles of the magnetic field and ionospheric plasma concentration, as measured by Pioneer Venus during three passages over the day side of Venus. The structure of the ionopause varies in a way that is correlated to the incident dynamic pressure of the solar wind, increasing from left to right. The altitude of the ionopause decreases considerably, from 420 to 250 km, while its thickness increases simultaneously from a few kilometers to more than fifty, probably due to an upper scattering of the plasma facilitated by low altitude collisions. Figure 3.21 also shows that the decrease of the altitude of the ionopause has a dramatic effect on the permeability of this barrier to the external magnetic field. When the ionopause is located at high altitude (left-hand side), only very localized maxima of the magnetic field that are a few kilometers wide remain in the ionospheric layer. These structures are spatially organized as magnetic flux "ropes". The formation of these Venusian flux ropes has been attributed to the development of a Kelvin-Helmholtz instability, produced by the flow of the magnetosheath plasma over the ionopause. When the ionopause is at low altitude (middle and right-hand side), the external magnetic field seems to penetrate the ionosphere and accumulate in its lower part, where a magnetic field layer is formed reaching a similar amplitude to the magnetosheath.

As in the cometary case, the accumulation of the interplanetary magnetic flux upstream of the planet results in the draping of magnetic field lines of the magnetosheath around the planet downstream of the flow. This produces a very low density plasma cavity, which was explored for the first time by Venus Express. The plasma density then rises progressively downstream of the flow by expansion of the magnetic field tubes, and by diffusion of the plasma along these field tubes. At a few planetary radii on the night side, the complete bending of the magnetic field tubes leads to the formation of two magnetic lobes, of opposite magnetic polarities, separated by a current layer, together forming a magnetotail.

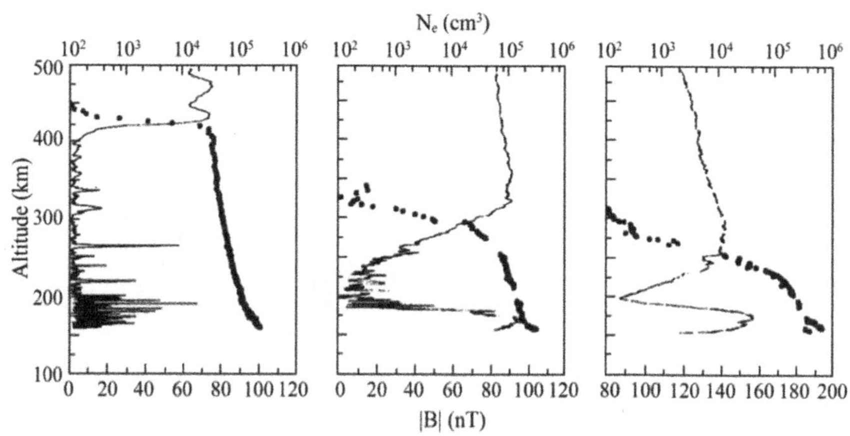

Figure 3.21. *Measurements of the electron density (black dots) and the magnetic field (thin lines), as a function of altitude, during three flybys over the day side of Venus by Pioneer Venus Orbiter. From left to right-hand side, the boxes show increasing values of the dynamic pressure of the solar wind (source: after Russel* et al. *1982)*

3.3.2.2.2. Mars

The ionized Martian environment has been much less explored than that of Venus, having mainly been surveyed by the Phobos Soviet mission, the Mars Global Surveyor (NASA), the Mars Express mission (ESA) and, since 2014, by NASA's Maven probe (Mars Atmosphere and Volatile Evolution), whose objective is to characterize the erosion of the Martian atmosphere.

The large structures of the Venus induced magnetosphere (upstream shock, magnetic stacking on the day side, magnetic lobes and current layer on the night side) have also been observed on Mars. In contrast, the Martian equivalent of the Venusian ionopause is more seldom observed (Brain *et al.* 2017). Mars Global Surveyor discovered pockets of magnetic field in the Martian crust. These residual magnetic sources are remnants of a past global magnetic field that probably existed for several hundred million years after the planet's formation (see section 4.1.3). The Mars Global Surveyor and, more recently, Maven probes have been able to make a fairly complete mapping of the magnetic field produced by these sources above the surface. The map presented in Figure 3.22 shows an inhomogeneous distribution of the radial magnetic field, with intense regions in the southern hemisphere, and the absence of intense sources in the impact basins and near the large volcanoes. This absence is interpreted as a local demagnetization of the crust, produced by impacts at the origin of these basins or by lava flows during volcanic eruptions.

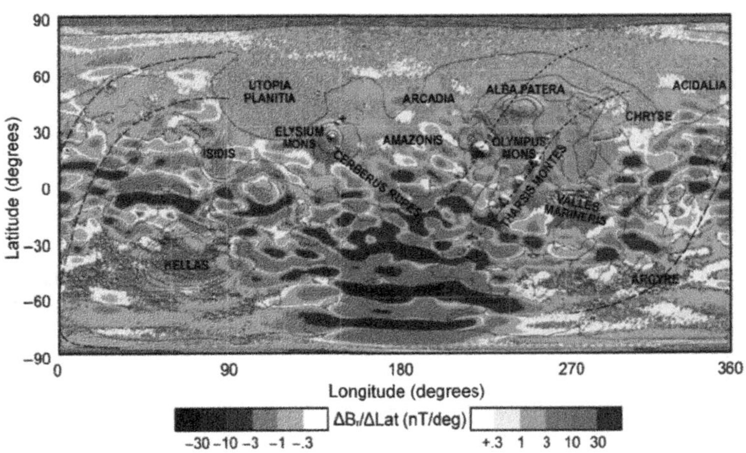

Figure 3.22. *Cylindrical projection map of the radial magnetic field of Mars, measured at 400 km altitude by the Mars Global Surveyor, plotted in 1°x1° steps in longitude and latitude, superimposed on a map of the Martian relief (source: from (Connerney* et al. *2005)). For a color version of this figure, see www.iste.co.uk/ encrenaz/solar1.zip*

Maven's first results show that the pockets of residual magnetic field (magnetic anomalies) modify the structure of the flow and the internal magnetospheric boundaries by creating local mini-magnetospheres, schematically shown in Figure 3.23. The current fundamental questions concerning the past dynamo involve finding out why it stopped and if the global magnetic field, which existed during the first few hundreds of millions of years after the formation of Mars, could have protected its atmosphere from erosion by the solar wind.

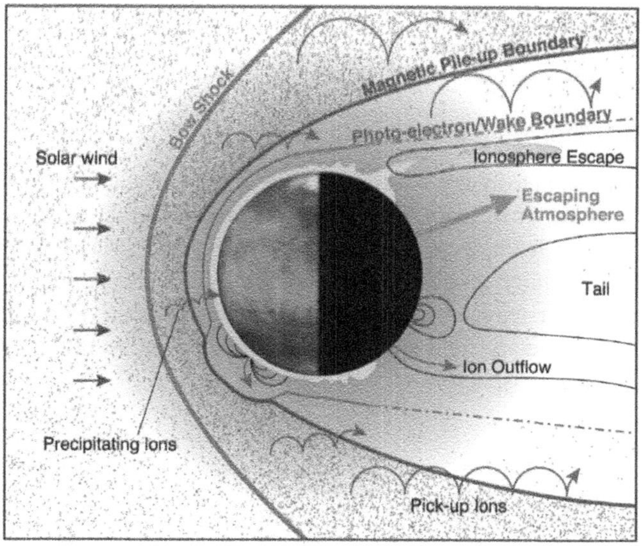

Figure 3.23. *The solar wind's interaction with the atmosphere, the ionosphere and Mars's residual magnetic field, indicating the main processes of atmospheric erosion (source: after F. Bagenal and S. Bartlett). For a color version of this figure, see www.iste.co.uk/encrenaz/solar1.zip*

3.3.3. *The case of magnetized planets*

For planets with an intrinsic magnetic field (Mercury, Earth, Jupiter, Saturn, Uranus and Neptune), the magnetic pressure in the nearby planetary environment exceeds that of the outer atmospheric layers. In this case, the magnetic field forms the main obstacle to the flow of the solar wind. It digs a cavity in the interplanetary medium, the magnetosphere, whose shape was calculated in the 1960s. Two surfaces of discontinuity are schematized in Figure 3.17d and Figure 4.15:

– the bow shock separates the unperturbed interplanetary medium from the solar wind, which is decelerated in the magnetosheath;

– the magnetopause is a tangential discontinuity separating the magnetosheath from the planetary magnetic cavity.

The external flow and shape of these boundaries have been well reproduced by models imposing magneto-hydrostatic equilibrium at the magnetopause, and by computing the external flow with hydrodynamic codes, leading to the concept of a closed magnetosphere, illustrated below in the Earth's reference case. The cases of Mercury and giant planet magnetospheres are developed in section 4.2.6 and section 5.4, respectively.

3.3.3.1. *The closed magnetosphere*

3.3.3.1.1. The magnetopause

The Earth's magnetosphere was discovered in the early 1960s, thanks to *in situ* measurements of the first space probes. These revealed that the Earth's magnetic field, which was far from decreasing regularly with geocentric distance, stopped at an abrupt discontinuity marking the limit of its extension into space: the magnetopause. For a magnetized planet, the magnetopause forms the boundary between the interplanetary field and matter (the solar wind) and the planetary field and matter (the magnetosphere).

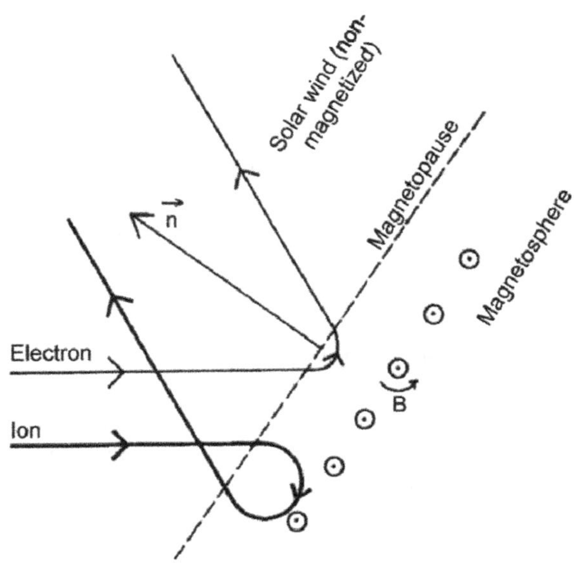

Figure 3.24. *The specular reflection of a charged particle of the solar wind (taken here as a non-magnetized jet of electric charges) on the magnetopause is the consequence of its Larmor gyration motion in the planetary magnetic field*

COMMENT ON FIGURE 3.24. – *This gyration motion has an opposite direction for ions and electrons, and the resulting charge separation produces an electric current that is tangential to the magnetopause and orthogonal to the magnetospheric field. This electric current isolates the solar wind from the planetary magnetic field (source: (Encrenaz* et al. *2003)).*

To determine the geometry of a magnetosphere, one considers the conservation of mass and momentum (including electromagnetic and gravitational fields) of the MHD conducting fluid circulating in the planetary environment during its motion. It can be shown that the equations governing the variations of a macroscopic quantity of density $\overline{\overline{\boldsymbol{D}}}$ (mass, momentum tensor) characterizing the moving fluid can be put into the following conservative form:

$$\partial\overline{\overline{\boldsymbol{D}}}/\partial t + \nabla.\boldsymbol{F} = 0 \qquad [3.16]$$

where $\boldsymbol{F}$ is the flux of this same quantity. At a surface of discontinuity S (of normal **n**), in a steady state ($\partial/\partial t = 0$), the jump conditions are written as:

$$\boldsymbol{F}.\mathbf{n}\ (\text{side } 1) = \boldsymbol{F}.\mathbf{n}\ (\text{side } 2) \qquad [3.17]$$

which means that the flow is continuous across the surface (S) between sides 1 and 2, respectively, on the left and right-hand side in Figure 3.24. Applied to the conservation of mass, equation [3.16] gives:

$$\rho\boldsymbol{V}.\mathbf{n}\ (1) = \rho\boldsymbol{V}.\mathbf{n}\ (2) \qquad [3.18]$$

where ρ is the density of the fluid and $\boldsymbol{V}$ is its bulk velocity. The flux of matter is conserved across the discontinuity surface (S).

To write the conservation of momentum simply, or in other words the balance of forces on this surface, we can consider the solar wind as a monokinetic jet of individual, non-magnetized, particles without collisions. We neglect the magnetic field and the thermal pressure of the plasma on side 1, which is justified by the supersonic character of the solar wind flow. If each particle of mass m undergoes specular reflection as it crosses the magnetopause under the effect of the planetary magnetic field, as shown in Figure 3.24, it yields an impulse $2m(\boldsymbol{V}.\mathbf{n})$. The impulse flux of the solar wind is thus $2\rho V^2\cos^2\chi$, χ being the angle between the velocity $\boldsymbol{V}$ and the normal **n** to the magnetopause.

In a steady state, Maxwell's equations impose:

$$\boldsymbol{\nabla}.\boldsymbol{B} = 0 \qquad [3.19]$$

$$\nabla \times E = 0 \qquad [3.20]$$

Therefore, B_N (normal component to the discontinuity) and E_T (tangential component to the discontinuity) are continuous.

On side 2 (magnetosphere), there is little matter but a magnetic field $\boldsymbol{B}$ produced by the planet. The fact that $B = 0$ in the solar wind imposes $B_N = 0$ at the discontinuity, and thus $B = B_T$: the planetary magnetic field is tangential at the magnetopause. The impulse flux on side 2 is therefore reduced to the magnetic pressure: $P = B_T^2/2\mu_o$.

The balance of the impulse (or pressure) flux on both sides of the discontinuity (S) thus leads to the very simple condition of magnetostatic equilibrium:

$$2\rho V^2 \cos^2\chi = B_T^2/2\mu_o \qquad [3.21]$$

whose resolution for $\chi = 0$ directly gives the position of the nose of the magnetopause in the subsolar direction. More generally, the expression [3.21] provides the global geometry of the magnetopause and the magnetic field distribution in the magnetospheric cavity, by solving the free boundary problem shown in Figure 3.25.

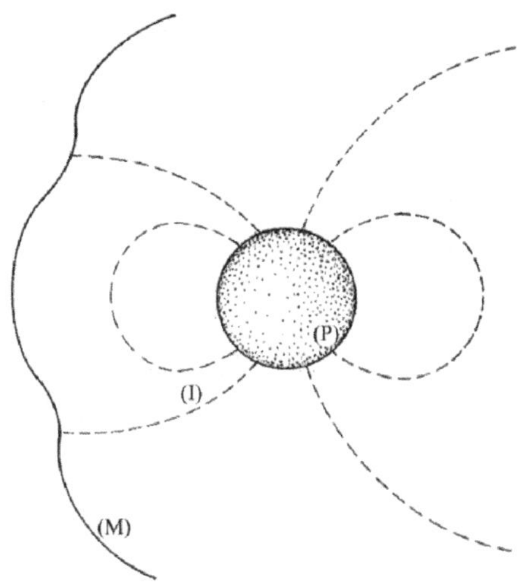

Figure 3.25. *Geometry of the free boundary problem, defining the shape of the magnetopause (M) associated with the magnetic flux source emerging from the planetary surface (P), with the dotted field lines occupying the space (I) (source: (Encrenaz* et al. *2003))*

Precisely, we want to find a function ψ defined in the domain (I) that separates the planetary surface (P) from a discontinuity surface (M), the magnetopause, such as:

$$\boldsymbol{B} = -\nabla\psi \qquad [3.22]$$

We neglect the effect of electric currents in the magnetic cavity, except for the internal currents of the planet, responsible for its intrinsic magnetic field.

B_N, the component of $\boldsymbol{B}$ that is normal to the surface (P), is set equal to the value given by the internal magnetic field source. Here, we choose the value of the planetary magnetic dipole for B_N:

$$B_N = B_e(1 + 3\cos^2\theta)^{1/2} \qquad [3.23]$$

with B_e being the intensity of the equatorial field on the planetary surface and θ the colatitude. On the surface (M), we also have:

$$B_T^2 = 4\mu_o\rho V^2 \cos^2\chi \qquad [3.24]$$

$$B_N = 0 \qquad [3.25]$$

the magnetopause being impermeable to the magnetic flux.

3.3.3.1.2. The Chapman-Ferraro model

As early as 1940, Chapman and Ferraro obtained a (wrong but approximate) solution to this problem through the method of images. They assumed a solar wind/magnetosphere interface that was planar and perpendicular to the Earth-Sun direction. To obtain a solution verifying equations [3.22] and [3.25] on the magnetospheric side of the interface (M), one simply needs to superimpose the mirror dipole to the interface on the Earth's dipole magnetic field. The resulting magnetic field lines in the planetary magnetic meridian plane, containing the Sun-planet axis, are shown in Figure 3.26.

For given characteristics of the solar wind, the position of the magnetopause is determined by imposing the condition of pressure equilibrium at the subsolar point S, at the intersection of the magnetopause and the Earth-Sun axis (this condition is not verified at any other point). The field strength at S is then twice that of the planetary dipole field alone:

$$B_S = 2\,B_e / L^3 \qquad [3.26]$$

where $L = R/R_P$ is the ratio of the planetocentric distance R from S and the planetary radius R_P. The magnitude of the dipole magnetic field at any point of vector radius R (expressed in R_P) and colatitude θ writes as:

$$B = (B_e / R^3)\,(1 + 3\cos^2\theta)^{1/2} \qquad [3.27]$$

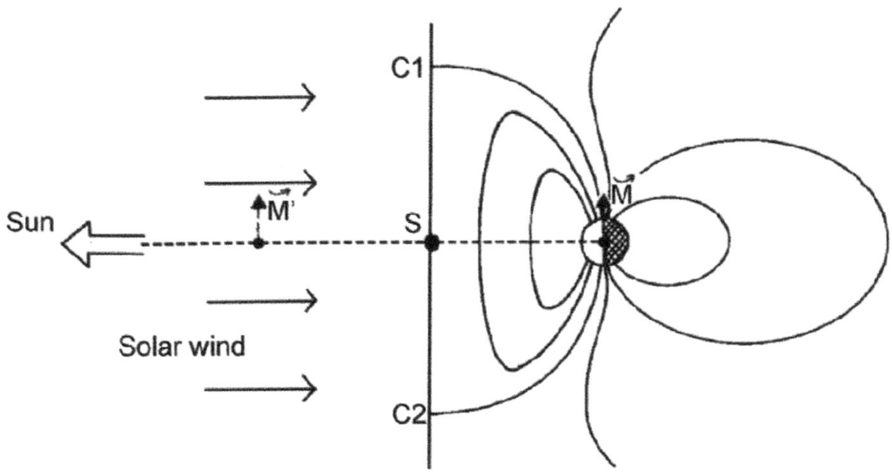

Figure 3.26. *Geometry of the magnetic field lines, resulting from the solar wind/magnetosphere interaction in the simplified Chapman-Ferraro model*

COMMENT ON FIGURE 3.26. – *The C1 and C2 points on the magnetopause along the axis of the M dipole correspond to neutral points where B = 0. The polar magnetospheric regions delineated by these points, and the space between the field lines oriented on the day side and on the night side, form the polar cusps (source: (Encrenaz* et al. *2003)).*

Another way to find this result is to express B_S as the sum of the contribution B_P of the planetary field and the contribution B_C of the electric currents flowing at the surface of the magnetopause, according to the Maxwell-Ampere equation (see Figure 3.24). For two infinitely close points located on both sides of the magnetopause, we obtain:

– magnetosphere side: $B = B_P + B_C$;

– solar wind side: $B' = B'_P + B'_C = 0$ with $B'_P = B_P$ and $B'_C = -B_C$.

We thus obtain the field in S (magnetosphere side): $B = 2B_P = 2B_e/L^3$.

Equation [3.24] with $\chi = 0$ then allows us to fix the planetocentric subsolar distance L_M of the magnetopause, expressed in planetary radii:

$$L_M = (B_e^2 \rho V^2 / \mu_o)^{1/6} \qquad [3.28]$$

For typical solar wind parameters at 1 au ($V = 400$ km s^{-1}, $N = 5$ cm^{-3}, with $\rho = Nm_p$), we find $L_M \sim 9$ R_T for the Earth's case, with R_T being the Earth's radius, in good agreement with the observations (see Table 1.6). The same calculation applied to the other magnetospheres also shows good agreement, with the exception of Jupiter (and, to a lesser extent, Saturn), where the expected value of $L_M \sim 60$ R_J with R_J being the Jovian radius, is lower than the value measured by the Jupiter exploration probes, which is about 90 R_J (see section 5.4). This discrepancy is due to the deformation of the Jovian magnetic field by the internal plasma sources that are specific to its magnetosphere, and whose pressure is not taken into account in the approach described above.

3.3.3.1.3. Hydrodynamic models

As the three-dimensional problem had no analytical solution, computer simulations have, since the 1960s, made it possible to determine the three-dimensional flow of the solar wind around the magnetospheric obstacle. These calculations used several simplifying hypotheses. Inspired by the Chapman-Ferraro solution discussed above, and by the expression of pressure equilibrium on the magnetopause given by the equation [3.21], it is assumed that, on the one hand, the solar wind pressure on the magnetopause is expressed by $K\rho V^2\cos^2\chi$, K being a parameter varying between 1 and 2 to simulate various conditions of particle reflection by the magnetopause (in the previous section, a specular reflection was considered with $K = 2$), and on the other hand, the B_T tangential component of the magnetic field at the magnetopause (on the magnetospheric side) is simply equal to twice the tangential component of the dipole field. Thanks to these two hypotheses, B_T at the magnetopause becomes a function of the position and of χ, and the following equation:

$$K\rho V^2 \cos^2\chi = B_T^2 / 2\mu_o \qquad [3.29]$$

can be solved as an ordinary differential equation giving (via χ) the shape of the magnetopause, even before the flow is known in detail. The flow is then calculated in a second step with a hydrodynamic code, the shape of the obstacle to be bypassed being now known.

Figure 3.27a shows the current lines, as solid curves, resulting from a simulation solving this calculation in a meridian plane. We see that the undisturbed flow of the solar wind is deflected around the obstacle upstream of the magnetopause, once past the bow shock. Along the Sun–Earth axis, the flow becomes immediately subsonic beyond the shock. Then, as indicated by the constant Mach number lines (long dashed lines) and the line showing the passage of the speed of sound (short dashed lines), the flow is again accelerated, during the circumvention of the obstacle, by the zone of strong pressures created behind the shock near the Sun–Earth line. The flow then becomes supersonic again very quickly, downstream of the shock, on the flanks of the magnetosphere.

Once the plasma flow is known, the magnetic field distribution in the magnetosheath can be calculated. Figure 3.27b plots the magnetic field lines (solid curves) and the current lines (dashed curves) for two different orientations of the interplanetary magnetic field, assumed to be uniform upstream of the bow shock. In both cases, the magnetic field lines are stacked, and thus the magnetic field is intensified, upstream of the magnetopause. As for the interactions seen in section 3.3.2, the field lines wrap around the magnetosphere. It should be noted that the evacuation of the accumulated magnetic flux occurs along the region that is orthogonal to the plane of the figure.

The calculation of the magnetic field is simply done by using the equation of the magnetic flux frozen in the plasma (equivalent to equation [3.9]):

$$\partial \boldsymbol{B}/\partial t + \boldsymbol{\nabla}\times(\boldsymbol{V}\times\boldsymbol{B}) = 0 \qquad [3.30]$$

which carries the interplanetary magnetic field of the undisturbed solar wind into the magnetosheath. For a velocity $\boldsymbol{V}$ along the Sun–Earth axis and an interplanetary magnetic field $\boldsymbol{B}$ perpendicular to $\boldsymbol{V}$, this equation reduces to $\boldsymbol{V}.\boldsymbol{B}$ = constant in the magnetosheath. The intensification of the magnetic field is then directly proportional to the flow deceleration, and the solar wind dynamic pressure $K\rho V^2\cos^2\chi$ is converted into an equivalent magnetic pressure $B^2/2\mu_o$, just upstream of the nose of the magnetopause.

Figure 3.28 compares the hydrodynamic simulations to the positions of the bow shock (black dots) and the Earth's magnetopause (circles), as measured by the IMP-1 satellite. The predicted shock and magnetopause positions for a given mean pressure value are plotted in solid lines. The selected measurements are picked up for comparable solar wind dynamic pressures. The agreement between observations and simulations is excellent, particularly on the day side. However, there is a systematic deviation which increases when moving away from the Earth on the night side. This discrepancy, although not significant, is due to the assumption that the magnetic field is completely frozen in the flow, a hypothesis made in the framework of the ideal MHD and not verified in these regions.

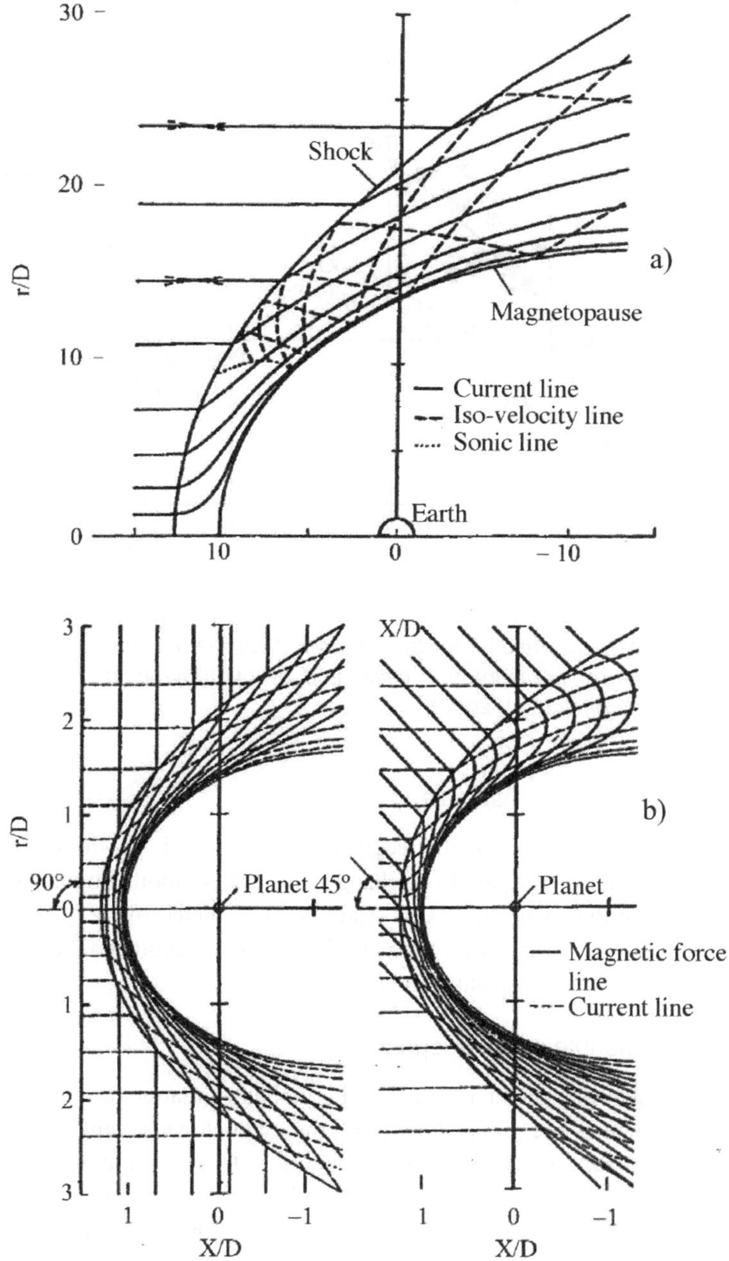

Figure 3.27. *Results of a hydrodynamic simulation of the solar wind's interaction with the Earth's magnetosphere, in a meridian plane (source: from (Spreiter* et al. *1966))*

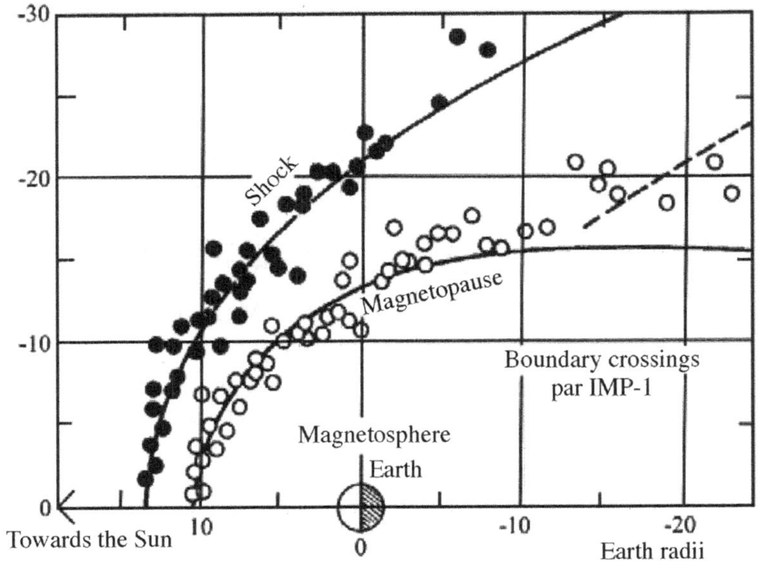

Figure 3.28. *Positions of the bow shock and of the terrestrial magnetopause (solid and hollow circles), as measured by the IMP-1 probe for the same range of interplanetary dynamic pressure and calculated by hydrodynamic simulation (solid lines) (source: (Encrenaz* et al. *2003))*

3.3.3.2. *The open magnetosphere*

In the ideal MHD framework, the magnetospheric cavity is, by definition, closed (and the magnetopause opaque) to exchanges of matter, momentum, energy and magnetic flux with the solar wind. This result is contradicted by daily observations of the control that the solar wind exerts on the terrestrial geomagnetic and auroral activity obtained both on the ground and in space.

3.3.3.2.1. A permeable magnetosphere

It is now known that about 1% of the interplanetary matter flux that intercepts the Earth's magnetosphere succeeds to cross the magnetopause, and constitutes an important magnetospheric plasma source, feeding in the auroral emissions and the radiation belts. Satellite measurements in the solar wind and in geostationary orbit have also shown that about 10% of the interplanetary electric field is transmitted into the magnetosphere. However, statistical studies, as well as the analysis of geophysical events, have shown that the geomagnetic activity is not equally influenced by all parameters of the solar wind. The component of the interplanetary magnetic field that is parallel to the z-axis of the planetary magnetic dipole, noted as B_Z, plays a preponderant role. This activity is all the more intense as the interplanetary magnetic field is more strongly oriented towards the south (in the case of the Earth), in other words, in the same direction as the planetary dipole. Magnetosphere compressions resulting from sudden increases of the dynamic pressure in the solar wind (called "magnetospheric storms") also cause intense auroral and geomagnetic activity. Lastly, any physical model of the solar wind/magnetosphere interaction must explain the very particular geometry of the Earth's auroral zones. Figure 3.29 shows a UV image of the Earth taken by the DE1 (Dynamics Explorer 1) satellite. On the left-hand side, it shows the sunlight reflected by the illuminated face of the Earth and, in the center, a circle of auroral emissions in the upper atmosphere, produced by the precipitation of energetic particles from the magnetosphere. This "auroral oval", which is brighter in the polar night side regions, is located at latitudes varying between 65° and 75° and is centered on a point offset by about 5° on the night side from the magnetic pole. Its radius and intensity increase with geomagnetic activity, and with the intensity of the B_Z component of the interplanetary magnetic field and/or of the dynamic pressure.

3.3.3.2.2. The Dungey model

To account for these observational facts, J. Dungey proposed an "open magnetosphere" model in 1961 (Dungey 1961). This model is based on the mechanism of magnetic reconnection to the magnetopause, which, by construction, excludes the ideal MHD. The principle is as follows: when the interplanetary and planetary magnetic fields on either side of the magnetopause have opposite directions, an unstable current sheet appears on the magnetopause, which flows perpendicular to the field lines. The development of this "shearing" instability can then lead to the reconnection of the interplanetary and planetary magnetic fields, rendering the magnetopause permeable to the solar wind. The terrestrial magnetic field at the nose of the magnetopause being, by definition, directed towards the north ($z > 0$), this situation occurs when the B_Z component of the interplanetary field is antiparallel ($z < 0$).

The magnetic topology and plasma flow resulting from Dungey's model are schematically shown in Figure 3.30, which shows the magnetic field lines in the magnetic meridian plane with the planet-Sun direction and the planetary pole axis. In the equatorial plane, a neutral line is formed, along which the magnetic field has a topology of neutral X points, which implements the physical connection between the solar and planetary magnetic fields. The trace of this neutral line in the plane of the figure corresponds to the N1 and N2 points. Three types of field lines appear:

1) the "closed" planetary field lines connecting the two magnetic poles, isolated from the solar wind;

2) the "open" interplanetary field lines, entirely immersed in the solar wind and the magnetosheath;

3) the "semi-open" field lines resulting from the reconnection phenomenon, connecting the interplanetary medium to the planet.

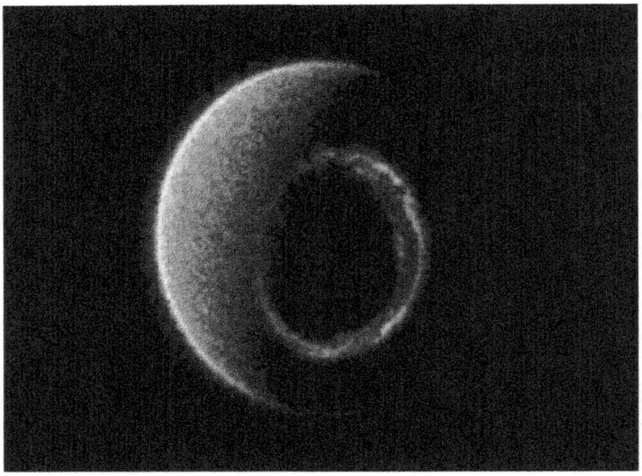

Figure 3.29. *UV observation of the aurorae of the Earth's northern hemisphere, taken by the DE1 satellite on November 23, 1981 at 10:48 UTC (source: NASA). For a color version of this figure, see www.iste.co.uk/encrenaz/solar1.zip*

At the level of the planetary upper atmosphere, the boundary between the type 3 and type 1 field lines fairly accurately coincides with the auroral oval (visible in Figure 3.29). This was evidenced by spatial measurements of electrons precipitating into the upper atmosphere. While inside each auroral oval, the precipitating electrons have the same characteristic energy as those encountered in the perturbed solar wind after the shock (the magnetosheath), those encountered at the same level than the auroral ovals, and outside of them, have much higher characteristic energies (from 1 to 10 keV). Thus, the auroral ovals seem to be the magnetic projection of the neutral line onto the Earth's upper atmosphere which, in Dungey's model, goes around the planet near the equatorial plane.

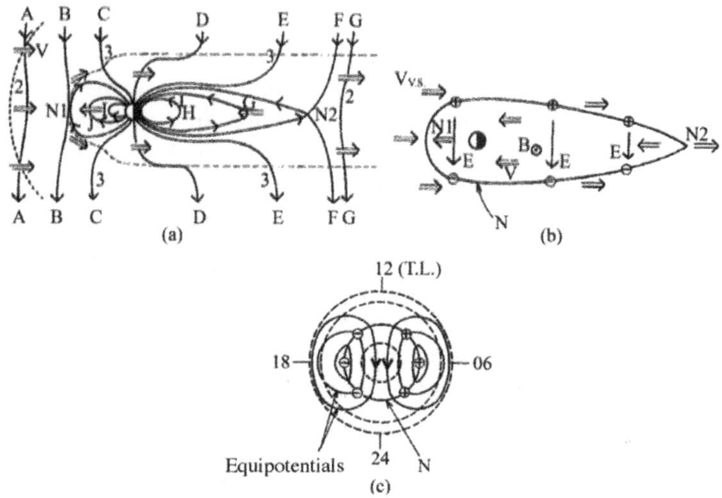

Figure 3.30. *Schematic diagram of Dungey's open magnetosphere model*

COMMENT ON FIGURE 3.30. – *(a) In a midday-midnight magnetic meridian plane, the interconnection of the interplanetary and terrestrial magnetic fields (solid lines) is shown for the case of a uniform southward-directed interplanetary field, with the bow shock in dotted line and the magnetopause in dashed line. This interconnection allows the day-night flow motion of the interplanetary plasma (steps A-G) to be transmitted to the semi-open field lines of the polar caps (noted as 3). Conservation of magnetic flux implies that the plasma on the closed field lines (noted as 2) flows inversely in a night-day motion (steps G-J). (b) In the equatorial plane, the neutral (or reconnection) line N continues, connecting the midday reconnection point N1 to the one located far in the tail at midnight, point N2. Plasma flows toward the Sun within the neutral line, driven by the solar wind-induced dawn-dusk electric field. c) Projected along the field lines onto the ionosphere, where circles of constant latitude are plotted in dashed, the N line corresponds to the polar cap/auroral zone boundary, along which the most intense aurorae appear. The plasma flow lines above these regions, which are also electric equipotentials, indicate that it flows toward the night (day) side inside (outside) the auroral circle (source: after (Lyons and Williams 1984)).*

3.3.3.3. *The circulation of magnetospheric plasma*

3.3.3.3.1. Convection motion

Type 3 field lines have an essential consequence for the solar wind/planet interaction: they allow the solar wind flow momentum to be partially transmitted into the magnetosphere. Indeed, the validity conditions for the frozen-in theorem are verified almost everywhere in the Dungey model, except in the immediate vicinity of the neutral line. We can thus understand the motion of the ambient plasma inside the magnetosphere, by following the type 3 field lines in their motion. In a steady state, these lines are driven by the solar wind on the night side (steps B-F). The circulation of the magnetospheric plasma thus follows the double arrows of Figure 3.30a. The conservation of magnetic flux then requires that, for the day side magnetic reconnection to be fed in by planetary magnetic flux, the circulation of plasma should be directed to the day side along type-1 closed field lines. This is indeed what is observed: in the equatorial plane shown in Figure 3.29b, the motion of the plasma is directed towards the Sun inside the neutral line, which, at least on the day side, draws the limits of the magnetospheric cavity.

At the level of the planetary ionospheric layers, of which Figure 3.29c represents a hemispheric polar projection, the motion of the plasma at high altitude (where the collisions of ions and electrons are negligible) can be deduced from the preceding figures by projection along the field lines: the plasma drifts towards the night inside the auroral circle, located at the ionospheric projection of the neutral line, and returns towards the day side outside this circle. Its circulation, whose current lines are represented in solid lines, thus forms two great vortices, which occupy the morning and evening sectors of the polar regions. This convection cell geometry is at the origin of the term "magnetospheric convection", denominating the circulation of magnetospheric plasma induced by the solar wind.

3.3.3.3.2. Corotation motion

Another important component of the circulation of magnetic flux tubes within a magnetosphere comes from the proper motion of the ionospheric conductor elements, through the planetary magnetic field. Each element is driven at the local velocity $\boldsymbol{V}$ of the upper atmosphere gas at the level of the conducting layer, or "ionospheric dynamo layer", which was described in section 3.2.1.2. This velocity, in the inertial frame of reference of the planet, writes as:

$$\boldsymbol{V} = \boldsymbol{\Omega} \times \boldsymbol{R} + \boldsymbol{V}_{\mathbf{n}} \tag{3.31}$$

where $\boldsymbol{\Omega}$ is the angular velocity vector of planetary rotation and $\boldsymbol{R}$ is the planetocentric vector radius of the ionospheric fluid element. $\boldsymbol{\Omega} \times \boldsymbol{R}$ thus represents the driving velocity of the fluid in the planetary rotation, known as the "corotation velocity". $\boldsymbol{V}_{\mathbf{n}}$ is the proper velocity of the atmospheric fluid relative to the planet, in other words, the neutral wind speed. Apart from in the immediate vicinity of the poles of rotation, the second term is generally insignificant for the calculation of the total circulation of magnetospheric plasma in the inertial frame of reference (it can, however, be added later as a corrective term).

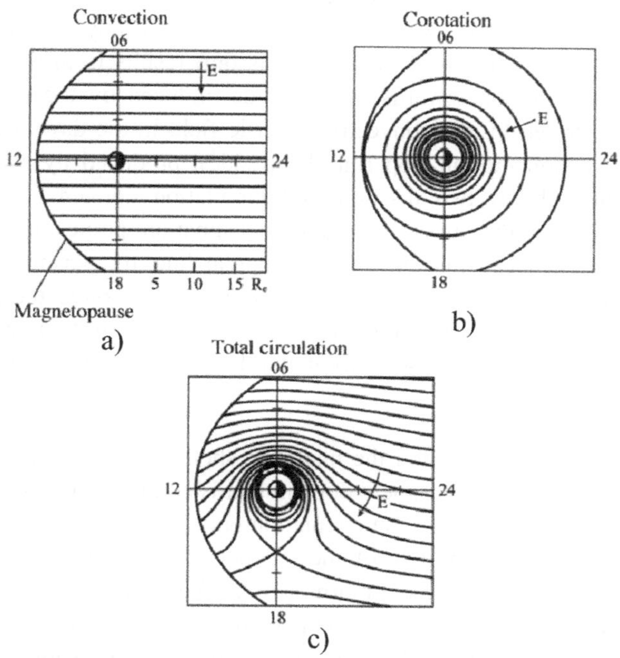

Figure 3.31. *Geometry of the electrostatic equipotential regions, which are also the flow lines of the cold plasma, in the magnetospheric equatorial plane*

COMMENT ON FIGURE 3.31. – *The total electric field, whose equipotential regions are represented in (c), results from the superposition of two electric field systems produced; one by the interaction with the solar wind, as represented in (a) by the electric field of convection; and the other one, by the rotation of the planetary ionospheric conductor, shown in (b) by the electric field of corotation (source: (Encrenaz* et al. *2003).*

3.3.3.3.3. Global motion

The effect of the rotational transport of the atmospheric fluid of the dynamo layer is simply to superimpose the corotational motion onto the solar wind-induced convection cells. This is done through the friction exerted by the rotating ionospheric conductor, at the footprints of each field line. Figure 3.30 must thus be updated to account for the total circulation of the plasma. At the ionospheric level, Figure 3.30c remains valid if we place ourselves in a frame of reference that is linked to the planetary rotation, for example, for an observer located at the planet surface. In the equatorial plane, however, the diagram in 3.30b must be modified as shown in Figure 3.31. This shows the current lines of the frozen-in plasma in the magnetic equatorial plane which are also, in a steady state, electrostatic equipotential regions. The total circulation (bottom) results from the superposition of the convection motion towards the day side, induced by the solar wind (top left-hand side), driven by the convective electric field in the dawn-dusk direction, and of the corotation motion (top right-hand side), produced by a radial electric field. The figure shows two disconnected circulation domains, separated by a singular current line that passes through a stagnation point located along the 06:00–18:00 axis in local time. Inside this singular line, the plasma frozen-in with the magnetic flux tubes travels along closed trajectories around the planet. These flux tubes are permanently closed and connected to the ionospheric plasma source; they are never connected to the solar wind and thus not (at first order) influenced by the interplanetary medium.

The singular current line, and the magnetic shell it defines, represents the true inner boundary of the magnetosphere domain that is controlled by the interplanetary medium. Figure 3.31 shows that the circulation outside this boundary goes from the magnetotail towards the day side of the magnetopause, bypassing the corotation region either by the evening or the morning side. We retrieve the motion of closed field lines on Figure 3.30a. The latter reminds us that these closed field lines, after being transported from the tail to the magnetopause, are subject to magnetic reconnection through crossing this boundary and then return to the tail of the magnetosphere above the two polar caps, in the form of field lines open onto the interplanetary medium.

These remarks allow us to identify the spatial extension of the region that is populated by the ionospheric plasma inside a magnetosphere, named the plasmasphere. Figure 3.32 shows a section of the magnetospheric cavity in the midday-midnight meridian plane. The shaded area represents the interior of the magnetic shell containing the singular current line: this is the region where the magnetic field tubes remain closed, connected to the ionospheric plasma source and are transported in corotation with the planet. As a result of this source always being active on the planetary side illuminated by the solar UV radiation, the flux tubes are filled by the ionospheric plasma until they reach a pressure balance with the source. The plasmasphere corresponds to a high concentration of cold plasma.

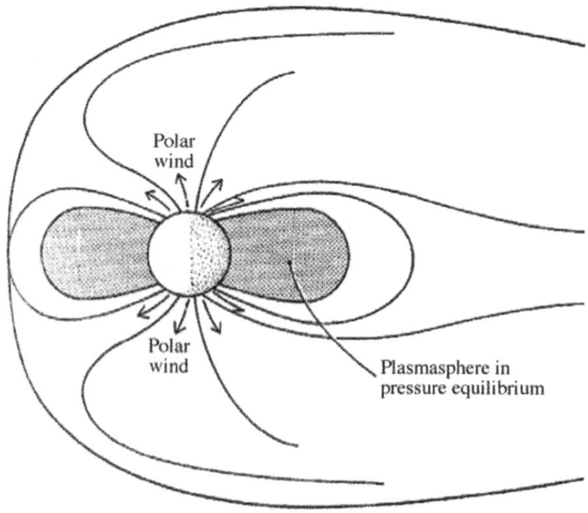

Figure 3.32. *The superposition of the convection and corotation motions of the magnetospheric flux tubes reveals two distinct regimes of flux tube filling by the ionospheric plasma reservoir*

COMMENT ON FIGURE 3.32. – *The inner grey part, the plasmasphere, is mainly populated by a cold and dense plasma, which is the extension of the ionospheric plasma in the magnetosphere. The high latitude flux tubes, above the polar caps, are open to the interplanetary medium, into which the ionospheric plasma that populated them escapes, forming the polar wind (source: after (Banks* et al. *1971)).*

Outside its boundary, the plasmapause, the flux tubes open periodically to the interplanetary medium during their transport across the poles. There, they lose a large part of their ionospheric plasma content, mainly composed of protons, which escape into the interplanetary medium by a nozzle effect, identical to that at the origin of the solar wind. This escape at supersonic speed, observed above the Earth's polar caps, has been named polar wind. Its periodic nature maintains the cold plasma concentration outside the plasmasphere, in the magnetospheric region strongly coupled to the interplanetary medium, at values at least one order of magnitude lower than those of the plasmasphere. However, at each opening to interplanetary space, the flux tubes outside the plasmapause are filled in with interplanetary plasma that diffuses along the magnetic flux lines through the magnetopause. This plasma is transported by magnetospheric convection to the equatorial plane of the magnetospheric tail, where it accumulates into a plasma reservoir known as the plasma sheet. After the flux tubes close across the night side of the neutral line (at point N2 in Figure 3.30a), this plasma is transported toward the Earth (steps F–H). During this motion, the adiabatic compression of the flux tubes yields kinetic energies exceeding the keV. It is the electrons of this hot plasma which, by precipitating in the planetary upper atmosphere along the high latitude magnetic field lines, give rise to the intense boreal and austral polar aurorae (the bright arcs on Figure 3.29), while the ions feed the external regions of the radiation belts (after recirculation and re-acceleration, they can reach energies exceeding 100 keV). This cycle of magnetospheric plasma circulation and intense auroral emissions, controlled by the polarity of the interplanetary magnetic field, is called a magnetospheric substorm. It has a total duration of about 12 hours and recurs periodically within a solar rotation.

3.3.3.3.4. Compared magnetospheres

The size of the plasmasphere varies considerably from one magnetosphere to another. It is calculated by determining the position of the stagnation point (Figure 3.31), where the electric fields of convection and corotation strictly balance each other. In the case of the Earth's magnetosphere, the relatively slow planetary rotation allows a large part of the magnetic cavity to be coupled to the solar wind, the plasmapause typically being located at a geocentric distance of 4 to 7 R_T. The Earth's magnetosphere is thus dominated (in volume) by convection.

In the case of Jupiter, characterized by a fast rotation (~ 10 hours) and an intense magnetic field, the same calculation places the plasmapause at or beyond the magnetopause. This implies that the whole magnetospheric cavity is dragged into the fast rotation of the planet. The influence of the interplanetary medium, probed by the observation of some magnetospheric phenomena such as auroral radio emissions, is probably limited to the immediate vicinity of the magnetopause and the magnetotail. The regions with closed magnetic topology of the Jovian magnetosphere are thus dominated by the internal plasma sources, which are the planetary ionosphere and the satellites (see section 5.4).

On the contrary, in the case of Mercury, the very slow rotation of the planet (sidereal period of 88 days) implies that the area dominated by corotation has a radius that is smaller than that of the planet (and is therefore non-existent), despite the probable existence of closed magnetic field lines near the magnetic equator and the planetary surface. The dynamics of Mercury's mini-magnetosphere is therefore entirely controlled by convection (see section 4.2.6).

The magnetospheres of Saturn, Uranus and Neptune are more or less intermediate cases between these two extremes (see section 5.4). With its magnetic axis perfectly aligned with its rotation axis, its modest magnetic field and a fast rotation of ~ 11 hours, the dynamics of Saturn's magnetosphere is governed by both the rotation and the solar wind. The case of Uranus is more atypical. During its flyby by Voyager 2 in 1986, during the solstice, the planetary rotation axis was almost exactly pointing in the direction of the Sun. The planetary magnetic field discovered by Voyager 2, unique in the Solar System, corresponded to a dipole inclined by -60° with respect to the rotation axis. In this solstice geometry, convection and corotation do not combine according to the scheme of Figure 3.31 because they act in orthogonal planes, and the plasma circulates on helical trajectories in the inertial frame. In 1986, it seemed that the flux associated with convection could freely penetrate up to the inner magnetosphere, without opposing corotation anywhere. Uranus' magnetosphere thus resembled the Earth's magnetosphere, but in precession around the planet-Sun axis at a rate of one revolution of ~ 17 hours, and probably subject to a greater influence of convection. During the 2007 equinox, its rotation axis was perpendicular to the direction of the Sun instead, its magnetic axis was successively almost aligned with, and then almost orthogonal to the solar wind flow twice per rotation, a much more dynamic situation than in the case of the Earth. This is the configuration that prevailed when Voyager 2 discovered Neptune's magnetosphere in 1989. With its magnetic field inclined by -47° with respect to its rotation axis itself pointing 29° from the ecliptic normal, the magnetic axis goes from a quasi-orthogonal configuration to a quasi-aligned configuration with the solar wind, once per rotation (~16 hours). This rapidly changing magnetospheric structure produces very complex plasma dynamics. Voyager 2 observations revealed an absence of dense plasmasphere, as well as irregularly shaped mid-latitude auroral regions, very different from the terrestrial typical auroral ovals.

3.3.4. *Planetary auroral processes*

Planetary magnetospheres are the source of intense electromagnetic radiation, produced in the vicinity of their magnetic poles through the precipitation of energetic charged particles into the atmosphere, which are essentially energetic electrons of a few keV, accelerated in the magnetosphere by different processes and then guided to the planet along the high latitude field lines: these are the polar aurorae. These emissions are produced by different physical mechanisms in the radio, optical (visible, UV, IR) and X-ray ranges, and are observed with on-board spectro-imagers, or space or ground-based telescopes. Their study provides a rich means of remotely probing the active regions, acceleration mechanisms, and dynamics of the terrestrial (Carlson and Egeland 1995) and planetary magnetospheres (Bhardwaj and Gladstone 2000; Badman *et al.* 2014).

Auroral emissions fall into two main categories: "atmospheric" (optical) aurorae produced by the upper atmospheric constituents excited by collision, and auroral "beams" (radio and X-ray), produced directly by the incident beam of energetic particles.

3.3.4.1. *The acceleration of energetic particles*

Terrestrial auroral emissions have been early associated with intense electric currents (see section 3.2.1.2), flowing along high-latitude magnetic field lines and connecting the solar wind, the magnetosphere and the ionospheric conductor. These are the Birkeland currents, visible in Figure 3.33, named after the Norwegian physicist who predicted their existence in the early 20th century. They are divided into two categories. Type I currents are produced by the solar wind/magnetosphere interaction and flow along planetary field lines, connected to the interplanetary medium on the inner boundary of the polar cap. The generator of this current system is the convective electric field applied to the planetary field lines. The type II currents are fed in by the ring current, produced on the inner edge of the equatorial plasma sheet by a drift motion, perpendicular to the field lines, of ions and electrons moving in opposite directions, which closes into the ionosphere on the morning and evening side along the field lines. These currents are therefore located at the boundary between the open and closed field lines, along the neutral line of Dungey's model, and develop dynamically during a substorm, which paces the circulation of plasma in the Earth's magnetosphere (Iijima and Potemra 1978). Their projection onto the ionosphere is shown in Figure 3.33b: the inner (outer) ring corresponds to the type I (type II) currents, and the upward and downward currents are shown in white and black, respectively. Their amplitude and position depend on the geomagnetic activity, controlled by the solar wind, which is quantified with geomagnetic indices. In Figure 3.33b, we can see that the current sheets widen and migrate towards the low latitudes when the geomagnetic activity increases (from left to right-hand side). These field-aligned currents induce quasi-static electric potential jumps, capable of accelerating cold plasma electrons and ions to sufficiently high energies so that they can precipitate into the atmosphere. This relationship was formalized in the 1970s by Knight's theory, which states that conservation of the current along a flux tube, coupled with decreasing plasma density toward increasing latitudes, requires acceleration of the remaining charge carriers at high latitudes (Knight 1973). Aurorae are mostly located at the footprint of the upward current layers, which accelerate electrons towards the planet. The auroral oval in Figure 3.29 therefore coincides with the white circle in Figure 3.33.

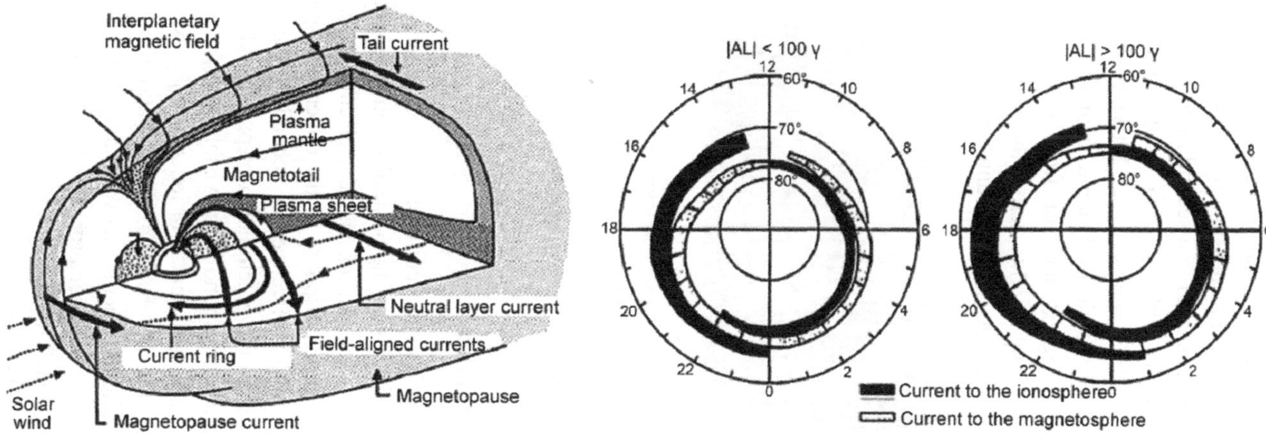

a) b)

Figure 3.33. *(a) Schematic representation of the Earth's magnetosphere. The main electric currents are indicated by bold arrows. Birkeland currents are aligned with high latitude magnetic field lines (source: from (Kivelson and Russell 1995)). (b) Projection of the amplitude of field-aligned currents onto the Earth's northern ionosphere, as measured by the Triad satellite for calm (left-hand side) and active (right-hand side) geomagnetic conditions (source: from (Iijima and Potemra 1978))*

The *in situ* exploration of the Earth's auroral regions confirmed the existence of quasi-static electric potential jumps, but also revealed that part of the energy particles measured locally presented characteristics that were not compatible with this mechanism. These have been associated to a second major acceleration process, not directly related to the interaction with the solar wind, based on the dynamic interaction between particles and Alfvén waves (Lysak 2003). Other types of intense field-aligned currents, generated by the electrodynamic coupling between the solar wind, the magnetosphere and the ionosphere, or from the electromotive force induced by the motion of a conducting satellite across the planetary magnetic field lines, have also been identified on Jupiter, Saturn, Uranus and Neptune.

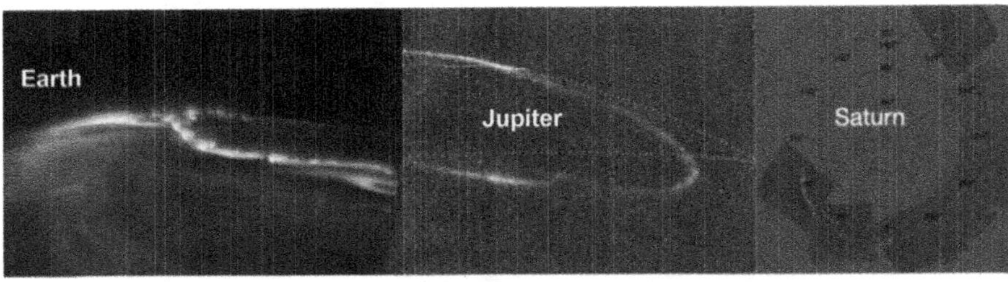

Figure 3.34. *High spatial resolution images of visible aurorae on Earth (atomic oxygen transitions), Jupiter and Saturn (atomic hydrogen lines), observed from the International Space Station, Galileo and Cassini probes, respectively (source: NASA, ESA). For a color version of this figure, see www.iste.co.uk/encrenaz/solar1.zip*

3.3.4.2. *Visible and ultraviolet aurorae*

"Atmospheric" aurorae result from the collision between precipitated particles, mainly electrons, and prominent species of the neutral or ionized upper atmosphere. They depend on the atmospheric composition, and the electronic transitions of the excited species probe the instantaneous response of the upper atmosphere to particle precipitations. The excitation is most effective for secondary electrons (of about 100 eV), generated by cascade from the precipitation of primary electrons (of a few keV at Earth). The Earth's aurorae correspond to the visible and ultraviolet transitions of atomic oxygen and molecular nitrogen, neutral or ionized. They extend from 90 km to more than 300 km in altitude and are observed, in the long-term, by a network of spectro-imagers on the ground (like the Themis network) and in space. The aurorae of the giant planets, discovered by the Voyager probes in the 1980s and then regularly observed with the HST, are dominated by visible and especially UV transitions (where the contrast with reflected sunlight is better) of atomic and molecular hydrogen (Grodent 2015). They are observed from several hundreds of kilometers (in the Jovian case, where the most energetic electrons reach several hundreds of keV) to several thousands of kilometers above the clouds. Aurorae have also been also identified on Ganymede by the HST in the UV lines of atomic oxygen. Figure 3.34 gives examples of planetary aurorae observed in the visible range, while UV aurorae are displayed in Figures 3.29 and 3.35. This spectral range benefits from a good angular resolution and allows one to accurately map the projection of active magnetosphere regions onto the atmosphere. Their spectral analysis provides a diagnostic of their precipitated power, of the excited species and of the energy of incident particles.

An example of the diagnostic brought by the regular observation of terrestrial UV aurorae is given by Figure 3.36, which shows a series of UV images taken by the Polar satellite covering the development phase of an auroral substorm. The accumulation of plasma in the tail of the magnetosphere, which results from the magnetic reconnection between planetary and interplanetary magnetic fields at the magnetopause, can be interrupted by a similar reconnection on the night side for stretched field lines (point N2 on Figure 3.30). The re-dipolarization of the field lines associated with an adiabatic compression of the plasma causes an influx of energetic particles into the auroral night side zones. Their precipitation into the atmosphere gives rise to the bright, discrete aurorae observed in Figure 3.36, the auroral oval being active: it widens and reaches lower latitudes, in a similar manner to the currents in Figure 3.33b.

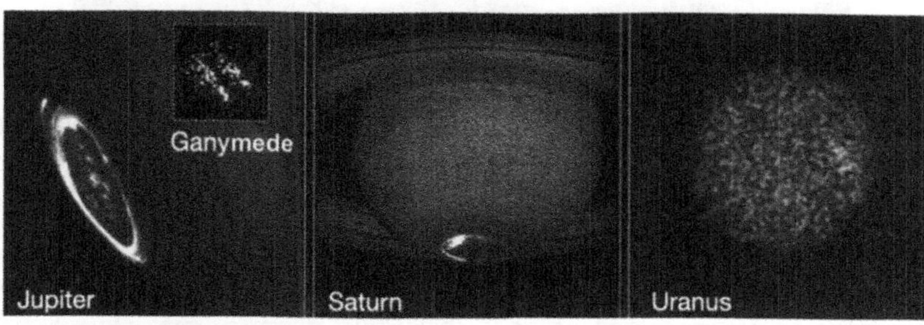

Figure 3.35. *Images of the UV aurorae of Jupiter, Saturn, Uranus (atomic and molecular hydrogen transitions) and Ganymede (atomic oxygen lines), taken with the HST. The Earth's equivalent is shown in Figure 3.29. The structure of the emissions and their intensity vary considerably from planet to planet (source: after (Lamy* et al. *2015)). For a color version of this figure, see www.iste.co.uk/encrenaz/solar1.zip*

3.3.4.3. *Infrared aurorae*

The H_3^+ ion produced by ionization of H_2 in the ionized upper atmosphere of the giant planets also emits photons, induced by collisions resulting from auroral precipitations, when the molecule is thermalized. These are ro-vibrational transitions observed in the near-infrared, which materialize the response of the ionized atmosphere with some latency, related to the lifetime of the H_3^+ ion and the conditions of its thermalization (Badman *et al.* 2014). H_3^+ was detected as early as the 1990s by ground-based telescopes on the disks of Jupiter, Saturn, and Uranus, which possess a high thermospheric temperature. Nevertheless, only Jupiter and Saturn have revealed auroral signatures associated with this ion, examples of which are shown in Figure 3.37. The near-infrared range also includes the quadrupole emission lines of H_2, which have also been detected in the auroral emissions of Jupiter and Saturn. Lastly, Jupiter has specifically revealed far-infrared auroral signatures from the tops of C_nH_n hydrocarbon layers, excited by deep precipitation (Bhardwaj and Gladstone 2000). The infrared spectral range also benefits from a good angular resolution, and allows the conditions on the emission site to be probed (temperature, density) in a complementary way to the visible/UV observations.

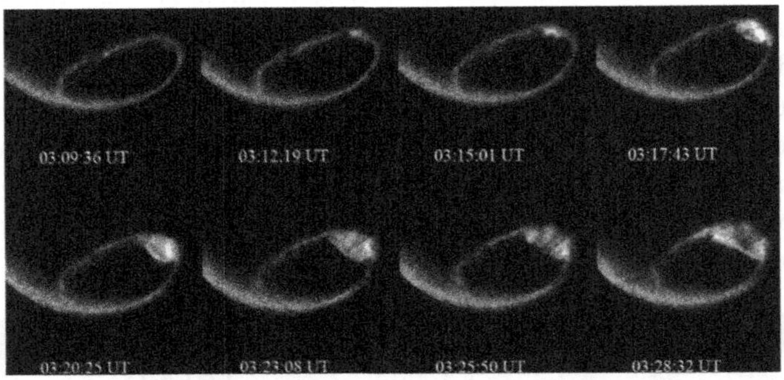

Figure 3.36. *Sequence of images of terrestrial UV aurorae taken by the Polar satellite on September 23, 1996, at three-minute intervals, illustrating the development of an auroral substorm on the midnight side (source: NASA, University of Iowa). For a color version of this figure, see www.iste.co.uk/encrenaz/solar1.zip*

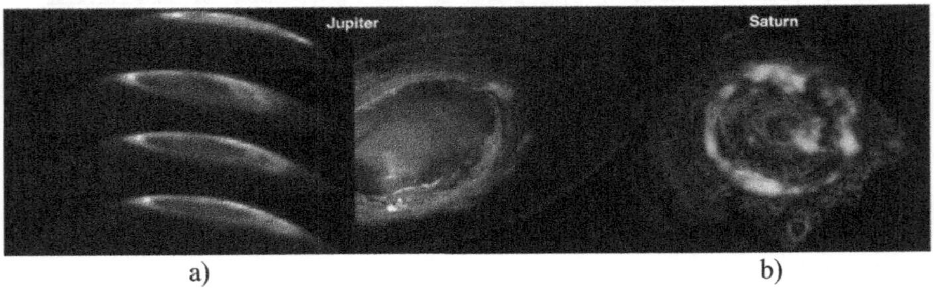

Figure 3.37. *Images of near-infrared auroral emissions from: (a) Jupiter, taken with the IRTF telescope and Juno probe and (b) Saturn, taken with the Cassini probe. These images reveal a great complexity of the auroral structure (source: after (Badman* et al. *2014), NASA). For a color version of this figure, see www.iste.co.uk/encrenaz/solar1.zip*

3.3.4.4. *X-ray aurorae*

The auroral emissions observed in the X-ray domain result from two very different mechanisms (Bhardwaj *et al.* 2007). The hard X-rays are produced by braking radiation from precipitated high energy electrons (*Bremsstrahlung*), while the soft X-rays are produced by charge exchange of very ionized heavy ions. To date, these emissions have only been observed on Earth and Jupiter, as shown in Figure 3.38. Current telescope sensitivity (Chandra, XMM-Newton) is probably too low to detect comparable emissions from Saturn or Uranus. In spite of the small number of observable photons in this spectral range, the combination of spectroscopy and imaging gives precise information on the nature of the observed radiation and the associated particles. Figure 3.38b shows that, at Jupiter, the energetic electrons responsible for the braking radiation coincide with the UV auroral oval, suggesting that the same population of electrons feeds in both emission processes, while the X-ray emissions inside the UV oval are attributed to charge exchange process, which involves strongly ionized heavy ions, probably resulting from the Jovian magnetosphere's interaction with the solar wind that may exist at high latitude.

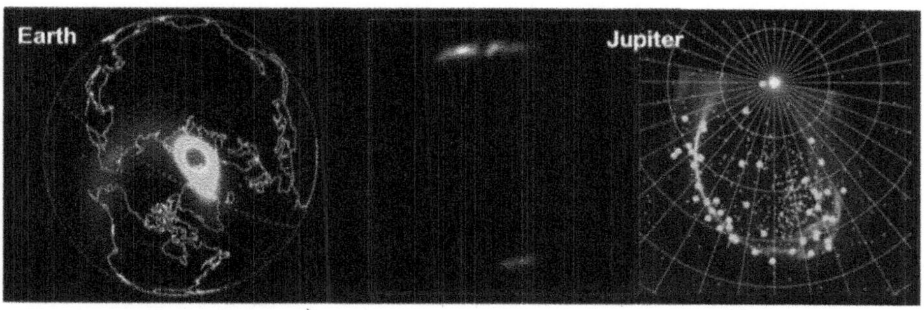

a) b)

Figure 3.38. *X-ray images of (a) the Earth aurorae, taken with the POLAR satellite (source: NASA) and (b) the Jupiter aurorae, taken with the Chandra telescope (yellow dots) (source: from (Branduardi-Raymont* et al. *2008)). For a color version of this figure, see www.iste.co.uk/encrenaz/solar1.zip*

COMMENT ON FIGURE 3.38. – *The Earth's emissions correspond to hard X-rays (Bremsstrahlung). Jovian emissions also include hard X-rays. On the right-hand side, the X-rays are compared to the UV aurorae observed with the HST on a polar projection. The hard X-rays (large dots) correspond to the main auroral oval, where very energetic electrons precipitate, while the soft X-rays (small dots) correspond to the interior of the oval, where highly ionized ions from the interaction with the solar wind may undergo charge exchange.*

3.3.4.5. *Auroral radio emissions*

We have known since 1955, and the rise of ground-based radio astronomy, that Jupiter is an intense radio emitter at decametric wavelengths (frequencies ranging from ~10 to 40 MHz). However, we had to wait for the spatial era to discover that the other planetary magnetospheres are also non-thermal radio sources at kilometric wavelengths (f > 1 MHz, unobservable from the ground) thanks to the Russian satellite Electron for the Earth in 1964, and the Voyager mission for Jupiter in 1979, Saturn in 1980, Uranus in 1986 and Neptune in 1989.

These radio emissions, shown in the time-frequency planes of Figure 3.39, are produced in the auroral regions, above the ionosphere and up to a few planetary radii away, at frequencies close to the local electron cyclotron frequency $f_{ce} = eB/2\pi m_e$, with e and m_e being the charge and mass of the electron, and B being the magnetic field strength. They have a much higher intensity than any other radio emissions produced in planetary environments. Their intensity, common characteristics and ubiquity give them a particular importance in the framework of magnetospheric physics.

Comparative study of observations obtained remotely over several decades by the Voyager (1977–1989), Ulysses (1992), Wind (≥ 1994), Galileo (1995–2002), Cassini (1999–2017), and Juno (≥ 2016) space probes, as well as observations of Jupiter's emissions with ground-based radio telescopes (since the 1960s), has led to a good understanding of their macroscopic properties. The radio sources are distributed along high-latitude magnetic field lines in regions where the f_{pe}/f_{ce} ratio (with f_{pe} being the electron plasma frequency, proportional to the square root of the electron density N_e) is much less than 1.

In spite of the low angular resolution prevailing at long wavelengths, sophisticated space measurements, such as those of the Cassini radio instrument, have made it possible to generate the first images of these emissions, as shown in Figure 3.40. The emitted radio intensity is very high and the instantaneous source is small (< 100 km perpendicular to the magnetic field), corresponding to a brightness temperature > 10^{15} K, and implying an efficiency of the order of 1% for the conversion of the kinetic energy of electrons into radio radiation. The waves are mainly emitted in the extraordinary mode (and marginally in the ordinary mode), which can propagate freely into space, with a 100% circular (or elliptical) polarization. They have a very anisotropic emission pattern, similar to a conical emission sheet with a large opening angle (30° to 90°) with respect to the local magnetic field, which induces a complex structuring in the time-frequency plane with "arcs" that can last several hours (Figure 3.39).

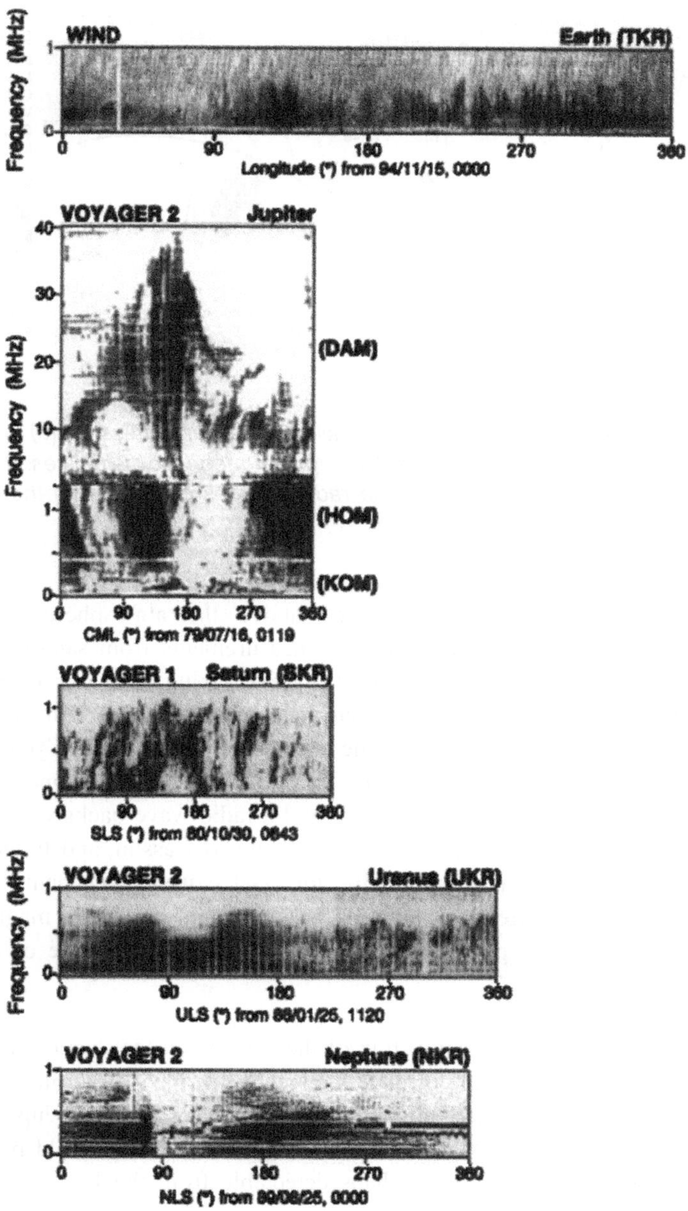

Figure 3.39. *Intensity of the auroral radio emissions of the Earth (as observed by Wind) and Jupiter, Saturn, Uranus and Neptune (as observed by Voyager) plotted as a function of frequency and longitude (or equivalently time) (source: from (Zarka 1998))*

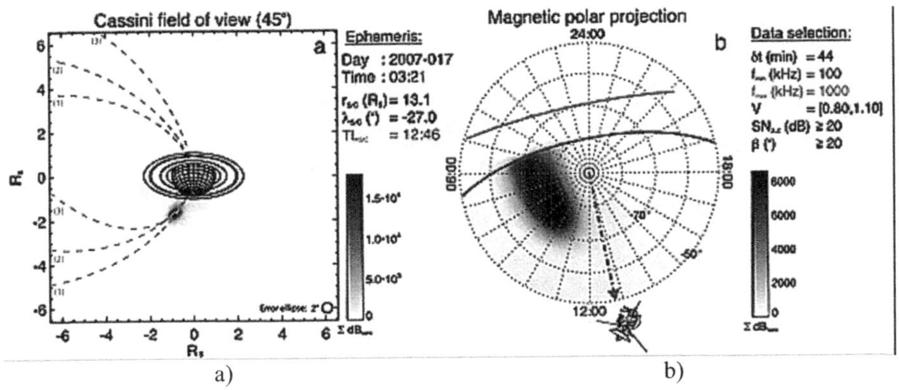

Figure 3.40. *Intensity maps of Saturn's kilometric radio emission (a) as observed from Cassini and (b) once projected along the magnetic field lines down to the ionosphere. The field lines carrying the radio sources coincide with the UV auroral oval (source: from (Lamy 2009))*

Since these radiations are produced above the atmosphere, an essential contribution has been provided by *in situ* measurements from satellites that have passed through the Earth's radio sources (by the Viking, Freya or FAST probes), from which the non-thermal and coherent mechanism at the origin of the Earth's kilometric emission was identified at the end of the 1970s as the Cyclotron Maser Instability or CMI (Wu and Lee 1979). This results from the resonant interaction between energetic electrons of a few keV and a radio wave background in regions where $f_{pe}/f_{ce} << 1$. The *in situ* exploration of Saturn by Cassini, and Jupiter by Juno, has confirmed the CMI as a potentially universal emission mechanism, although operating under different conditions depending on the magnetospheric characteristics (different populations of source electrons, presence or absence of auroral cavities, etc.).

Generally speaking, the energy flux of the solar wind that has been intercepted by a magnetosphere seems to correlate well with the average auroral radio power that it emits. The extrapolation of this law to cases of "hot Jupiters" (giant exoplanets that are very close to their star) suggests that they could produce very intense auroral radio emissions, perhaps detectable from the Earth (Zarka 2007). Although the search for exoplanets at radio wavelengths is still in progress, the radio detection of about 20 more massive objects at high frequencies (stars, brown dwarfs) has already been attributed to the CMI. The observation of these extrasolar targets is a major objective of current (Lofar, NenuFAR) or future (SKA) radio telescopes. In addition to this, the radio environment of Mercury is soon to be explored for the first time by the Bepi-Colombo magnetospheric orbiter. The weak planetary magnetic field (B_{max} < 0,01 G) corresponds to a maximum electron cyclotron frequency f_{ce} < 30 kHz. If auroral radio emissions are produced by Mercury, they will certainly be trapped in the magnetosphere by the surrounding solar wind, and will only be detectable *in situ* near their emission region.

3.4. Acknowledgements

The author would like to thank J.-Y. Chaufray for his careful proofreading and his suggestions about the interaction between the solar wind and Mars/Venus.

3.5. References

Alfvén, H. (1957). On the theory of comet tails. *Tellus*, 9(1), 92–96.

Anderson, B.J., Johnson, C.L., Korth, H., Winslow, R.M., Borovsky, J.E., Purucker, M.E., Slavin, J.A., Solomon, S.C., Zuber, M.T., McNutt Jr., R.L. (2012). Low-degree structure in Mercury's planetary magnetic field. *Journal of Geophysical Research*, 117(E00L12).

Badman, S.V., Branduardi-Raymont, G., Galand, M., Hess, S.L.G., Krupp, N., Lamy, L., Melin, H., Tao, C. (2014). Auroral processes at the giant planets: Energy deposition, emission mechanisms, morphology and spectra. *Space Science Reviews*, 187, 99–179.

Banks, P.M., Nagy, A.F., Axford, W.I. (1971). Dynamical behavior of thermal protons in the midlatitude ionosphere and magnetosphere. *Planetary and Space Science*, 19, 1053–1067.

Behannon, K.W. (1978). Heliocentric distance dependence of the IMF. *Reviews of Geophysics and Space Physics*, 16, 125–145.

Bhardwaj, A. and Gladstone, G.R. (2000). Auroral emissions of the giant planets. *Reviews of Geophysics*, 38(3), 295–353.

Bhardwaj, A., Elsner, R.F., Gladstone, G.R., Cravens, T.E., Lisse, C.M., Dennerl, K., Branduardi-Raymont, G., Wargelin, B.J., Waite, J.H., Roberton, I. *et al.* (2007). X-rays from solar system bodies. *Planetary and Space Science*, 55(9), 1135–1189.

Brain, D.A., Barabash, S., Bougher, S.W., Duru, F., Jakosky, B.M., Modolo, R. (2017). Solar wind interaction and atmospheric escape. In *The Atmosphere and Climate of Mars*, Haberle, R.M., Clancy, R.T., Forget, F., Smith, M.D. (eds). Cambridge University Press, Cambridge.

Brandt, J.-C. and Mendis, D.A. (1979). The interaction of the solar wind with comets. In *Solar System Plasma Physics. Volume II*, Kennel, C.F., Lanzerotti, L.J., Parker, E.N. (eds). North-Holland Publishing Company, Amsterdam.

Branduardi-Raymont, G., Elsner, R.F., Galand, M., Grodent, D., Cravens, T.E., Ford, P., Gladstone, G.R., Waite Jr., J.H., (2008). Spectral morphology of the X-ray emission from Jupiter's aurorae. *Journal of Geophysical Research*, 113, A02202.

Carlson, H.C. and Egeland, A. (1995). The aurora and the auroral ionosphere. In *Introduction to Space Physics*, Kivelson, M. and Russel, C. (eds). Cambridge University Press, Cambridge.

Connerney, J.E.P. and Acuna, M.H. (1991). The magnetic field of Neptune. *Journal of Geophysical Research*, 96, 19023–19042.

Connerney, J.E.P., Acuña, M.H., Ness, N.F., Kletetschka, G., Mitchell, D.L., Lin, R.P., Reme, H. (2005). Tectonic implications of Mars crustal magnetism. *Proceedings of the National Academy of Sciences of the United States of America*, 102(42), 14970–14975.

Connerney, J.E.P., Kotsiaros, S., Oliversen, R.J., Espley, J.R., Joergensen, J.L., Joergensen, P.S., Merayo, J.M.G., Herceg, M., Bloxham, J., Moore, K.M., Bolton, S.J., Levin, S.M. (2018). A new model of Jupiter's magnetic field from Juno's first nine orbits. *Geophysical Research Letters*, 45, 2590–2596.

Dougherty, M.K., Cao, H., Khurana, K.K. Hunt, G.J., Provan, G., Kellock, S., Burton, M.E. Burk, T.A., Bunce, E.J., Cowley, S.W.H., *et al.* (2018). Saturn's magnetic field revealed by the Cassini Grand Finale. *Science*, 362(6410).

Dungey, J.W. (1961). Interplanetary magnetic field and the auroral zones. *Physical Review Letters*, 6(2), 47–48.

Encrenaz, T. *et al.* (2003). Chapter 5. In *Le Système solaire*, Encrenaz, T., Barucci, M.-A., Bibring, J.-P., Blanc, M., Roques, F., Zarka, P. (eds). EDP Sciences/CNRS Editions, Paris.

Encrenaz, T., Bibring, J.-P., Blanc, M., Barucci, M.-A., Roques, F., Zarka, P. (2004). *The Solar System*, Springer-Verlag, Berlin.

Futaana, Y., Stenberg Wieser, G., Barabash, S., Luhnmann, J.G. (2017). Solar wind interaction and impact on the Venus atmosphere. *Venus III, Space Science Reviews*, 212, 1453–1509.

Glassmeier, K.-H. (2017). Interaction of the solar wind with comets: A Rosetta perspective. *Philosophical Transactions A*, 375(2097).

Grodent, D. (2015). A brief review of ultraviolet auroral emissions on giant planets. *Space Science Reviews*, 187(1–4), 23–50.

Gurnett, D.A. and Kurth, W.S. (2019). Plasma densities beyond the heliopause from the Voyager 1 & 2 plasma wave instruments. *Nature Astronomy*, 3, 1024–1028.

Gurnett, D.A., Kurth, W.S., Allendorf, S.C., Poynter, R.L. (1993). Radio emission from the heliopause triggered by an interplanetary shock. *Science*, 262(5131), 199–203.

Herbert, F. (2009). Aurora and magnetic field of Uranus. *Journal of Geophysical Research*, 119(A11206).

Hundhausen, A.J. (1972). *Coronal Expansion and Solar Wind.* Springer, New York.

Iijima, T. and Potemra, T.A. (1978). Large scale characteristics of field-aligned currents associated with substorms. *Journal of Geophysical Research*, 83(A2), 599–615.

Kivelson, M.G. and Russell, C.T. (1995). *Introduction to Space Physics*. Cambridge University Press, Melbourne.

Knight, S. (1973). Parallel electric fields. *Planetary and Space Science*, 21, 741–750.

Lamy, L. (2009). An auroral oval at the footprint of Saturn's kilometric radio sources, co-located with the UV aurorae. *Journal of Geophysical Research*, 114(A10212).

Lamy, L., Prangé, R., Henry, F., Le Sidaner, P. (2015). The Auroral Planetary Imaging and Spectroscopy (APIS) service. *Astronomy and Computing*, 11, 138–145.

Lepping, R.P. (1986). *Magnetic Configurations of Planetary Obstacles*. Cépaduès, Toulouse.

Lyons, L.R. and Williams, D.J. (1984). Quantitative aspects of magnetospheric physics. In *Geophysics and Astrophysics Monographs*. D. Reidel Publishing Company, Dordrecht.

Lysak, R.L. and Song, Y. (2003). Kinetic theory of the Alfven wave acceleration of auroral electrons. In *Geophysics and Astrophysics Monographs*. D. Reidel Publishing Company, Dordrecht.

McComas, D.J., Ebert, R.W., Elliott, H.A., Goldstein, B.E., Gosling, J.T., Schwadron, N.A., Skoug, R.M. (2008). Weaker solar wind from the polar coronal holes and the whole Sun. *Geophysical Research Letters*, 35(18).

Richmond, A.D. (1983). Thermospheric dynamics and electrodynamics. In *Solar-Terrestrial Physics*, Carovillano, R.L. and Forbes, J.M. (eds). D. Reidel Publishing Company, Dordrecht.

Russell, C.T. *et al.* (1982). In *Comets*, Wilkening, L.L. (ed.). University of Arizona Press, Tucson.

Saunders, M.A., Russell, C.T., Luhmann, J.G. (1986). Interactions with planetary ionospheres and atmospheres: A Review. In *Étude comparative des systèmes magnétosphériques*. Cépaduès, Toulouse.

Spreiter, J.R., Summers, A.L., Alksne, A.Y. (1966). Hydrodynamic flow around the magnetosphere. *Planetary and Space Science*, 14(3), 223.

Steinberg, J.-L. (1986). Le Vent solaire. In *Actes du colloque "Technologie de l'environnement spatial*". Éditions du CNES, Paris.

Thébault, E., Finlay, C.C., Beggan, C.D., Alken, P., Aubert, J., Barrois, O., Bertrand, F., Bondar, T., Boness, A., Brocco, L. *et al.* (2015). International geomagnetic reference field: The 12th generation. *Earth, Planets & Space*, 67(79).

Wu, C.S. and Lee, L.C. (1979). A theory of the terrestrial kilometric radiation. *The Astrophysical Journal*, 230, 621–626.

Zarka, P. (1998). Auroral radio emissions at the outer planets: Observations and theories. *Journal of Geophysical Research*, 103(E9), 20159–20194.

Zarka, P. (2007). Plasma interactions of exoplanets with their parent star and associated radio emissions. *Planetary and Space Science*, 55, 598–617.

4

Telluric Planets

Thérèse ENCRENAZ, Marcello FULCHIGNONI and Laurent LAMY
LESIA, Paris Observatory, PSL University, Paris, France

4.1. The exploration of the telluric planets

4.1.1. *From antiquity to the space age*

Visible to the naked eye, telluric planets have been observed since antiquity, and the study of their trajectories has served as the basis for the cosmogonic models that have succeeded each other, from the geocentric models of the Greeks to the Copernican revolution. The observation of telluric planets as physical objects began at the beginning of the 17th century with Galileo, who discovered the reliefs on the Moon and the phases of Venus. The transits of Venus were observed and used in 1639 by the English astronomer Jeremiah Horrock, who deduced a first estimate (to within 30%) of the astronomical unit. The first map of Mars was drawn by Christiaan Huygens, who deduced the rotation period of the planet. Jean-Dominique Cassini used simultaneous observations of Mars in Paris and Cayenne to refine the value of the astronomical unit; in 1769, a new value was obtained with 0.1% accuracy, thanks to the transit of Venus; during the previous transit, eight years earlier, the Russian astronomer Michael Lomonosov was able to deduce the probable existence of an atmosphere around Venus. A century later, the American astronomer Asaph Hall discovered Mars's satellites, Phobos and Deimos. In the 18th and 19th centuries, telescopic observations of the planets were systematically conducted; photography only replaced drawings in the 20th century.

In 1877, the Italian astronomer Giovanni Schiaparelli announced the discovery of straight lines on the surface of Mars, often grouped in pairs: he named them *canali* (channels). These were quickly interpreted by several enthusiastic astronomers, including Camille Flammarion and Percival Lowell, as the signature of an intelligent civilization fighting against drought. However, other astronomers remained skeptical, including Eugène Antoniadi, who, at the beginning of the 20th century, obtained high quality images showing that the canals observed by Schiaparelli were nothing more than an optical illusion; the myth of the Martian canals lasted until the advent of the space age. Schiaparelli, in the early 1880s, also obtained the first images of the illuminated part of Mercury. The 20th century saw the advent of spectroscopy, which allows the identification of gases present in planetary atmospheres. In the early 1960s, CO_2 was identified as a major component on Mars and Venus, and it was shown that the level of H_2O was very low. CO_2 ice was also identified as a constituent of ice caps. Simultaneously, the development of infrared astronomy allowed the first temperature measurements to be taken: the surface of Mars was colder than that of the Earth by a few tens of kelvins (K), and that of the clouds of Venus (opaque to visible radiation) was close to 0 K; the surface of the day side of Mercury is close to 700 K. In 1958, the first measurements of Venus in the radio domain brought an enormous surprise: the surface temperature of Venus was very high. This was the first clue that would lead astronomers to consider the hypothesis of an extreme greenhouse effect (see Box 4.1).

4.1.2. *The beginning of the space age*

The beginnings of space exploration are marked by the competition between the United States and the Soviet Union. After the Soviets' launch of the first satellite, Sputnik, in 1958, then the first man in space in 1962, the Americans took the lead with the Apollo program of manned lunar exploration and the sending of the first man to the Moon on July 21, 1969. At the same time, the Soviets also brought back lunar samples by means of rover missions, albeit in much smaller quantities. The analysis of the samples, through providing an absolute dating scale, has provided valuable information regarding the age of the Solar System.

Meanwhile, rover space exploration of Mars and Venus began in the 1960s, both in the United States and the Soviet Union. It is marked, on both sides, by multiple failures. In 1962, the American probe Mariner 2 successfully flew over Venus and confirmed its surface temperature (730 K or 457°C). In 1964, Mariner 4 flew over Mars and sent back the first photographs of its soil, putting an end to the myth of Martian canals. The next, most important steps were the successes of the Mariner 9 orbiter in 1972 and the Viking mission in 1975. Consisting of two identical orbiters and two descent probes, the Viking mission was an unprecedented scientific and technological success. However, the search for life turned out to be negative, and this disappointment resulted in the American space exploration program stopping for about 20 years. The Soviet Union, on its side, focused on Venus with the Venera program. In 1975, the Venera 9 and 10 probes sent back the first images of Venus's soil. In 1979, NASA sent the Pioneer Venus mission (see Figure 4.1), equipped with an orbiter and four descent probes, which measured the atmospheric composition down to an altitude of 12 km.

a) b) c)

Figure 4.1. *(a) Venus, imaged by Pioneer Venus; (b) the Earth and (c) Mars, imaged by one of the Viking probes; the images are not to scale (source: NASA). For a color version of this figure, see www.iste.co.uk/encrenaz/solar1.zip*

Between 1991 and 1994, the American orbiter Magellan carried out a mapping of Venus's surface by radar; the latter is covered with relatively recent volcanoes, dating back a few hundred million years at most.

4.1.3. *The return to Mars*

In the 1990s, NASA turned its attention back to Mars. Indeed, images from the Viking probes showed possible traces of fluvial flow that could imply that liquid water once flowed on the surface of Mars. The new strategy was to highlight such sites to search for traces of fossil life. In spite of several failures, NASA succeeded once again thanks in particular to the Mars Global Surveyor orbiter, launched in 1996, and Mars Odyssey, launched in 2001. Mars Global Surveyor has made two remarkable discoveries:

– the evidence of a residual magnetic field in the oldest terrains of the southern hemisphere, proof of the existence of an internal dynamo (and thus of an intrinsic magnetic field) at the beginning of the planet's history;

– the discovery of a "shoreline" of constant altitude over more than a thousand kilometers, which could testify to the presence of a boreal ocean some three billion years ago (see section 4.3.3.6).

As for the Mars Odyssey orbiter, thanks to its gamma ray spectrometer, it discovered the presence of hydrogen atoms attributed to the water molecule under the Martian poles, trapped in the form of ice or permafrost.

In 2003, NASA landed two rovers on the surface of Mars, Spirit and Opportunity, which operated successfully for many years (more than 15 years for Opportunity). The same year, the European Space Agency (ESA) launched its first planetary mission, the Mars Express orbiter, which is still in operation today. Its infrared spectrometer mapped the Martian mineralogy and revealed the presence of clays in the oldest terrains, confirming that liquid water did once flow in abundance, in the planet's distant past. Mars Express also observed Martian aurorae for the first time, revealing emissions of atmospheric species like CO, produced in the UV range by localized precipitations of energetic particles coinciding with pockets of residual magnetic field. NASA then successfully launched the Mars Reconnaissance Orbiter in 2005, followed by the Phoenix descent module in 2007, near the North Pole, and then the Curiosity rover in 2012, an ambitious laboratory aimed at searching for traces of past life or, failing that, signs of a habitable environment. Curiosity discovered stratified sedimentary rocks that show the presence of past liquid water and a habitable environment three billion years ago (see Figure 4.2). In addition, Curiosity reported the detection of sporadic methane emissions, which intrigued researchers, as the presence of CH_4 is completely unexpected in the CO_2-dominated Martian atmosphere. Other detections of CH_4 had been reported during the previous decade, but they remain controversial; the problem of the existence of methane and its possible origin remains, to this day, an open-ended question.

Martian exploration continued with the Maven orbiter, launched in 2013 and dedicated to the measurement of atmospheric escape, with the objective of determining what the primitive atmosphere of the planet may have been. In 2016, a new mission was launched in partnership between Europe and Russia: Trace Gas Orbiter (TGO), the first part of the ExoMars mission. Its objective is to search for atmospheric gases and, more precisely, methane. The next steps will be Mars 2020 for NASA and ExoMars 2022 for Europe and Russia.

The objective for the coming decades is the return of Martian samples, a program that will be conducted in international cooperation. Mars 2020 (operational) should be the first part of this program, as its main purpose will be to select and collect samples. Other vehicles, developed by ESA and NASA with other partners, will be in charge of recovering the samples, taking them to Martian orbit and then returning them to Earth.

Figure 4.2. *Discovery of a stratified terrain near Mount Sharp by the Curiosity rover. The presence of strata testifies to the presence of a lake in the past. The neutral pH, the low salinity, the presence of the elements C, H, N, O, P, S, and the presence of Fe and S in different states of oxidation have made it possible to define this environment as "habitable" in the past (source: NASA). For a color version of this figure, see www.iste.co.uk/encrenaz/solar1.zip*

4.1.4. *The return to Venus*

Ten years after the success of the Magellan mission, Europe launched a mission to Venus: the Venus Express orbiter, operating in polar orbit between 2006 and 2014, was mainly dedicated to the study of the planet's atmospheric circulation and its super-rotation, which accelerates the atmosphere with a period of four Earth days at the cloud top level. Venus Express discovered the south pole vortex, similar to the north pole vortex previously observed by Pioneer Venus, as well as high-altitude oxygen and carbon dioxide emissions on the night side; a signature of a subsolar-antisolar circulation (see section 4.3.3.6 and Figure 4.3). On the night side and in near-infrared, the VIRTIS-M imaging spectrometer observed variations in emissivity at a few points on the surface, a probable indication of recent volcanism (a few tens of thousands of years, or even less).

In 2010, the Japanese agency JAXA launched the Akatsuki mission, intended to be placed in equatorial orbit around Venus. Despite the initial failure of the insertion, the probe was recovered in 2015 and provided spectacular images of cloud turbulence. It also, for the first time, indicated the presence of gravity waves linked to the topography in the lower troposphere. Several projects of exploration of Venus are under study at ESA, NASA and in Russia for a launch in the 2020s or 2030s.

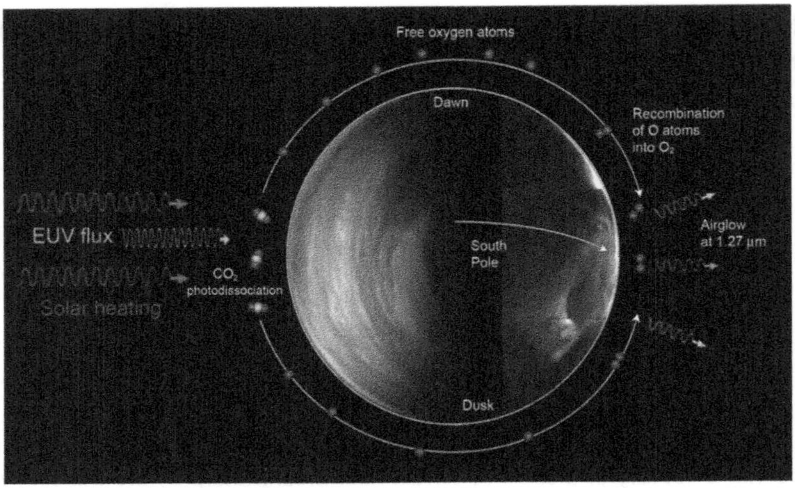

Figure 4.3. *The subsolar → antisolar atmospheric circulation that prevails in the Venus mesosphere, evidenced by Venus Express. Oxygen atoms, coming from the dissociation of CO_2 on the day side, are transported to the night side where they recombine to form excited oxygen O_2, detected at 2.7 µm by its fluorescence radiation (source: ESA/VIRTIS/Venus Express). For a color version of this figure, see www.iste.co.uk/encrenaz/solar1.zip*

4.1.5. *Observations from the ground*

Although our knowledge of the telluric planets owes much to space exploration, it is also based on ground observation, in particular thanks to high spectral resolution spectroscopy, which is difficult to achieve on board a space probe because of the complexity of spectrometers. This is how carbon monoxide CO was detected on Mars and Venus, first in the millimetric range in 1967, then in the near-infrared by Fourier spectroscopy, as well as HCl and HF on Venus. In 1988, the HDO molecule was detected by high resolution spectroscopy from the ground, allowing the first measurement of the D/H ratio on Mars (see section 4.3.3.7). In the 1990s, millimeter heterodyne spectroscopy allowed the analysis of the thermal structure and winds on Mars and Venus, thanks to the analysis of CO lines; these studies continued in the sub-millimeter range, particularly with the SMA and Alma interferometers (see Figure 4.4). In 2003, hydrogen peroxide (H_2O_2) was detected on Mars in the millimeter range, followed by the infrared range, and then the detection of methane (CH_4) was identified from the ground and the Mars Express and Curiosity space probes (this result remains controversial).

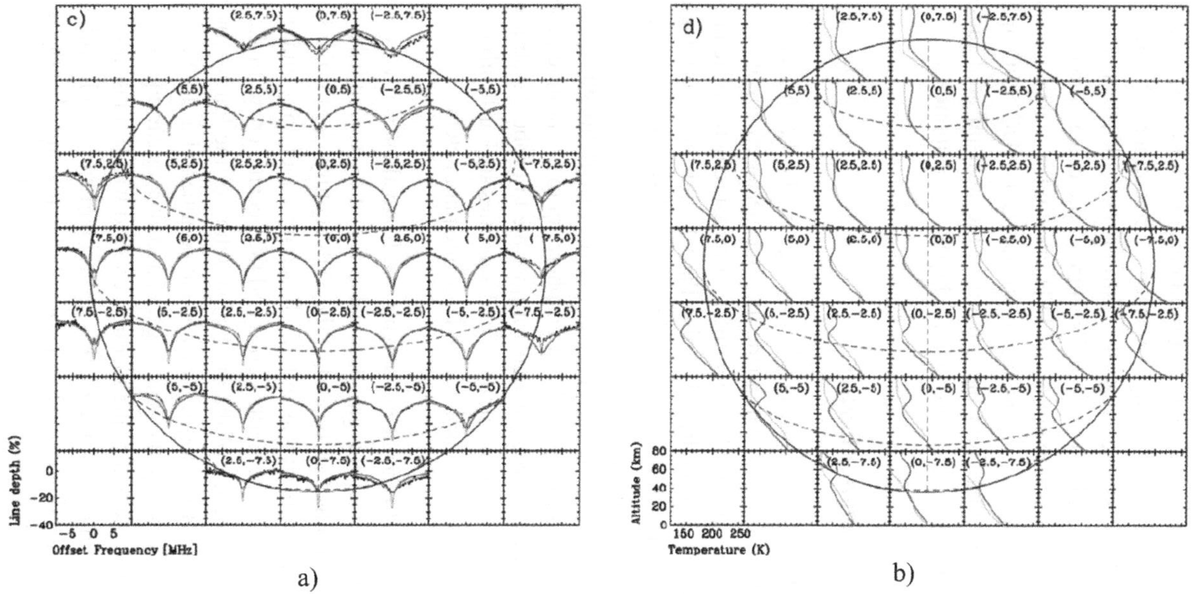

Figure 4.4. *Mapping of Venus in the CO line (1–0) at 115 GHz, with the IRAM interferometer (Institute of millimeter radioastronomy) at Plateau de Bure in September 2003. (a) Profiles of the CO line. The spectral resolution power is 10^6. IRAM spectra are in black; predictions of a general circulation model (GCM) are in red; spectra corresponding to the best agreement and used to deduce the thermal profile are in green. (b) Map of thermal profiles, deduced by inversion of the spectra. Profiles predicted by the GCM are in red; profiles deduced from IRAM measurements are in green (source: from (Moreno* et al. *2009)). For a color version of this figure, see www.iste.co.uk/encrenaz/solar1.zip*

In the case of Venus, infrared spectroscopy from the ground allowed us, in 1989, to detect emissions in the near-infrared range, coming from the low troposphere and even from the surface of Venus. These thermal emissions were observed in some spectral windows located between the CO_2 absorption bands, and were detectable on the night side thanks to the very high surface temperature of the planet; subsequently, the Venus Express and Akatsuki instruments took advantage of this characteristic. In 1990, high-resolution infrared spectroscopic observations in one of these windows allowed the measurement of the D/H ratio in the Venus troposphere (see section 4.3.3.7). Some minor constituents of Venus (CO, HDO, H_2O and SO_2) were mapped by spectro-imaging from the ground in the near-infrared range and thermal domains, complementing the more localized space measurements.

4.1.6. *The exploration of planet Earth*

As in the case of Mars and Venus, our knowledge of the Earth's atmosphere stretches back to the time of Galileo, who made the first measurement of air density and invented the first thermometers. We owe the first measurements of temperature and pressure to Evangelista Torricelli and Blaise Pascal, and the determination of the atmospheric composition to Antoine Lavoisier. The stratosphere was discovered in 1899 thanks to the sending of balloons, and the ozone layer in 1913 by spectroscopy from the ground. An international network of radio-survey stations was set up for systematic monitoring of the atmosphere up to an altitude of 40 km. With the advent of the space age, rockets made it possible to probe the stratosphere and the mesosphere up to 150 km in altitude, and satellites in Earth orbit made it possible to permanently monitor atmospheric and oceanic parameters.

While the radius of the Earth has been known since Eratosthenes, it was not until Newton, and the measurement of the gravitational constant by Cavendish, that the mass of the Earth was determined at the end of the 18th century. As for the study of the internal structure of the Earth, it was initially based on mineralogy and geology, which made it possible to probe the Earth's crust and the oceanic crust. In the course of the 20th century, our knowledge of the internal structure of the Earth took a leap forward thanks to two discoveries: seismic waves and continental drift. Networks of seismometers were set up to study the waves generated during earthquakes, which made it possible to determine whether the medium crossed was solid or liquid. This is how the presence of a solid core in the Earth's central part and then liquid, up to a height of 3,400 km, was determined: it is in this conductive medium that the dynamo that generates the magnetic field is formed. The latter has been known since the invention of the compass, by the Chinese, and has been systematically measured by ground stations since the work of Carl Friedrich Gauss. At higher altitudes, the first space probes revealed a large-scale magnetic cavity around the Earth, with complex properties, the magnetosphere. As for the phenomenon of continental drift, it was first proposed by Alfred Wegener in 1912 on the basis of various arguments, including the similarity of the coastline on each side of the South Atlantic and in the Indian Ocean (Wegener 1966). However, this hypothesis was initially not well received, for lack of a satisfactory physical justification. This was provided in the 1960s, when the discovery of oceanic ridges demonstrated the existence of convective motion in the upper mantle.

4.1.7. *Global Climatic Models*

Global Climatic Models (GCMs) aim to predict the climatic evolution of a planet on a global scale (see section 6.3.2, Volume 2). They use a three-dimensional grid (latitude, longitude, altitude) and, for each grid cell, they take into account the basic equations of hydrostatic equilibrium, the law of perfect gases, the conservation of energy, mass and momentum, and photochemistry. GCMs have been developed since the 1970s to model the Earth's climate and predict its evolution, and have greatly benefited from the development of super-computers.

The first Martian climate models appeared in the 1990s. These models have been perfected over time and reproduce, with great accuracy, the evolution of the different atmospheric parameters (temperature, pressure, winds, relative abundance of the different constituents) according to latitude, longitude, altitude and the solar longitude L_s, which describes the seasonal cycle of Mars. These models have also been used to model the Martian paleoclimate, by assuming different values of the obliquity of the planet. Global climatic models for Venus are currently under development.

4.1.8. *The electromagnetic spectrum of telluric planets*

Most of the information we have on the composition of the surface and the atmosphere of telluric planets comes from the observation of their electromagnetic spectrum, from X-rays and ultraviolet to radio waves. The infrared and radio sub-millimeter spectra show signatures of neutral molecules (vibration-rotation bands and rotation bands), while the ultraviolet range mainly probes the upper atmosphere and gives us information on the dissociation products of molecules (radicals, atoms and ions; see section 4.3.3.4).

Like for all Solar System objects, the electromagnetic spectrum of the telluric planets is characterized by a reflected solar component that peaks at 0.5 μm and a thermal component that dominates beyond a few micrometers (see section 1.3.1). The Earth also presents intense intrinsic emissions (airglow, and especially aurorae), produced by the precipitation of charged energetic particles near the magnetic poles and observed in the low frequency, visible, UV and X-ray ranges. The spectrum of objects without an atmosphere (Mercury, Moon) can be roughly represented in the visible range by that of a black body at 5,770 K (temperature of the solar photosphere) multiplied by the albedo of the object, and in the mid-infrared range by that of a black body at the equilibrium temperature of the object (see section 1.3.1). These continuous spectra are modulated by spectral signatures, which give us information on the mineralogical composition of the surfaces. In the case of planets with an atmosphere, the reflected component is also modulated by the spectral signatures of atmospheric constituents, observed in absorption in front of the solar continuum. The thermal component is sensitive to the temperature of the atmosphere and the surface; if the temperature of the atmosphere is lower than that of the surface (which is generally the case), the spectral signatures appear in absorption (see Figure 4.5); on the other hand, they can appear in emission in the polar regions or in the event of temperature inversion in the atmosphere (as in the case of the Earth).

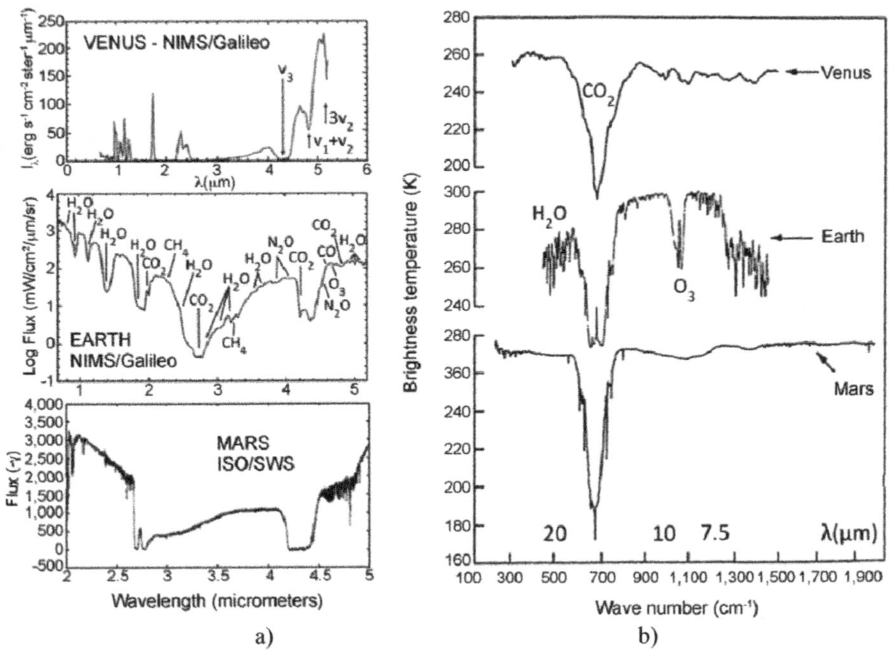

Figure 4.5. *The spectrum of telluric planets with an atmosphere*

COMMENT ON FIGURE 4.5. – *(a) The 1–5 µm region. The spectrum of Venus corresponds to the night side. The heat flux from the lower troposphere is observed in the windows between the CO_2 absorption bands. This spectrum was taken by the Galileo probe that flew over Venus in December 1989 (Carlson* et al. *1991). The spectrum of the Earth was similarly taken during the Earth flyby in December 1990 (Drossart* et al. *1993); that of Mars was taken by the SWS instrument of the ISO satellite (Lellouch* et al. *2000). In the Earth and Mars spectra, the reflected solar component dominates at low wavelengths, and the thermal component beyond 4 µm. b) The thermal spectra of Venus, Earth and Mars, obtained by the Voyager probes between 100 cm^{-1} (100 µm) and 2,000 cm^{-1} (5 µm). The strong CO_2 band at 660 cm^{-1} (15 µm) dominates the three spectra. On the Earth spectrum, the O_3 ozone band is clearly identifiable at 1,040 cm^{-1} (9.6 µm) (source: from (Hanel* et al. *2003)).*

4.2. Objects without an atmosphere: Mercury, the Moon[1]

The Moon has been, and continues to be, the target for a great number of space expeditions. In parallel to the 24 automatic Soviet missions carried out between 1959 and 1976, which allowed the analysis of the ground and the return of some samples, the American Apollo program led to the implementation of 29 missions from 1961 to 1969 (9 Ranger, 5 Orbiter, 5 Surveyors and the first 10 missions of the Apollo program) in preparation for the six manned flights with astronauts landing between July 1969 (Apollo 11) and December 1972 (Apollo 17). A failure imposed a premature return of Apollo 13, without it having reached the Moon. The scientific, cultural, economic and strategic impact of this program was considerable. Hundreds of scientific teams around the world worked, and for some, still work, on the data and samples (almost 400 kg) brought back from these missions.

Following the Apollo and Luna space programs, space exploration of the Moon ceased for a period of nearly 20 years. However, in spite of the immense progress made during the 1970s, much remained to be done in the study of our satellite. In particular, only the regions close to the equator had been flown over and mapped with a good resolution, in terms of mineralogy and chemical composition. This is the reason why a new program of space exploration of the Moon was undertaken, on a more modest scale, during the 1990s: 25 new missions were programmed until the first quarter of the 21st century, with the arrival of the Japanese (JAXA), Indian (ISRO) and Chinese (CNSA) space agencies on the Moon exploration scene. The observations of the missions already carried out have made it possible to complete

1 This section is adapted from J.-P. Bibring's text in the book *Le Système solaire* (Encrenaz *et al.* 2003, Chapter 6).

the mapping of the whole lunar surface with measurements of altimetry, gravimetry, magnetic field and composition (elements and minerals). They have also highlighted the presence of water ice in polar regions, permanently shaded from solar radiation; as in the case of Mercury, these ice deposits were created by impacts of small water-rich bodies (comets or asteroids).

Because of the synchronous rotation of the Moon, a part of the lunar surface remains inaccessible to observation from Earth. The first image of the "hidden side" of the Moon was taken by the Soviet probe Luna 3 in October 1959. The latter proved to be different from the visible side, in that it did not have vast dark "seas", which contrast with the visible side's clearer surrounding grounds. The entire lunar surface is riddled with meteorite impact craters, which attest to the absence of any geological activity and violent erosion for several billion years.

From the perspective of its motion in the sky, the Moon should obviously be considered as the Earth's satellite. However, its essential characteristics do not seem to result from its proximity to the Earth. It is therefore appropriate to treat the Moon as a planetary body, in the same way as the four other telluric planets, as its macroscopic properties make it similar to this family. The Earth–Moon system (see section 4.2.6.3) can be closely considered as a double system, as is the case of the dwarf planet Pluto and its moon Charon.

The Moon and Mercury are the two smallest objects of the family of telluric planets: they are of comparable size and without an atmosphere, both have old surfaces, and are characterized by very cratered terrains that alternate with smoother regions (see Figure 4.6). The difference between these two bodies concerns their internal structure. The Moon, which is very poor in iron, has a core that, if it exists, must be very small and has no magnetic field; Mercury is the richest planet in iron, has a metallic core that constitutes more than 80% of its mass and has a magnetic field of its own. This indicates that the evolutions of these two bodies were marked by similar exogenous processes, but by very different endogenous processes.

Mercury is an almost perfect sphere, with an equatorial radius of 2,439 km. It is the smallest of the four telluric planets, barely larger than the Moon. Its mean density of 5.427 g/cm^3 (see Table 1.2) is close, however, to that of the Earth and Venus, and is clearly higher than that of Mars and the Moon. This leads to a very low escape velocity of 4.3 km/s (see Table 1.2). A review of our knowledge of Mercury before the Messenger probe's exploration can be found in (Vilas *et al.* 1989).

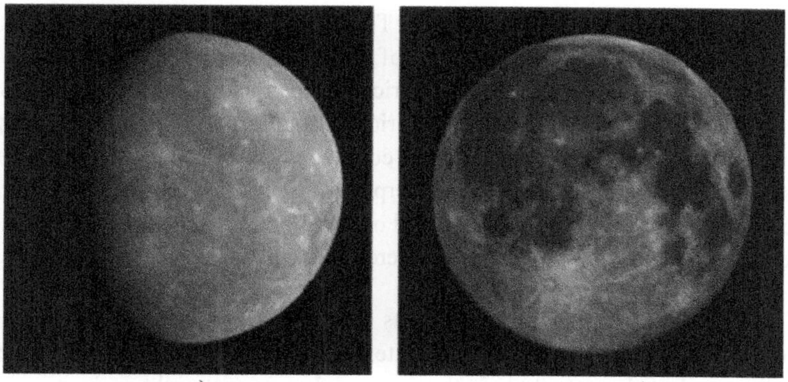

a) b)

Figure 4.6. *(a) Mercury, imaged by the Messenger probe (source: NASA); (b) the Moon (source: Wikimedia Commons/G.H. Revera). The scale is not respected*

The appearance of Mercury's surface, riddled with impact craters, seems to attest to the absence of internal activity for billions of years. The daytime temperature of Mercury is very high, which, together with a low escape velocity, has ruled out the possibility of it having a dense atmosphere. Mercury thus produces the appearance of an extinct planet, whose internal structure is similar to that of the Earth and whose surface bears a great resemblance to that of the Moon; for this reason, Mercury has been defined as "a mini-Earth dressed as the Moon".

Mercury is the least explored telluric planet in terms of space missions: the extreme conditions of its environment, due to its proximity to the Sun, require very advanced technologies and navigation techniques to ensure the survival of the probe, and the success of the mission. NASA has dedicated two space missions to the study of Mercury: Mariner 10 (1974–1975) and Messenger (2011–2015). The European and Japanese space agencies (ESA and JAXA) launched the joint BepiColombo mission in 2018, which includes two spacecrafts: the European Mercury Planetary Orbiter (MPO) and the Japanese Mercury Magnetospheric Orbiter (MMO). When BepiColombo arrives at Mercury in late 2025, both spacecrafts will collect data during the nominal one-year mission, with a possible, even likely, extension.

4.2.1. *Orbital parameters and macroscopic characteristics*

Mercury's orbit, whose semi-major axis is 0.466 au, is strongly eccentric: $e = 0.206$ (see Table 1.1). It is the most elliptical of the planetary trajectories. Thus, its perihelion is 0.33 au, and its aphelion 0.56 au. Another specificity of Mercury's orbit is its 7° inclination on the ecliptic. The period of sidereal revolution is 87.97 Earth days, for an average orbital speed of 48 km/s. Mercury rotates on itself in the prograde motion, around an axis that is strictly perpendicular to the plane of its orbit, in 58.65 Earth days. We notice that its period of rotation is equal to 2/3 of its period of revolution, which reflects a dynamic coupling between the motions of rotation and sidereal revolution. This can be interpreted as the result of the rotation of the planet slowing down, during or at the end of accretion, under the gravitational effect of the Sun. The duration of one day on Mercury is equal to 176 Earth days.

The absence of an atmosphere results in very strong variations of the surface temperature, which can reach approximately 700 K at the sub-solar point near the perihelion, and go down to 90 K in the equatorial regions at night.

The Moon completes an orbit around the Earth in about 27.3 days, at a mean distance of about 385,000 km from the center of the Earth, at a mean velocity of 1,023 m/s. It differs from most of the natural satellites of the other planets in that its orbit is closer to the ecliptic plane than to the Earth's equator (see Table 1.3). The Moon's orbit is clearly elliptical, with a mean eccentricity of 0.0549. This causes a variation in the angular velocity and apparent size of the Moon for an Earth observer. The orientation of the orbit is not stable in space, but undergoes a precession effect over time. The line of the apsis, connecting the perigee (the closest point of the orbit to the Earth) and the apogee (the farthest point), rotates slowly in the same direction as the Moon itself (prograde motion), making a complete revolution in 3,232.6054 days (about 8.85 years). It also rotates on itself in the same time that it orbits around the Earth, which explains why the same face is always shown. From any point on its surface, except in the polar regions, a lunar day (from sunrise to sunset) lasts 29.53 Earth days. This is called the "synodic lunar month".

The Moon has no atmosphere, its surface is not protected from the heat of the Sun; thus, its temperature goes from about 400 K during the day to less than 100 K during the night. In some craters, it can drop to 23 K.

4.2.2. *Exospheres*

Although Mercury's gravity field is not sufficient for the planet to maintain a stable atmosphere, there is a very tenuous and transient atmosphere, with a surface pressure in the nanobar range, consisting of hydrogen, helium and oxygen atoms. These elements, identified by the UV spectrometer of Mariner 10, are the result of the interaction of the solar wind with the surface of Mercury. The latter is also responsible for the sodium and potassium emissions, detected by spectroscopy, that were visible from the Earth in 1985–1986.

Mercury's tenuous atmosphere scatters sunlight, and the brightness of the emission is proportional to the abundance of its constituents. The scattered sunlight gives the sodium a bright orange glow, which varies with the season according to the Mercury-Sun distance. During some parts of a Mercury year, the radiation pressure resulting from the scattering process is strong enough to strip off a large part of the atmosphere and form a long glowing tail.

The lunar exosphere, between the vacuum of space and the surface of the Moon, is an atmosphere that is so thin that atoms rarely collide. This environment, mainly composed of neutral atoms and molecules, is created by the interaction of solar radiation, radioactive decay and meteorite bombardment, counterbalanced by losses in space and return to the surface. The sputtering by solar H^+ and He^{++} ions ejects lunar material into the exosphere; a fraction of this material is formed of ionized species. Direct information about the composition of the planet's surface is carried by sputtered secondary ions, which are observed by ion mass spectrometers. Measurements by the Wind, Selene, AMPTE, and LADEE missions have detected photoionized neutral elements and secondary ions in the exosphere on the Moon, including H_2^+, He^+, C^+, Na^+, O^+, Al^+, Si^+, K^+, CO^+, Ar^+, Fe^+ and Ca^+.

The main components of the thin lunar exosphere are helium, neon and argon. The presence of neon had been a subject of speculation since Apollo, and the Lunar Atmosphere and Dust Environment Explorer (LADEE) mission, operated by NASA in lunar orbit in 2013–2014, finally confirmed it by systematically measuring these gases; we now know that neon comes from the solar wind. Argon comes largely from the natural transformation of a radioactive form of potassium, found in the rocks of all terrestrial planets. As potassium in the Moon decays, the resulting argon rises to the lunar surface and becomes part of the exosphere. The helium in the lunar exosphere is closely related to the contribution of the solar wind. The residence time of each helium atom on the Moon is only a few days. For neon and argon, which are more massive, this time is longer.

4.2.3. *Internal structure*

We cannot directly examine the interior of the bodies of the Solar System (even that of the Earth), so we therefore refer to numerical models. A model contains the fundamental laws of physics and calculates what the conditions are at a particular point inside the planet. To perform these calculations, a model requires a series of input parameters (density versus radius, rotation speed, temperature, magnetic field, chemical composition along the radius). The model is tested by comparing the boundary conditions, measured at the surface of the planet (mean density, total mass, temperature, magnetic field intensity, crustal thickness, seismic activity), with their values, obtained by this model.

The interiors of the Moon and Mercury are very different from those of other telluric planets. The Moon is extremely poor in iron and its core, if it has one, should be very small. Mercury's density, given its small radius, is particularly high, almost equal to that of the Earth, while the latter undergoes a significant compression factor due to the high pressures exerted at its center. In the absence of a compression factor, the mean density of the Earth would only be 4.4 kg/m^3. This implies that there is a significantly higher proportion of heavy elements found within Mercury than in the Earth and other telluric planets.

4.2.3.1. *Internal structure of Mercury*

Mercury's very high mean density (5.55 g/cm^3) implies that nearly 60% of its mass is iron. Static models suggest that the iron core occupies three quarters of the planet's radius and that the remaining 600 km constitute the rocky mantle, of which the outer 200 km are the lithosphere. Such a small mantle, compared to that of the Earth, may be the result of the impact of a large planetesimal in the final stages of Mercury's formation, which would have ejected most of its mantle, leaving an iron-rich planet. When the large core began to cool, it contracted, causing the collapse of the outer rigid crust, giving rise to escarpments that are characteristic of the surface of Mercury. The fact that the surface rocks do not contain basaltic materials that are rich in iron and titanium implies that the deep volcanic activity stopped very early on during the formation of Mercury and, consequently, that the cooling of the iron core would have been so slow (especially if it also contains iron sulfide FeS) that it could still be partially molten. The interior of Mercury is therefore still active due to the presence of the liquid core, which sustains the magnetic field of the planet.

The spin and gravity of the planet give information on the structure of the core. Messenger's radio observations have been used to determine the gravitational anomalies and the location of its rotation pole, and to understand the orientation of the planet. Mercury rotates much more slowly than the Earth, with the length of a day being about 58 Earth days: very small variations in the way an object rotates are used to reveal clues about its internal structure. In 2007, radar observations from Earth revealed slight variations (librations) in Mercury's rotation, which proved that Mercury's core must be made of molten metal. However, observations of the rotation rate alone were not sufficient to give a clear measure of the nature of the inner core. Gravity is a powerful tool for examining the deep interior of a planet, because it depends on the density structure of the planet. This structure can be reconstructed from the small variations observed in the orbit of a spacecraft, as it gets closer and closer to the surface. In the final phases of the mission, the Messenger spacecraft flew about 200 kilometers above the surface and less than 100 kilometers in its last year. These final low-altitude orbits provided the data to determine Mercury's internal structure. The results showed that Mercury must have a central solid iron core of about 2,000 kilometers, which constitutes half of Mercury's total core mass.

4.2.3.2. *Internal structure of the Moon*

The current model of the Moon's interior (see Figure 4.7) is, aside from that of the Earth, the most accurate compared to those of other planets and satellites, because it was obtained thanks to the large amount of data collected during the numerous lunar exploration missions: data from seismometers deposited at different sites on the surface by the Apollo missions, gravity and altimetry maps, rotational field measurements, and the composition of the surface materials.

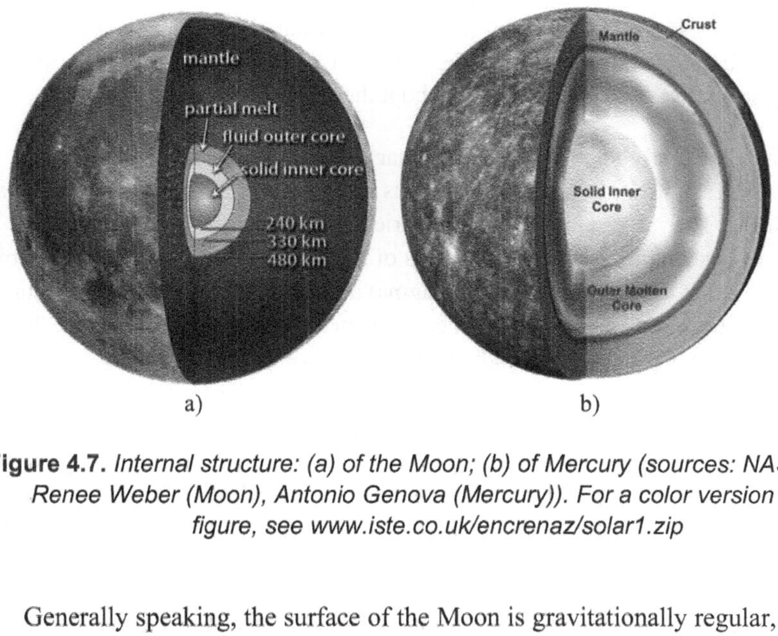

Figure 4.7. *Internal structure: (a) of the Moon; (b) of Mercury (sources: NASA/MSFC/ Renee Weber (Moon), Antonio Genova (Mercury)). For a color version of this figure, see www.iste.co.uk/encrenaz/solar1.zip*

Generally speaking, the surface of the Moon is gravitationally regular, indicating that the crustal elements buried at shallow depths are in equilibrium with the mantle (isostatic compensation). However, some of the large basins on the visible side show important gravitational anomalies called *mascons* (from the fusion of the words *mass* and *concentration*), generated by the lifting of the mantle material as a consequence of the impact that formed each basin and by the lava that subsequently filled it. These gravitational differences, together with the analysis of seismic measurements of lunar tremors originating near the surface, have provided a first estimate of the thickness of the lunar crust. Assuming a constant density of 2.85 g/cm^3 for the crust, the hidden face of the Moon would be on average thicker (68 km, with a maximum of 107 km under the topographic heights) than the visible face (60 km); below the seas, the crust is generally much thinner (0 km under Mare Crisium, 4 km under Mare Orientalis, 20 km under the Aitken basin). The information provided by seismography also shows that the source of the moonquakes is located at a depth of ~ 1,000 km. There is a slight inversion of the speed of seismic waves around 300–500 km, which could indicate the presence of a chemical gradient between the upper mantle (mainly composed of olivine) and the intermediate mantle (composed of olivine and pyroxene).

Recent reinterpretation of seismic data from the Apollo missions, as well as data from the Clementine and Grail missions, suggest the presence of a solid inner core with a 300 to 400 km radius, a liquid outer core and a partially molten zone at the base of the lower mantle. New gravimetric measurements suggest that the crust is thinner than previously thought, with an average thickness of 30–40 km. The rocks collected on the Moon come from the crust and not the mantle, but some volcanic glasses (the green glass spherules collected during the Apollo 15 mission) are very "primitive" (chemically speaking) and could inform us about the composition of the mantle.

From the chemical composition of the lunar rocks collected by the astronauts, it appears that at least 50% of the entire Moon has gone through a molten state: a "lunar magma ocean". The large anorthositic crust (rich in plagioclase feldspar) requires a mechanism to concentrate significant amounts of aluminum in the crust. Incompatible elements (those that tend to remain in the magma) are highly concentrated on the lunar surface, and these rocks are very old. The sequence of crystallization in the lunar magma ocean can give us some indication of the composition inside the Moon. Olivine and pyroxene were the first minerals to crystallize; they were dense, sank to the bottom and accumulated in a thick and dense zone: these are known as accumulated rocks. The molten magma was thus depleted in Fe and Mg, and became rich in Ca and Al, which caused the crystallization of plagioclases. These plagioclases were less dense than the magma and floated upwards to form the crust of anorthositic rocks of the Moon, 4.46 billion years ago (Gyr).

The small amount of magma remaining was highly enriched in elements that are not found in olivine, pyroxene or plagioclase: potassium (K), rare earth elements, as well as Th, U, Ba, Zr, etc., and phosphorus (P). These elements that form the KREEP rocks imply large amounts of initial magma: they date back to about 4.36 Gyr. Thus, the melting was large-scale and crystallization was early and rapid. The oceanic magma hypothesis can therefore explain the thick anorthositic crust, the different types of basalts and the KREEP rocks (for potassium K, rare earths REE and phosphorus

P), and is consistent with the ages of the lunar samples. Isotopic measurements of the Hf-W radiogenic system show that the core of the Moon was formed about 30 million years after the formation of the Solar System. Other isotopes show that most of the magma ocean was crystallized about 4.4 Gyr ago, and that the final solidification of the KREEP rocks took place 4.36 Gyr ago: everything happened very early in the Solar System! A consequence of the sequence of mineral crystallization, and the cooling of the lunar magma ocean, is that very dense minerals tended to form last, especially from the last melting points containing a lot of titanium: thus, the upper part of the mantle has a denser layer than the material below. It has been suggested that this combination was not stable and that the dense, Ti-rich material must have fallen back to the bottom of the mantle ("overturning the magma ocean"). This idea is important, because these Ti-rich molten masses also contained many energy-producing elements that could have maintained volcanic activity for a long time.

4.2.4. *The surfaces of Mercury and the Moon*

The surface of Mercury, dominated by impact craters, is very similar to the surface of the Moon, but there is a certain degree of diversity, related to the different evolutionary history of the two bodies. Mercury is brighter (mean albedo is 0.14) than the visible side of the Moon (mean albedo is 0.11) and does not show any contrast between the maria and the highlands. The result of Mercury's greater surface gravity is that its craters are shallower than similar sized craters on the Moon, and the secondary craters and ejecta are closer to the primary craters. The smooth intercrater plains that characterize much of Mercury's northern hemisphere resemble the maria on the Moon in some respects, but their composition indicates that Mercury lacks basaltic material that is rich in heavy elements such as iron and titanium.

4.2.4.1. *Mercury's surface*

Large areas of smooth plains lie between the heavily cratered regions of Mercury's surface. The northern polar regions of the planet are largely covered by these plains, which may be volcanic basalt deposits, several kilometers thick.

The annular basin of Caloris (~ 1,300 km in diameter) is the main structure observed on Mercury, similar to the large impact basins of the Moon. The mountains surrounding the basin are 2 km high. After the impact, the basin was filled with lava: ridges and fractures formed as the volcanic rock contracted and stretched under its own weight. The mouth of a small volcano was observed on the inner edge of the Caloris basin, which testifies to an internal activity of the planet during its primordial history.

On the opposite side of Caloris is a chaotic formation of boulders and hills, called "weird terrains". These terrains were probably created by the earthquake caused by the shock waves following the impact that produced the basin, which, propagating in opposite directions, met at the antipodes of Caloris and interacted destructively.

The Rembrandt basin (~ 715 km in diameter), Mercury's youngest basin, is surrounded by blocks excavated by the impact; inside, we can observe a hummocky terrain, with smooth plains of volcanic origin, characterized by a system of wrinkles (contraction zones) and depressions (extension zones), both radial or concentric. The concentric ridges form an almost complete ring with a diameter of about 375 km. The ripples, which form a particular radial structure, and the troughs are mostly inside this ring. The Rembrandt basin is crossed by a large lobe-shaped escarpment, running from the southwest to the north. It is about 600 km long and belongs to the global system of escarpments covering the whole surface of Mercury. These escarpments are several hundred kilometers long and their height varies from a few hundred meters to a few kilometers (see Figure 4.8). They indicate that Mercury's crust contracted as the planet's interior cooled. They are long, winding cliffs called lobate escarpments that are the surface expression of thrust faults, where the crust was broken along an inclined plane and pushed upward; gravity then forced the crust to fit into a smaller interior. The escarpments suggest an overall decrease in Mercury's radius by 1 to 2 km.

Figure 4.8. *One of the largest lobate escarpments on Mercury (Discovery Rupes), which is about 550 kilometers long and up to 1.5 kilometers high. Note that the walls and bottom of two impact craters in the center of the image have been deformed by the fault that formed the escarpment (source: NASA). For a color version of this figure, see www.iste.co.uk/encrenaz/solar1.zip*

The most striking surface features are the small haloed depressions called "hollows". Clusters of hollows, ranging from one kilometer to hundreds of kilometers in diameter, are found all over the planet's surface, usually within impact craters. They look as if the surface has been eaten away, leaving flat-bottomed depressions about 50 meters deep. Bright, bluish materials on their floors, and in a halo around them, had already made them stand out as bright spots (called "BCFD": Bright Crater Floor Deposits) in Mariner 10 images. The hollows appear to be very recent: their margins are sharp and they are not covered by craters. The way they cluster to form large and relatively flat areas indicates that they are not volcanic vents. It is assumed that they were formed by the loss of volatile matter from the surface of the planet. Given the very high temperatures reached during Mercury's day, the volatile elements near the surface sublimated, passing directly from the solid state to the gaseous state, and were then lost in the exosphere. The distribution of troughs on the surface of the planet seems to confirm this. The surface regions that receive more heat from the Sun generally have more troughs than the colder areas. In addition, troughs in these cooler areas are more frequently associated with volcano vents than troughs elsewhere, which may mean that heating of the interior may also provide the energy for volatile loss in areas of lower solar heat. Craters with troughs are almost always superimposed on dark substrates containing volatile-rich low reflectance materials (LRM), which may be the result of high volatile content throughout Mercury, or the result of the concentration of volatile content in a specific substrate: indeed, X-ray and γ-ray spectrometers of the Messenger probe have detected ten times more sulfur than in terrestrial rocks, as well as a higher potassium/thorium ratio than on Mars. Even though it is possible that internal processes have concentrated volatile material on the surface of Mercury and that the overall concentration of volatiles on the planet is lower, Mercury is not being depleted in volatile elements.

There has been a lot of volcanic activity on Mercury: lava plains cover 27% of the globe, and volcanic pipes surrounded by deposits of reddish material, that are present everywhere, seem to be the result of widespread explosive volcanism. Volcanism on Mercury has been active for a long time: the lava plains are 3.5 billion years old, and effusive and explosive eruptions seem to have occurred for at least 2 billion years and probably even up to a billion years ago. However, it is logical to think that Mercury would cool more rapidly than a larger planet, leaving less energy for magma melting. The cooling resulted in a global contraction and thickening of the lithosphere, which hindered the ascent of magma. Yet, the presence of volatile substances must have reduced the melting temperature of the rock: the more volatile substances there were in the crust and mantle of Mercury, the easier it would have been to melt them and bring them up. The widespread presence of explosive volcanism is an indicator of the presence of volatile substances: volcanism becomes explosive when the volatile substances in the magma expand and tear the magma into fragments which, as they fall, create pyroclastic deposits. On Mercury, 150 places have been observed showing these bright, red deposits with diffuse margins around the central pipes; the largest of them is in the northeast of the Rachmaninov basin, and extends up to 130 km from the eruptive center.

4.2.4.2. *The Moon's surface*

When viewed through a telescope, the Moon appears to be covered with impact craters of all sizes. The most visible features on the Moon's surface – those that can be seen with the naked eye and that appear to show a "man on the Moon" – are vast expanses of darker lava flows. Centuries ago, early observers believed that the Moon had continents and oceans, and that it was possibly home to life. They called the dark areas "seas" (*maria*, in Latin). Their names, *Mare Nubium* (sea of clouds), *Mare Tranquillitatis* (sea of tranquility), etc., are still used today. However, the areas between the seas, the lands (*terrae*, in Latin), are no longer named as such. Thousands of individual craters have been named in honor of great scientists and philosophers (among the most important craters are those bearing the names of Plato, Copernicus, Tycho, Kepler and Galileo).

The lunar reliefs' resemblance with those of the Earth is superficial: indeed, the characteristics of the lunar craters and mountains are very different from their terrestrial counterparts. The relative lack of internal activity on the Moon, as well as the absence of air and water, allows access to most of its geological history, unlike anything we know on Earth. The solidification age of lunar rock samples brought back to Earth by Apollo astronauts ranged from about 3.3 to 4.4 billion years, which is significantly older than most terrestrial rocks. Both the Earth and the Moon formed between 4.5 and 4.6 billion years ago: most of the Moon's crust (83%) is made up of silicate rocks called anorthosites; these regions constitute the lunar highlands. These rocks have a relatively low density and initially solidified on the Moon as they cooled. As a result of forming so early in lunar history (between 4.1 and 4.4 billion years ago), the highlands are heavily cratered and bear the marks of all the interplanetary debris impacts that have hit them since. The mountains on the lunar *Terrae* have low, rounded profiles that resemble the older, eroded mountains of Earth (see Figure 4.9). As there is no atmosphere or water on the Moon, there was no wind, water, or ice to erode them and form the sharp cliffs and peaks that are seen on Earth. Their smooth features are attributed to progressive erosion by meteorite impact. The *Maria* are much less cratered than the highlands and only cover 17% of the lunar surface, mostly on the side of the Moon that faces Earth.

Figure 4.9. *Beside the seas covered with basalt, the surface of the Moon presents uneven reliefs with faults and ridges (here, the Apennines chain) (source: NASA)*

Today, we know that the *Maria* are primarily composed of dark-colored basalt (volcanic lava), deposited during volcanic eruptions billions of years ago. Eventually, these lava flows filled in part huge depressions called impact basins, which were produced by collisions of large pieces of material with the Moon at a relatively early point in its history. The basalt on the Moon is very similar in composition to the crust beneath the Earth's oceans or to the lavas of many terrestrial volcanoes. The youngest of the lunar impact basins is *Mare Orientale*. Volcanic activity may have begun very early in the Moon's history, although most of the evidence from the first half-billion years is lost. What we do know is that the major volcanism of the lunar seas, which involved the upwelling of lava from a depth of more than 100 km below the surface, ended about 3.3 billion years ago. After that, the interior of the Moon cooled and volcanic activity was limited to local phenomena. The primary forces that then modified the surface came from the exterior, not the interior.

The surface of the Moon is covered by a fine soil called regolith (a combination of the words *regular* and *lithology*), formed by tiny fragments of broken rocks (microbreccias), minerals and vitreous spherules, which is the result of meteorite showers of all sizes that has affected the surface of the Moon since its formation. Each cratering event, be it large or small, breaks the rock of the lunar surface and scatters the fragments. Billions of years of impact have reduced much of the surface layer to dust or sand-sized particles.

In the absence of air, the lunar surface experiences much higher temperatures than the Earth's surface, even though the Earth is at nearly the same distance from the Sun. Around noon, when the Sun is highest in the sky, the temperature of the dark lunar soil rises above water's boiling point. During the long lunar night (which, like the lunar day, lasts two Earth weeks), the temperature drops to about 100 K. The extreme cooling not only results from the absence of air, but also from the porous nature of the regolith, which cools more rapidly than solid rock.

To summarize, the Moon, like the Earth, formed about 4.5 billion years ago. The heavily cratered highlands of the Moon are made of rocks that are over 4 billion years old. The darker volcanic plains of the seas mainly formed between 3.3 and 3.8 billion years ago.

4.2.4.3. *Polar residues*

There are regions near the poles of Mercury and the Moon which, because of the absence of body obliquity, are permanently protected from solar radiation (see Figure 4.10). The temperature is then permanently lower than 120 K, allowing the condensation of water on long time scales.

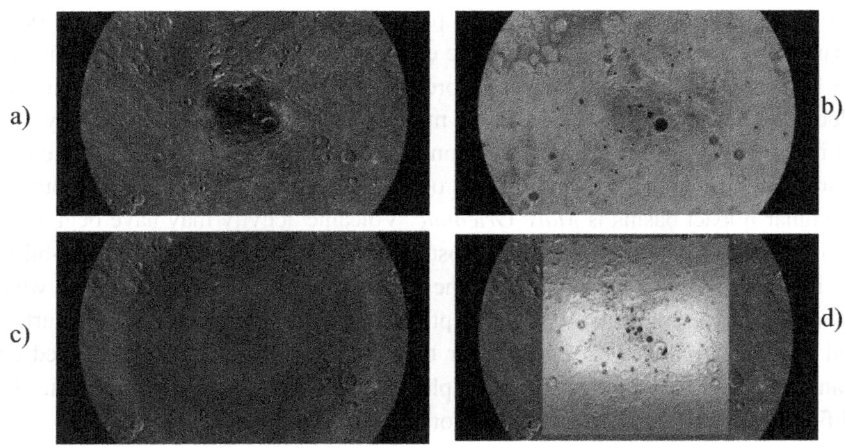

Figure 4.10. *The north polar region of Mercury, with 0° longitude at the bottom and extending to 65° north latitude. For a color version of this figure, see www.iste.co.uk/encrenaz/solar1.zip*

COMMENT ON FIGURE 4.10. – *(a) Radar image taken from the Arecibo radio telescope, superimposed on a mosaic of Messenger images, which, for the first time, make it possible to identify craters with bright deposits. (b) Topography of the region (purple: – 5 km, red: + 5 km). Lighting models derived from the topography show that bright deposits are located in areas of permanent shadow. (c) Longitudinal average of the neutron flux over the polar region: the decrease of the neutron flux at higher latitudes testifies to the presence of hydrogen (in water ice) in Mercury's polar region. d) Mercury's maximum surface temperature, modeled from the topography (violet: 50 K, red: > 550 K). The permanently shaded craters near Mercury's north pole have thermal environments that allow water ice to be stable in these craters, either at the surface or a few dozen centimeters below the surface (source: NASA).*

Radar data, taken in 1991, showed that there were regions of very high reflectivity at the poles of Mercury, which could be covered with water ice. This observational result was confirmed by data collected during the Messenger mission: water could come from meteorite bombardments from comets or asteroids that were rich in water, and could be trapped at the bottom of Mercury's polar craters.

Ice deposits have also been observed on the Moon (see Figure 4.11). Most of the water ice is found in the shadow of craters near the poles where, due to the very low inclination of the Moon's rotation axis, sunlight never reaches the bottom of these structures and temperatures never exceed 120 K. Data from the NASA M3 instrument on board the Chandrayaan-1 probe, launched in 2008 by the Indian Space Research Organization (ISRO), have identified the presence of ice on the Moon's surface. M3 (Moon Mineralogy Mapper) was particularly well-equipped to confirm the presence of solid ice. Between December 2008 and August 2009, M3 collected infrared spectra that made it possible to differentiate between liquid water, steam and solid ice.

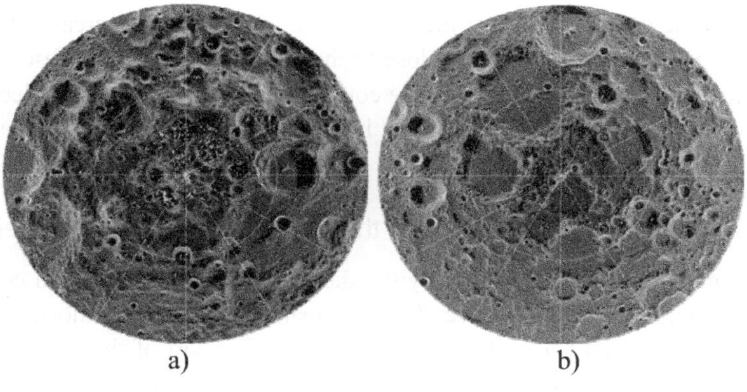

Figure 4.11. *Distribution of surface ice (a) at the south pole and (b) at the north pole of the Moon, detected by NASA's Moon Mineralogy Mapper Instrument. For a color version of this figure, see www.iste.co.uk/encrenaz/solar1.zip*

COMMENT ON FIGURE 4.11. – *The blue represents areas of ice, drawn on an image of the lunar surface, where the gray scale corresponds to the surface temperature (the darker shades representing the coldest areas and the lighter shades indicating the warmest areas). Ice is concentrated in the darkest and coldest areas, in the shadow of craters (source: NASA).*

On October 9, 2009, an impactor from the NASA LCROSS mission penetrated the bottom of the Cabeus crater, creating two plumes of ice-containing material. The first was made of steam and fine dust, and the second of heavier material. Data from the satellite's infrared spectrometer provided accurate information on the presence of water in the material ejected by the impact. An additional confirmation came from an emission line attributed to hydroxyl OH, detected in the visible-ultraviolet range by the spectrometer.

4.2.4.4. *Laboratory analysis of lunar samples*

We briefly describe below the scientific fields that have benefited from the laboratory analysis of lunar samples.

4.2.4.4.1. Dating of the main lunar metamorphic events

The date of formation of the Moon is slightly later than that of the Solar System as a whole, that is $-4.55\ 10^9$ years. The differentiation of the feldspar crust dates back from the very beginning of the Moon's history. During the following hundreds of millions of years, the Moon underwent a strong meteorite bombardment, until $-3.8\ 10^9$ years. On the visible face, the large basins formed by these impacts were later filled by deep basaltic effusions, giving rise to the lunar seas. This volcanism, however, did not take place on the hidden face, probably because of the greater thickness of the crust. The rate of meteorite bombardment then decreased from -3.8 to $-3.5\ 10^9$ years, reaching a value close to the present value. In the absence of any internal metamorphic activity, these meteorite bombardments would indeed be the source of the main modifications of the lunar surface during the last three billion years.

4.2.4.4.2. Comparative mineralogy of the different lunar soils analyzed

Lunar mineralogy is dominated by the dichotomy between seas and continents. These represent 80% of the surface. They are mainly made of anorthositic feldspars that are rich in Al and Ca, and constitute the emerging part of the lunar crust. The seas are basaltic and come from lava from the mantle. They are almost entirely located on the visible face, where they represent 35% of the surface. The feldspars are plagioclases, poor in Ca. The pyroxenes contain lamellae of pigeonite and augite that are a few hundred angstroms thick, indicating very rapid crystallization kinetics. The overall mineralogical composition of the Moon is quite close to that of the Earth's mantle, with respect to the major elements Si, Mg and Fe. Volatile elements are considerably depleted on the Moon, while there could be an overabundance of refractories. This is particularly the case of titanium, which appears in ilmenite, a double oxide of iron and titanium, with the formula $FeTiO_3$. Lastly, an essential difference with the Earth is the absence of water in all lunar minerals.

4.2.4.4.3. History of grain exposure

The incessant impacts of meteorites of all sizes, by the crushing of rocks on the surface, have formed a layer of debris a few meters thick, known as the regolith. The smallest meteorites, which are also the most numerous, reduce the size of the grains of the regolith, transforming them into a very fine dust. Under the power of the impacts, some of the grains are melted and, in the form of droplets, give rise to a great diversity of partially or totally vitreous materials: "spherules", "vitreous aggregates", "splatters" on the lunar rocks. The craters formed in the regolith by meteorite impact frequently turn over this regolith, which, in turn, exposes the grains on the surface of the regolith, in direct contact with space. The smaller the grain, the less time it remains on the surface: a grain of 1 μm only remains a few thousand years, whereas for a grain of 100 μm, this exposure time is counted in hundreds of thousands of years. The successive stacking of layers of debris from neighboring craters forms a stratified structure, which allows us to follow the evolution of interplanetary energetic particle flows: no atmospheric or magnetic shield protects the Moon, which has been constantly irradiated by the solar wind and cosmic radiation. The solar wind is, by far, the most intense of these particle fluxes: currently, the flux exceeds 10^8 protons cm^2 s^{-1}, and the speed varies from 350 to 650 km/s, which corresponds to an energy close to a keV/nucleon. Particles of this energy only penetrate the first few hundred angstroms of the grains, only irradiating the surface grains. The analysis of grains, taken at different levels of the lunar cores, indicates that the isotopic composition of nitrogen in the solar wind would have changed over time, which reflects an evolution of the solar corona. Nitrogen would have been progressively enriched in ^{15}N isotope over two billion years. The average energy of the solar wind ions also seems to have increased during this period. The solar cosmic radiation, a million times less intense, but with an energy a thousand times higher (1 MeV/nucleon), penetrates up to a few dozen microns into the regolith. The "heavy ions" (iron group) form traces of irradiation at the end of their path, which can be observed under a microscope. The number of these traces makes it possible to measure the duration of grain exposure on the surface of the regolith. This also allows us to determine the energy spectrum and relative abundances of the major groups of elements in the cosmic radiation. The solar cosmic radiation seems to have changed little, both in its chemical composition and in its energy spectrum over the last two billion years. Lastly, the galactic cosmic radiation, which is 100 times less abundant than the previous one, is energetic enough (1 GeV/nucleon) to penetrate deeper than one meter into the regolith and induce nuclear reactions on the atoms constituting the grains. These effects make it possible to trace the history of the stacking of the successive strata that can be observed in each lunar soil core.

The combined effects of the meteorite bombardment and the flux of charged particles cause a progressive alteration of the macroscopic properties of the surface, or aging, which is characterized by a maturity index. The permanent sandblasting by the smallest meteorites progressively blunts the relief. A newly formed impact crater of 100 meters, with sharp edges, is eroded in a few hundred million years. Aging is also reflected in the change of visual characteristics of the surface. The light-dark contrasts observed on the Moon's surface have several origins. First, the seas have a different mineralogical composition than the continents, which makes their albedo twice as low on average. Aging also produces a progressive darkening of the seas and continents, and a decrease in contrast.

The analysis of lunar samples, even of very small mass (less than a milligram), sometimes going as far as the study of individual grains, thus provides access to global properties of planetary bodies, or even of the Solar System or the interplanetary medium as a whole.

4.2.5. *The origin of Mercury and the Moon*

4.2.5.1. *Mercury's origin*

The lack of Mercury's heavy surface elements, such as iron and titanium (as well as its high density), could be explained by the assumption that the planet's mantle was stripped by a giant impact early on in its history. Such an impact would have removed the more volatile elements, turning the planet into a ball of hot magma, dominated by extremely refractory material. The planet cooled by radiating into space, while heavy material migrated to the center, leaving Mercury's crust relatively devoid of heavy elements. The lava that covered or created the smooth plains must have originated near the crust, unlike the lunar seas, which were flooded by basalt from deep beneath the Moon's surface.

4.2.5.2. *The Moon's origin*

Following the space exploration of the Moon in the 1970s, three models were proposed to account for the Moon's origin (Canup and Righter 2000):

– fission of part of the Earth's mantle by a rapidly rotating liquid Earth (2 hours). This model was based on similarities in density (3.3 g/cm^3) and in major element composition. In particular, it explained the low iron content by the concentration of this element in the Earth's core at the time of fission. However, an analysis considering the distribution of siderophile elements and refractory elements imposed serious constraints on this model. Moreover, there were dynamic objections to this model; it seemed particularly difficult to account for the ratio of the masses 1/81 and the 5° angle between the lunar orbit and the ecliptic;

– capture by the Earth of a Moon formed elsewhere in the Solar System. This model was suggested by the study of the evolution of the lunar orbit under the effect of the tides, which indicates a greater proximity of the Moon in the past. The main objection was, above all, statistical: it is a very unlikely event. Moreover, the initial orbit should have been very close to the Earth orbit, with a semi-major axis between 0.95 and 1.05 au. It was therefore difficult to understand how an object formed at the same heliocentric distance as the Earth, would have such a different global composition, without a large metallic core;

– accretion in orbit around the Earth. This model had the advantage of not requiring any cataclysmic event. On the other hand, it did not solve the objection presented for the capture model, concerning the global composition: one could expect that the two objects, the Earth and the Moon, formed in the same place of the solar cavity, would have accreted materials that were similar in composition.

Two important new theoretical developments occurred in the early 1980s. Firstly, numerical simulations of chaotic dynamics showed that planetary orbits could be stable for millions of years, then evolve very rapidly due to perturbations exerted by giant planets; this result showed that collisions between massive bodies could have occurred at the beginning of the history of the Solar System. Secondly, three-dimensional hydrodynamic simulations showed that the result of the lateral impact of a body the size of Mars with the Earth could be the orbital ejection of part of the mantles of the two bodies concerned, accompanied by the fusion of the two nuclei that were rich in heavy elements. This scenario is compatible with the initial orbit of the Moon, strongly elliptical and inclined, such as the models deduce from the current measurements of the lunar recession rate (4 cm/year; see section 4.2.6.7). It also has the advantage of taking the results of numerical models of planetary formation into account, which seem to indicate the formation of at least twenty planetary bodies of the size of the telluric planets: they make the existence of major impacts necessary between these bodies, in order to reduce their number to that of the telluric planets observed today.

4.2.5.3. *The Earth–Moon system*

Located at 60 Earth radii and with a radius equal to more than a quarter of the Earth's radius, the Moon constitutes a real double system with the Earth (Figure 4.12); the only other known example in the Solar System is the Pluto-Charon couple. Currently, the average distance between the Earth and the Moon is 384,402 km. This distance increases with time because of the tidal effect, at a rate of about 4 cm/year at the present time (a similar effect causes the progressive distancing of Deimos from Mars, and the approach of Phobos). Assuming that the dissipated energy has remained constant in the past, we calculate that the Moon was at its closest to the Earth about two billion years ago. We should find traces of the considerable tidal effects of this time in the lunar rocks, the youngest of which solidified three billion years ago, but this is not the case. In fact, the dissipated energy has no reason to have remained constant: we know that it is essentially dissipated in the Earth's oceans, while the distribution of the Earth's continents has varied significantly during its history. There is therefore no objection to the fact that the progressive distancing of the Earth from the Moon started at the very beginning of the history of the Solar System.

Figure 4.12. *The Earth–Moon system, observed from the Galileo probe during its flyby of the Earth in 1990 (source: NASA). For a color version of this figure, see www.iste.co.uk/encrenaz/solar1.zip*

The tidal effect is the reason why the Moon always presents the same face to the Earth: it has slowed down the rotation of the Moon, until the periods of rotation and revolution are equal, as is the case for all satellite systems close to their planet (see section 1.1.5). Since the eccentricity of the orbit of the Moon around the Earth is not zero (0.054), the orbital velocity is not constant. Consequently, the rotational motion is periodically slightly ahead or behind the revolutionary motion. The relative oscillation of the trajectory of the Moon with respect to the Earth produces the libration in longitude, which is accompanied by a libration in latitude, because of the non-zero inclination (5.8°) of the rotation axis on the ecliptic. It allows the observation of a little more than half (59%) of the lunar surface from Earth.

The Sun and the planets perturb the orbit of the Moon. Let us mention the advance of the lunar perigee, which results from the rotation of the line of the apsis, with a period of 8.85 years. The line of the nodes is subject to a retrograde motion, with a period of 18.6 years.

Several lunar months can be defined depending on whether the Moon's position is relative to the Earth or to the stars. The sidereal month corresponds to the return of the Moon in the same configuration, with respect to a fixed reference point, and is 27.32 mean solar days. The duration between two identical lunar phases is a synodic month (29.53 days), which is longer than the previous one because of the orbital motion of the Earth.

4.2.6. *Mercury's magnetosphere*

Along with the Earth, Mercury is the only other telluric planet with a large-scale magnetosphere. Their global structure and the physical processes they host are very similar in nature. However, Mercury's closer proximity to the Sun, combined with its specific properties (slow rotation, weak magnetic field), produce a unique magnetosphere in the Solar System, with a reduced size and subject to dynamical phenomena varying at much smaller spatial and temporal scales than around the Earth. Our knowledge of Mercury's magnetosphere is based on its exploration by the Mariner 10 (three flybys in the 1970s) and by the Messenger probe (with a series of close polar orbits performed between 2011 and 2015), whose results have been the subject of recent reviews (Korth *et al.* 2014, 2017).

4.2.6.1. *Structure*

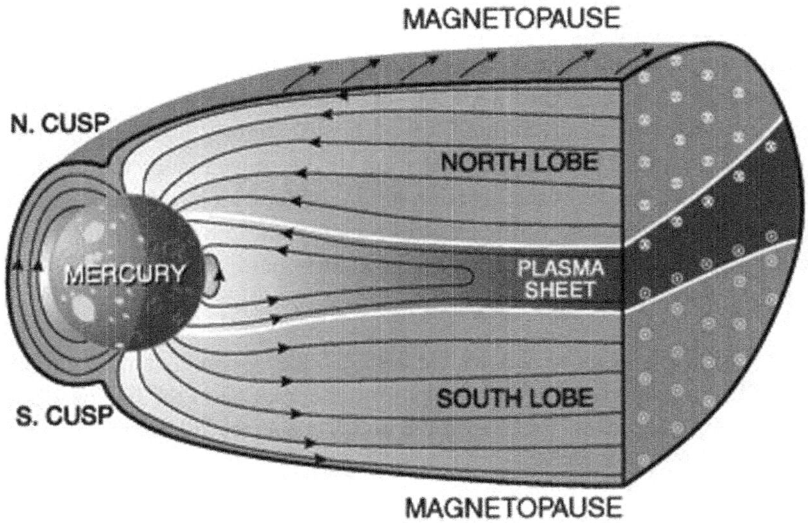

Figure 4.13. *Diagram of Mercury's magnetosphere (source: adapted from (Slavin* et al. *2010)). For a color version of this figure, see www.iste.co.uk/encrenaz/solar1.zip*

Mercury's intrinsic magnetic field is weak, with a magnetic moment 1,000 times smaller than the Earth, while the dynamic pressure of the solar wind along Mercury's elliptical orbit is about 5 to 15 times greater than at 1 au. The resulting magnetosphere is 20 times smaller than that of the Earth, with a typical sub-solar distance from the magnetopause of about 1,000 km (0.45 planetary radius). This mini-magnetosphere, the smallest of the Solar System, is very sensitive to the solar wind. The latter represents the main source of magnetospheric plasma, seconded by the exosphere. Mercury's magnetic field is almost aligned with its rotation axis and shifted northward by about 480 km (0.2 planetary radius). This asymmetry implies that the polar cap, the polar magnetic region that can interact with the solar wind, is about four times larger in the south than in the north. During episodes of high solar activity, the magnetopause can be compressed enough to let the solar wind reach the surface (and contribute to sustaining its transient exosphere by sputtering), or it can be maintained thanks to the appearance of induction currents that are produced in the planetary interior by the magnetospheric compression.

4.2.6.2. *Dynamics*

Due to Mercury's slow rotation (with a period of 58 days), the magnetospheric dynamics is only governed by the interaction with the solar wind, and the plasma is transported by the solar convection mechanism according to the Dungey cycle (see Chapter 3). The rate of magnetic reconnection between the planetary and interplanetary magnetic fields on the day side is 10 times higher than Earth's, and causes the solar wind to enter the magnetosphere near the polar cusps during brief episodes, with an average duration of 2–3 seconds (vs 1–2 minutes in the case of the Earth). During these episodes, favored by an interplanetary magnetic field oriented towards the south and often produced in a cascade of hundreds of events, the plasma and magnetic field are transported towards the tail of the magnetosphere in a few minutes (vs 1 hour in the case of the Earth), along an equatorial plasma sheet. This accumulation favors episodes of magnetic reconnection on the night side, which release magnetic islands (or plasmoids) that are evacuating plasma in the antisolar direction, and inject accelerated electrons on the closed field lines towards the planet, which re-dipolarize. The Mercurial substorm activity is thus much more dynamic than that of the Earth's.

These transient injections of energetic electrons from the tail of the magnetosphere, with high energies up to 100–200 keV, take place at the boundary between the open and closed field lines, corresponding to invariant latitudes of +50° and -20°, in both hemispheres. At Earth, these particle precipitations in the atmosphere are responsible for the auroras, organized along two magnetic circumpolar ovals. In the absence of a dense atmosphere, similar auroral emissions have not been observed on Mercury, but Messenger has revealed the existence of localized X-ray fluorescence emissions produced by the precipitation of 1–10 keV electrons at the surface, drawing part of an oval in each hemisphere. The Earth's auroral processes are associated with Birkeland electric currents, aligned with the magnetic field and closing in the ionosphere, called region 1 (on the open field lines of lower latitude) and region 2 (on the closed field lines of higher latitude). Only the former have been observed on Mercury, carrying between 20 and 40 kA (100 times less than at Earth). At the surface, the regolith is insulating and most of these currents close in the conductive planetary interior. The absence of region 2 currents has been attributed to the absence of a large-scale Earth-like ring current, flowing at the equator of the magnetosphere.

4.2.6.3. *Bepi-Colombo's contribution*

The exploratory flybys of Mariner 10 and the first orbital measurements of Messenger have laid the foundations of our understanding of Mercury's magnetosphere. These flybys have also raised questions that the Bepi-Colombo mission, scheduled to arrive in orbit around 2025, will answer. Bepi-Colombo will include a dedicated magnetospheric orbiter, associated with a nearby planetary orbiter. These two modules will thus provide simultaneous measurements in the solar wind and the magnetosphere, and will make it possible to address the details of their interaction. Bepi-Colombo will also, for the first time, explore the radio and plasma-wave environment of Mercury's magnetosphere, make high-resolution particle measurements to study the microphysics of magnetic reconnection and the Kelvin-Helmholtz process, and probe the exospheric species with a mass spectrometer in order to diagnose the exosphere/magnetosphere interaction.

4.3. Objects with an atmosphere (Venus, Earth, Mars)

The three telluric planets with an atmosphere – Venus, Earth and Mars – have radically different surface conditions. On Venus, the ground pressure reaches 94 10^5 Pa, while it is, on average, only 6 hPa on Mars; the ground temperature of Venus is 730 K, while it is on average 220 K on Mars. As a result of its low mass, as well as the strong obliquity of the planet, Mars's atmosphere reacts very quickly to the variations of insolation, which results in strong local and seasonal variations of the pressure and temperature. The Earth, whose orbit is located between those of Venus and Mars, presents intermediate surface conditions, with a mean temperature of 288 K at ground level and a mean pressure of 10^5 Pa at ground level.

However, the three planets were formed under relatively similar physico-chemical conditions, as shown by their primitive atmospheric composition, which was rich in carbon dioxide and water vapor, with a small fraction of molecular nitrogen. How did these three planets evolve to the extreme conditions we observe today? This is one of planetology's major questions today.

The satellite environment of the three planets is very different. While Venus has no satellite, and the Moon forms the equivalent of a double system with the Earth, Mars has two very small satellites, Phobos and Deimos (see Table 1.2). Due to the tidal forces exerted by Mars, they are in synchronous rotation around the planet, like the Moon around the Earth (see section 4.3.4).

Reviews of the state of our knowledge of Venus since the beginning of the space age can be found in (Hunten *et al.* 1983; Bougher *et al.* 1997; Taylor 2014; Bézard *et al.* 2020). Reviews related to the physical properties of Mars and Earth can be found in (Kieffer *et al.* 1992) and (Lunine 1999), respectively.

4.3.1. *The interior and the magnetic field*[2]

4.3.1.1. *Venus*

Venus is very similar to Earth in size and mean density, suggesting similar interior structures for both planets. This argument is strengthened by Soviet lander observations, which show the mean density of the Venus crust to be 2.7 to 2.9 g/cm^3, as expected for matter condensed from differentiated materials. The mean uncompressed planetary density is slightly lower than that of Earth (~3%). This small difference may, in fact, be related to a major difference between the two planets: Venus, unlike Earth, does not have an intrinsic magnetic field. Can the core be completely solid? The slightly lower mean density of Venus compared to the Earth suggests the absence of some abundant heavy elements. On the basis of theories about the origin of the Solar System, we can assume that Venus must contain less sulfur than the Earth, because it was formed in a hotter region of the protosolar nebula. If this is true, the core of Venus contains less iron sulfide than the Earth's core. As iron sulfide lowers the melting temperature of iron, the absence of this alloy may lead to a completely solid core. Conversely, the difference in planetary densities may be the result of stochastic perturbations of differentiated planetesimals. It is also possible that the core is liquid, but that there is no convection. On Earth, a major source of energy driving convection in the liquid outer core comes from the phase transition between the solid inner core and the liquid outer core. Solidification of the core material releases energy and stimulates convection in the Earth's core. If this phase boundary is absent on Venus, there may be no convection, even if the core is completely liquid. Measurements of the Venusian gravitational field, estimated from the Doppler monitoring of the Magellan and Pioneer Venus Orbiter spacecraft data, appear to support the hypothesis that the core of Venus is liquid. An alternative explanation for the lack of convection in a liquid core would be if the mantle were hotter than the core, a situation that would inhibit convection in the core instead of triggering it. Why might the mantle be hotter than the core? Global-scale tectonic activity on Earth (with volcanism as a complement), removes most of the internal heat: its absence on Venus implies that the interior heats up more. If radioactive heating occurs in the mantle rather than the core, this can lead to a hot mantle covering a relatively cooler core.

The lowlands and highlands of Venus are sometimes compared to the seabed and continents of Earth, although the composition and detailed topography of the Venusian regions are quite different from those on Earth. The biggest difference between the two planets is the absence of planet-wide tectonic activity on Venus.

2 This section is adapted from J.-P. Bibring's text from the book *The Solar System* (Encrenaz *et al.* 2003, Chapter 6).

Although the mantle of Venus is probably very similar to that of the Earth, the higher surface temperature of Venus suggests that the thickness of the lithosphere is about half that of the Earth. The thinner lithosphere may be too buoyant for tectonic plate activity; in other words, if plates exist, they may not be able to move by sub-conduction. On the Earth's plates, tectonic activity is a major reason for heat loss. In the absence of plate movement, hot spots and volcanic activity may be more prominent. It is generally thought, however, that the current surface heat flux on Venus is less than the current internal heat production, such that the interior heats up. The mantle could thus be hotter than the core it overlies.

With a relatively thin lithosphere, one would expect the local topography of the planet to be isostatically compensated. However, gravity field measurements show a strong correlation between local topography and gravity highs and lows, suggesting that Venus's lithosphere may be thicker than theoretical expectations, and/or that Venus does not have a zone of low viscosity, the asthenosphere, in its upper mantle. What produces this low viscosity zone in the Earth's mantle? Recycling of water in the upper mantle of subduction zones is a plausible cause. Since Venus lacks water, it has no low viscosity zone and therefore no plate tectonics. Numerical models such as those leading to plate tectonic convection on Earth show that, under Venusian conditions, the lithospheric plate can undergo complete resurfacing every $\sim 10^8$ years or so.

On the basis of the arguments developed above, the following model of the internal structure of Venus could be proposed. In its center, Venus would have a core with a 2,900 kilometers radius, which would be constituted of an external core of liquid iron and nickel, and an internal core composed of solid iron and nickel. This core would be surrounded by a silicate mantle about 3,000 km thick, hotter than the liquid core and in direct contact with the lithosphere, whose thickness is estimated between 10 and 30 km.

The space exploration of Venus has shown that the planet is devoid of a magnetic field. Two reasons can be given to explain this absence:

– the inner core, made of heavy elements, could be solid and not liquid (see above) because of a lower abundance of iron sulfide, which is itself linked to its heliocentric distance of formation lower than that of the Earth;

– the core could be liquid, but free of convection, which could occur if the temperature of the mantle is higher than that of the core, as we have mentioned above.

4.3.1.2. *The Earth*

The density of the Earth (5.5 g/cm^3), much higher than that of the silicate rocks of its surface (3 g/cm^3), is a first indication of internal differentiation. To study the deep layers, we use seismology. The pressure waves emitted during earthquakes have two components, transverse (S) and longitudinal (P), which make it possible to define discontinuities within the globe. The globe is composed of several layers (see Figure 4.14). The central core, with a radius of 3,500 km, is surrounded by the mantle, which is 2,900 km thick, and topped by a crust of a few dozen kilometers. The core is thought to be mostly iron, with a nickel alloy. Seismological measurements indicate that the core consists of a central solid nucleus with a radius of 1,250 km, surrounded by a liquid envelope. The mantle is essentially made of ferromagnesian silicates. The crust, of lower density, has two constituents: granite and oceanic floors of basalt. The temperature increases from 288 K at the surface (equilibrium temperature with solar radiation, taking into account the greenhouse effect) to more than 3,000 K at the point between the mantle and the core, and perhaps even 5,000 K at the center, where, according to models, the pressure is several megabars.

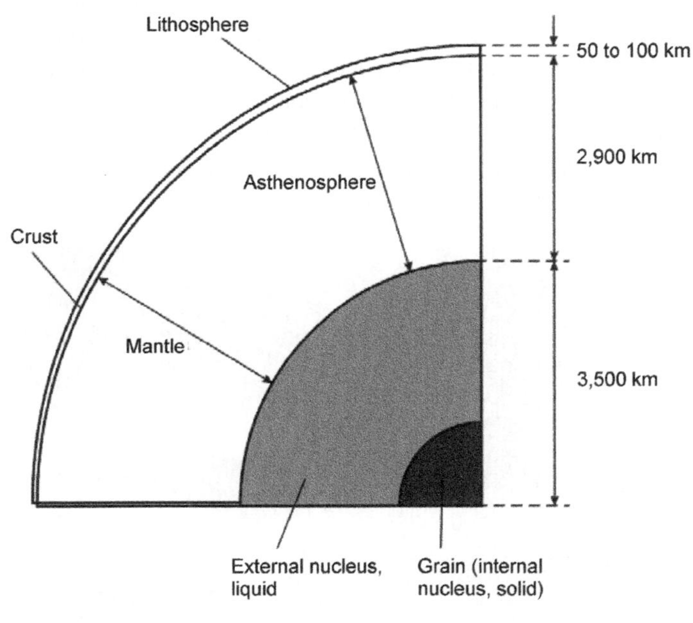

Figure 4.14. *Schematic representation of the internal structure of the Earth (source: (Encrenaz* et al. *2003))*

The Earth has an intrinsic magnetic field, the strongest among the telluric planets. This field originates from the outer liquid core. This fluid, set in motion by the Earth's rotation or by thermal convection, generates electric currents along closed lines. The resulting magnetic field has the structure of a dipole. It has an axis of symmetry, shifted by 11° with respect to the Earth's axis of rotation. The magnetic dipole moment of the Earth is 8 10^{15} Tm^3. The intensity is 0.3 G at the equator and 0.6 G at the poles.

A remarkable property of the Earth's magnetic field is the reversal of its direction over the last few million years. This phenomenon is not well understood. It could be due to a progressive rotation of the magnetic axis or to a variation of the field strength. These reversals define epochs, of a typical duration of about one million years, that have followed one after the other over the last million years. These changes are found in magnetic minerals crystallized at different times, and allow us to follow continental drifts and the expansion of ocean floors.

The Earth's magnetic field generates a magnetosphere, whose structure reflects the solar wind's interaction with a magnetized object with an atmosphere (see Chapter 3). The supersonic flow of the solar wind leads to the formation of a stationary shock wave ahead of the magnetopause, where the terrestrial and interplanetary magnetic pressures balance. The magnetopause is located at about ten Earth radii in the direction of the Sun. A geomagnetic tail develops in the antisolar direction, up to about 80 Earth radii. The magnetosphere shows the characteristic structures of a dipolar field, with Van Allen belts that constitute trapping zones and polar cusps at high latitudes (see Figure 4.15).

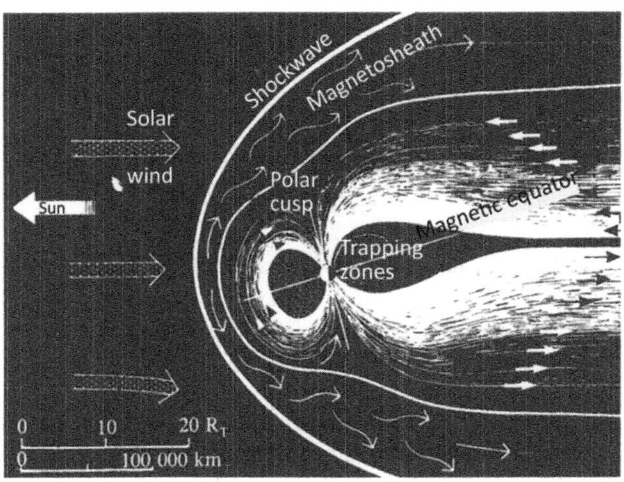

Figure 4.15. *The Earth's magnetosphere (source: from (Encrenaz* et al. *2003))*

4.3.1.3. *Mars*

Mars is of intermediate size, between that of the Moon and the Earth. Its average density is 3,930 kg m^{-3}, slightly higher than that expected for a planet composed entirely of chondritic material. Mars's low surface temperature suggests a relatively thick lithosphere, which is consistent with the observed absence of tectonic plate activity. A large part of the initial heat loss was probably due to volcanic activity, as suggested by the presence of several large volcanic shields. These large volcanoes further suggest that the erupted magma had a low viscosity, as measured for FeO-rich lavas (25% in mass). It can be shown that the magma rose to a depth of about 200 km, which is in the planet's mantle, a region that may correspond to the Earth's asthenosphere. The mantle of Mars may thus be enriched in FeO compared to its surface, but even its surface is rich in iron oxides. Mars's red color has been attributed to relatively large amounts of Fe_2O_3, which, as measured by the Viking and Mars Pathfinder landers, is about 18% in mass of the surface minerals. Such quantities of iron oxides have not been seen on any other planet.

A map of the global topography of Mars reveals a drop of about 5 km between the northern and southern hemispheres (see Figure 4.20). After adjusting the reference ellipsoid as best as possible, which is located about 3 km south of the center of mass, most of the hemispheric asymmetries disappear. A global gravity map shows that the deviations from the Martian geoid are strongly correlated with the topography, indicating that the topographic features are not fully isostatically compensated. This may be due to the thickness of the rigid lithosphere. The Tharsis region, for example, has a high net gravity of 1,000 mGal, whereas the Hellas Basin has a low gravity.

Unlike Mercury, Venus and the Moon, the gravitational moments of Mars are largely representative of the hydrodynamic response of a planet to rotation, except for a slight effect due to the presence of the Tharsis region. The measured moment of inertia is $I/MR^2 = 0{,}377 \pm 0{,}001$, or after correction for the Tharsis region $I/MR^2 = 0{,}365$; it is completely determined by the response of the planet in hydrodynamic balance to the rotation. The interior structure of Mars is derived from its mean density and moment of inertia. The core of the planet could be like the Earth's core: solid Ni-Fe, with a density of 7,000–8,000 kg m^{-3}. Alternatively, the core may consist of a liquid Fe-FeS mixture, in which case the uncompressed density is 6,000 kg m^{-3}, and the core radius is larger than that of a denser solid core. At present, no model explains the formation of the liquid core with the observed absence of an internal magnetic field. The uncompressed density of the mantle is 3,550 kg m^{-3}, which is significantly greater than that of the Earth (3,340 kg m^{-3}), indicating a greater abundance of iron. Estimates based on models fitting the observed parameters, such as I/MR^2, produce mass abundances between 16% and 21% for FeO, more than twice the value in the

Earth's mantle (7.8%). Analysis of Mars meteorites yields an absolute FeO mass concentration of 18% for the Martian mantle, a value that is within the range of values above. Although the iron content is much higher than in the Earth's mantle, it is still much lower than in the chondritic material (the normalized value for chondrites is 0.39). In addition to a higher concentration of Fe compared to Earth, Mars also has higher abundances of relatively volatile elements, such as Na, P, K and Rb. The planet probably accumulated material containing more volatile elements than the planetesimals that formed the Earth, leading to greater oxidation of metals, thus higher concentrations of FeO and other siderophile elements compared to the Earth. Separation of an FeS phase depletes Fe compared to the chondritic material (or the solar value of Fe), and also depletes some siderophile elements that have strong chalcophilic affinities (such as Co, Ni and Cu).

The mean density of Mars is much lower than that of the Earth. Although different degrees of internal compression must be taken into account, this probably reflects a lower overall iron content on Mars: about 25%, compared to 33% on Earth. This lower content is accompanied by a higher concentration in the outer layers (crust, mantle). Measurements of the moment of inertia indicate a value of 0.366, which implies the presence of a high central density; let us be reminded that the moment of inertia of a hollow sphere is 1, that of a homogeneous sphere is 0.4, and that it leans towards 0 when the mass tends to concentrate near the axis of rotation.

The results of the observations available today have made it possible to define two possible configurations for the interior of Mars: according to the first, Mars has a small dense core, essentially made up of iron (representing 15% of the planetary mass for a radius of 1,300 km); according to the second, the central core, made up of iron and sulfur, is larger (with a radius of 2,000 km) and less dense; it represents 25% of the planetary mass. It is also known that Mars has no measurable magnetic field of its own, which implies that the interior of the planet is not convective. A crustal and local magnetic field has, however, been measured by Mars Global Surveyor, particularly in the oldest terrains of the southern hemisphere. It could have been induced by a past magnetic field, due to the presence of a dynamo effect at the beginning of the planet's history.

The thickness of the mantle, which is mainly composed of silicates (notably olivine, the formula of which is $(Mg,Fe)SiO_4$), is estimated at 1,500–2,100 km, the estimates varying according to assumptions on the characteristics of the core. By and large, the mantle of Mars appears to be richer in iron and less rich in aluminum than the Earth. Lastly, the crust, of lower density, could have a thickness varying between 30 to 100 km, with strong spatial variations. It seems to be mainly made of hydrated silicates.

Three models of the evolution of the internal structure of Mars are plausible to account for the data collected so far.

According to the first model, a sulfur-rich metallic core, which, after the formation of Mars was totally liquid and convective, would have generated a global magnetic field. Convection also animated the mantle, but no plate tectonics took place and thus an unfragmented crust covered the whole planet. As the planet cooled, the convection of the liquid core gradually decreased until it stopped, thus extinguishing the magnetic field. In this configuration, the mantle could have remained the seat of convection motions, or could have solidified over time.

In the second model, the structures of the mantle and the crust are the same as in the first model, but the core, which was poor in sulfur, consisted of a liquid outer part and a solid inner part (grain). Convection in the liquid part would have set up a global magnetic field. The grain, formed by crystallization of the initial metallic liquid, following the cooling of the planet, continued to grow rapidly, leaving a liquid layer that was too thin to sustain sufficient convection to maintain a magnetic field.

In the third model, the crust of Mars was dominated by plate tectonics, which were displaced by convective motions in the mantle, and the core was composed of a liquid convective outer part, and a solid inner part. The cooling of the planetary interior, favored by the greater efficiency of heat transfer in the presence of a lithosphere cut into plates, led to the end of convective motions in the mantle and consequently the end of plate tectonics. The convection of the liquid part of the outer core thus disappeared, as well as the global magnetic field.

The final result is the same for all models: the stop of the internal dynamo led to the disappearance of the magnetic field, which protected the planet. Due to erosion by the solar wind, the atmosphere escaped in space, the temperature dropped, the water became solid and a process of definitive desertification took place.

Until now, the only data on the internal structure of Mars came from the seismometers of the Viking missions, which were insufficiently sensitive to be exploitable. The InSight mission, launched in 2018 by NASA and equipped with a broadband seismometer, has, since early 2019, allowed the monitoring of Martian earthquakes. The latter seem to be more numerous than expected, but of lower amplitude. The measurements will continue in the years to come.

Mars does not have an intrinsic magnetic field. However, as previously mentioned, the Mars Global Surveyor probe, launched in 1996, has revealed a crustal magnetic field in the oldest rocks of the southern hemisphere. This fossil field testifies to the past existence of a dynamo, generated by convection motions within a liquid core. Mars's core then cooled down over a few hundred million years, putting an end to the convection and to the Martian dynamo.

4.3.2. *The surface*[3]

4.3.2.1. *Venus*

The first radar images of the surface of Venus date from 1974–1975, and were taken by the Goldstone and Arecibo stations, in the S band of 2.32 GHz (λ = 12.6 cm). The ground resolution of these observations (a few kilometers) made it possible to measure the contrasts related to differences in albedo (and thus roughness), and not in altitudes. Many circular formations were thus wrongly attributed to impact craters. Later, radar data from the Pioneer Venus missions (launched in 1978), Venera 15 and 16 (launched in 1983), and ultimately Magellan (launched in 1989), achieved almost complete coverage (98%) of the planet with a resolution of 120 to 300 m on the ground, and 200 m in altitude. The majority of the surface (70%) is made up of plains, from which a few high reliefs stand out, corresponding to about 10% of the surface, the remaining 20% being of negative altitude with respect to the reference level (see Figure 4.16). The plains are made up of five large areas (*Planitia*), in which some regions (*Regios*) are higher. These low-lying plains were formed by fluid lava, and are characterized by a high density of fractures and narrow ridges that form when the crust is subjected to compressive and dilation forces. They also contain numerous circular structures, mostly of volcanic origin.

The reliefs are mainly composed of two plateaus: Ishtar Terra and Aphrodite Terra. The first one contains Mount Maxwell, the highest point of Venus, with an altitude of 12 km. These vast plateaus, comparable in size to that of a terrestrial continent, dominate the neighboring plains by high cliffs. Beta Regio, a smaller plateau, located in the middle of a plain, has a summit altitude of 4,000 m.

The surface of Venus is formed by both volcanic and tectonic structures, which present very different characteristics to those of the Earth (see Figure 4.17).

Volcanoes have been classified into several major categories:

– the largest structures reach altitudes of 3–5 km above the surrounding surface. There are more than 150 of them, most of them in the highlands. They present large flows of fluid lava, composed of rocks similar to terrestrial basalts. Many of these

---

3 This section is adapted from J.-P. Bibring's text from the book *Le Système solaire* (Encrenaz *et al.* 2003, Chapter 6).

flows are several hundred kilometers long. The mobility of the lava could be increased by the very high surface temperature of the planet;

– the structures of intermediate size (20–100 km) have various dome forms; about 300 have been recorded. These volcanoes were all formed from very thick, viscous lavas and have very different morphological characteristics. Pancake volcanoes have pancake-shaped domes; they are up to 100 times larger than the lava domes found on Earth, with a very flat top about 1,000 meters high. The "tick-like" volcanoes are similar in appearance to pancake volcanoes and are found on the plains, far from the largest shields on Venus. In view of their resemblance to pancake domes, these "ticks" could be the result of structural modifications of these domes;

– volcanoes, which are rarer, present fan-shaped flows, mostly located in the southern plains. They are very similar to the most viscous terrestrial lava flows, which could indicate the existence of granitic or rhyolitic lavas on Venus;

– small volcanoes (about 550) are distributed over the entire surface of the planet.

Figure 4.16. *View of the Venus hemisphere, centered at 180 degrees east longitude, obtained by assembling 10 years of radar observations by the Magellan, Venera and Pioneer Venus missions (resolution 3 km). The false colors represent the altitude (scale giving the local radius from 6,048 to 6,062 km), the blue regions being the lowest (source: NASA/Interkosmos). For a color version of this figure, see www.iste.co.uk/encrenaz/solar1.zip*

The volcanoes in the first two groups indicate the presence of successive eruptions and thus the presence of stable reservoirs of magma; in the case of the large volcanoes, the average eruption rate has been estimated at 1.7 10^{-2} km^3 of magma per year. The resurfacing rate appears to have decreased significantly over the history of Venus, with an average value of about 4 km^2 per year at the time of the formation of the plains and 0.5 km^2 per year over the last 300 million years. The lava, analyzed from the examination of volcanic channels, seems to be particularly fluid. It could be mainly composed of sulfur or alkali-carbonate compounds, which would have resulted from the interaction of the molten crust with the atmosphere of Venus, rich in CO_2, SO_2 and halogens.

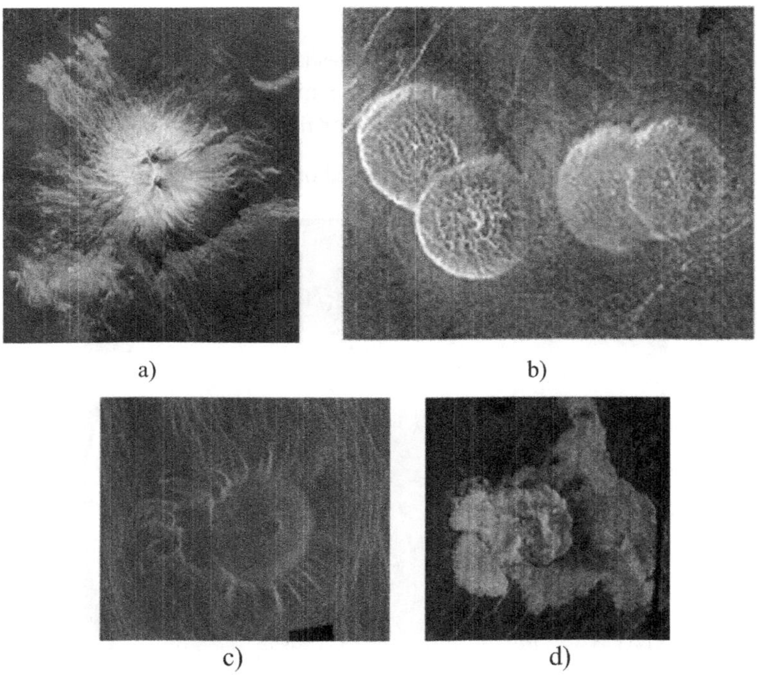

Figure 4.17. *Radar images of Venus volcanoes. (a) The Sapas Mons volcano. The radial aspect of the volcano is caused by the long lava flows extending from the two craters to the summit and numerous eruptions on the flanks (width of the image: 617 km); (b) "pancake" volcanoes; (c) "tick" volcanoes and; (d) volcanoes with a fan-shaped flow (source: NASA/Messenger). For a color version of this figure, see www.iste.co.uk/encrenaz/solar1.zip*

There are various tectonic structures observed on the surface of Venus. Some, like Maxwell Montes, are large, up to 1,000 km (mountains, valleys and cliffs). The high altitudes generally appear very bright in the radar images of Venus. This may be due to a thin layer of metal-rich material that only forms above a certain altitude. Other structures, generally smaller than 300 km in diameter, seem to be associated with volcanic activity. This is the case of "coronae", circular systems of fractures that form under the surge of hot material below the crust that deforms the surface, allowing the release of internal heat. The coronae surrounding some volcanoes are often accompanied by large lava flows.

The surface of Venus has very few impact craters: due to the very high atmospheric pressure, the less massive impact objects are completely burned up during their passage through the atmosphere, while the larger ones break up as they fall. We can thus find areas where craters are very close to each other or even overlap. The dark terrain around these craters is covered with a layer of fine material, pulverized by the impact. The craters on Venus differ in some respects from those found on other planets and satellites. The thick atmosphere prevents material thrown up by impacts from falling far away, so Venus craters do not have long lines of ejected material or chains of secondary craters. Probably due to the high surface temperature, impacts can melt and sink large amounts of material, forming large flows.

The craters are distributed in a more or less uniform way on the whole surface. The surface is thus globally very young, and was the object of a recent general resurfacing. It is thought that this resurfacing activity ceased somewhere between -300 and -800 million years ago.

4.3.2.2. *The Earth*

Among the telluric planets, the Earth has a particular tectonic activity, the plate tectonics: this is expressed by the drift of the continents, which move in relation to each other by a few cm/year. Proposed for the first time in 1912 by Alfred Wegener (1880–1930) and later improved (Wegener 1966), the theory of continental drift, at first greeted with skepticism, has been confirmed many times over the years, the most important confirmation being undoubtedly the identification of oceanic ridges; it has been unanimously accepted since the late 1960s. Continental drift is the consequence of a more global activity of the Earth's lithosphere. The lithosphere is divided into about ten large plates, delineated by ridges, through which material from the mantle (the ocean floor) escapes, and subduction trenches, through which this material returns to the mantle (see Figure 4.18). The seafloor cycle is 100 to 200 million years.

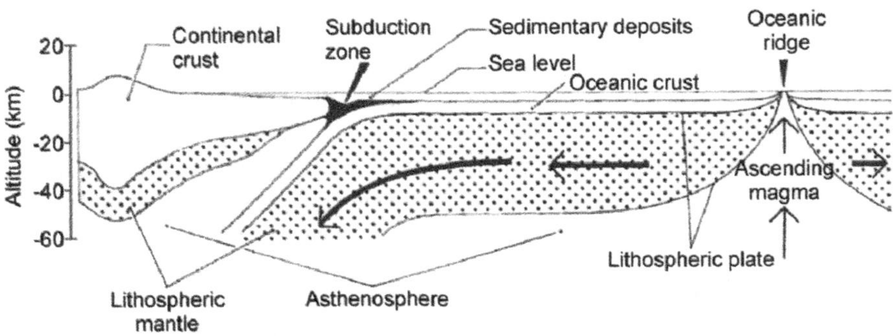

Figure 4.18. *Schematic representation of a portion of the Earth's lithosphere, showing the escape of material from the mantle at the oceanic ridge level, and its burial by subduction at the trench level (source: after (Encrenaz* et al. *2003))*

The continents, of lower density, are unsinkable, and therefore much older; the oldest rocks in Greenland are more than 3 billion years old. A consequence of plate tectonics is orogeny: mountain ranges are formed when two plates collide, one of which carrying a continent. Oceanic ridges and subduction zones are the site of strong volcanic activity.

The energy capable of setting plate tectonics in motion comes from the radioactive decay of long-lived isotopes in the mantle: uranium (^{235}U and ^{238}U, lifetime of 7.1 10^8 years), thorium (^{232}Th, lifetime of 4.51 10^9 years) and potassium (^{40}K, lifetime of 1.28 10^9 years). It could also come from the solidification of the liquid outer core.

4.3.2.3. *Mars*

Mars is flatter than the Earth: the polar radius is 18 km smaller than the equatorial radius, which corresponds to a ratio of 1/192 compared to 1/298 in the Earth's case. Moreover, the altitudes present very strong contrasts (from -6 km at the bottom of the Hellas basin to more than 20 km at the top of Olympus Mons) and we have seen that there is a strong north-south dissymmetry.

The polar caps of Mars have been known since the first telescopic observations of Mars. Of high albedo, they form successively at the north and south poles according to the seasonal cycle. The perennial caps of the two poles (see Figure 4.19) are mainly composed of a combination of CO_2 and H_2O ice, as well as sediments. Frozen carbon dioxide accumulates there in winter to form a relatively thin seasonal layer of a few dozen centimeters thick. This phenomenon plays an essential role in the seasonal atmospheric cycle, since 30% of the atmospheric mass of Mars is alternately condensed at the winter pole. The northern polar cap is about 1,000 km in diameter at its maximum and contains about 1.6 million km^3 of ice which, if distributed evenly over the cap, would be 2 km thick (about half the thickness of the Greenland ice cap). At the south pole, a permanent CO_2 cap, 420 km in diameter and a few dozen meters thick, rests on a substrate of water ice. The total volume of ice in the south polar cap, plus adjacent layered deposits, has also been estimated at 1.6 million km^3. Both ice caps show spiral troughs, which are the result of katabatic winds that rotate due to the Coriolis effect.

Mars has some of the largest impact basins in the Solar System (see Figure 4.20). They were formed when asteroids collided with Mars, ejecting material over a large area of the planet. In the south are the Hellas Basin (about 9 km deep and 2,300 km in diameter) and Argyre Planitia (deeper than 5 km, with a diameter of 1,800 km). In the northern hemisphere, we can identify the smooth-bottomed remains of other large ancient basins: Utopia Planitia (D = 3,000 km) and Isidis Planitia (D = 1,500 km).

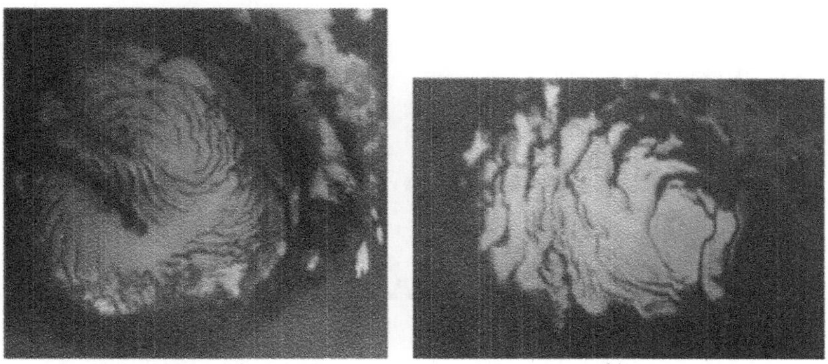

a) b)

Figure 4.19. *The perennial polar caps of Mars: (a) the northern cap; (b) the southern cap (source: NASA). For a color version of this figure, see www.iste.co.uk/encrenaz/solar1.zip*

The southern hemisphere of Mars is dominated by the highlands, an ancient and highly cratered terrain, a few kilometers above the planet's average level. The northern hemisphere is dominated by low-lying volcanic plains. At the equator is the Tharsis Plateau, about 10 km above sea level, which extends for about 8,000 km. There are three giant volcanoes lined up (Ascraeus Mons, Pavonis Mons and Arsia Mons), as well as Olympus Mons, the largest volcano in the Solar System, which peaks at more than 20 km in altitude. All these volcanoes have low slopes and present a morphology similar to the shield volcanoes of Venus and, on Earth, the volcanoes of Hawaii. The rising of the Tharsis Plateau dates from 4.1 to 3.3 billion years ago, according to the impact crater count rate. The volcanic complex only appeared later, 3 to 2.5 billion years ago. The large size of these volcanoes implies the presence of a lithosphere thick enough to support them. Their high altitude is probably due to the absence of plate tectonics: the volcanoes thus remained stable, above the magmatic zones, and the lava effusions lasted throughout the local magmatic activity. It seems that these effusions continued until recently, for around a few hundred million years. Other smaller volcanoes, the *paterae*, all located in the northern hemisphere, have a pancake shape, a central caldera and flanks with channels. Older than the volcanoes of Tharsis, they would have been at the heart of hydrothermal explosive volcanism, the result of the meeting between the magma chamber and the underground water. This explosive stage would have been followed by an effusive phase. To date, there is no trace of active volcanism on Mars.

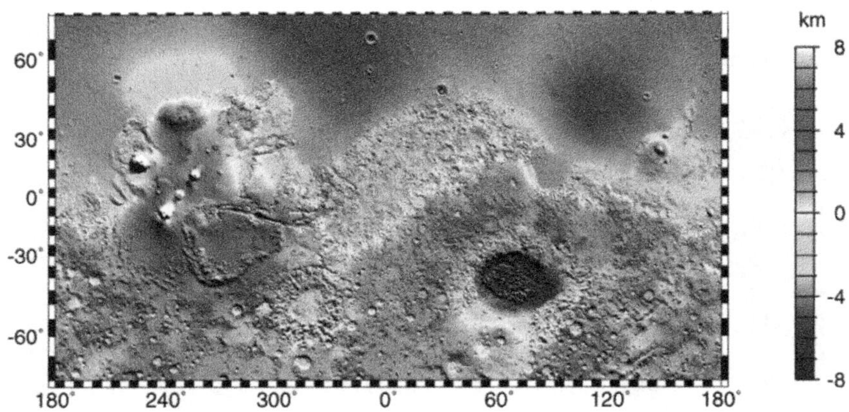

Figure 4.20. *Map of Mars altimetry, obtained in 1999 by the radar altimeter of the Mars Global Surveyor mission. For a color version of this figure, see www.iste.co.uk/encrenaz/solar1.zip*

COMMENT ON FIGURE 4.20. – *On the left (centered at 250° longitude and 0° latitude), we can see the highlands of Tharsis (in red) and its giant volcanoes. The Valles Marineris canyon system extends a little below the equator between 240° and 320° longitude. The Argyre basin (longitude 310°, latitude 45°) is at the bottom (in green) and the largest impact basins are on the right (in blue) (source: NASA).*

Mars has a gigantic network of canyons, organized around Valles Marineris, which runs along the surface of Mars over a length of 5,000 km, with a depth of up to 7 km and an average width of 200 km. These canyons seem to result from a fluvial type of erosion.

Today, Mars is a cold, dry desert with an atmosphere so thin that any surface water evaporates or freezes. However, there are signs of a warmer, wetter past: sulfur-rich salts and other deposits that could only form in a water-rich environment. Three kinds of valleys reminiscent of dry river beds can be distinguished on the Martian surface (see Figure 4.21):

– valley networks, associated with ancient terrains, which suggest fluvial erosion;

– outflow channels, suggesting massive floods during cataclysmic events;

– steep valleys with flat bottoms, suggesting debris flow erosion.

All of this evidence suggests the presence of liquid water during the planet's past history (see also section 4.3.3.7). Much of the original water remains frozen in the polar ice caps and beneath the surface in permafrost.

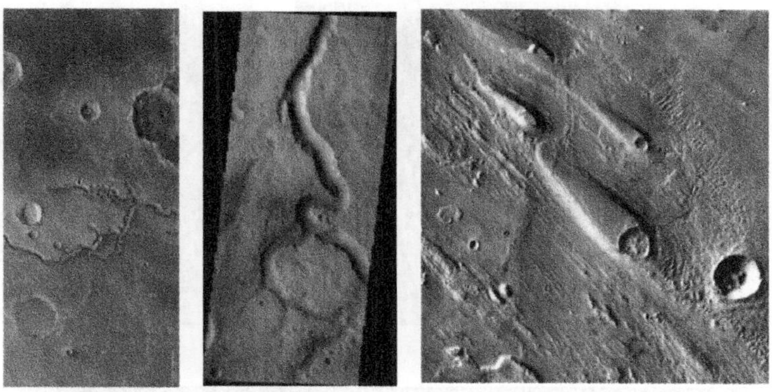

a) b) c)

Figure 4.21. *Examples of (a) a valley network with Nirgal Vallis; (b) a steep valley with Scamandre Vallis; (c) an outflow channel with Ares Vallis (source: NASA). For a color version of this figure, see www.iste.co.uk/encrenaz/solar1.zip*

Moreover, altimetry suggests the possible presence of a liquid ocean that would have covered all the low-lying plains of the northern hemisphere, at the beginning of the planet's history. It is possible that the initial accretion of the planet was followed by an internal outgassing, resulting in a dense and warm atmosphere, particularly rich in water vapor. As the surface cooled, this water would have condensed in the form of torrential rains, strongly eroding the surface, and then penetrating the ground by percolation; the process would have led to the progressive growth of a thick permafrost. The subsequent melting of these masses of ice, sometimes stimulated by volcanic activity, would have led to violent floods, at the origin of the canyons.

In addition to the large geological formations – impact craters, volcanoes, canyons – there is a whole series of complex morphological structures on the surface of Mars: faults, landslides, valleys. A tectonic has thus taken place on Mars, but different from the Earth's plate tectonics. At the beginning of its history, Mars had a liquid core in which motions generated a magnetic field, of which the Martian rocks in the southern highlands bear the evidence. The absence of magnetic features in the large impact basins indicates that the magnetic field died out before they formed, more than three billion years ago. As the liquid core cooled, the tectonic forces associated with convective motions of material in the mantle diminished, and today, ground motions along faults and "Mars quakes" are probably much less frequent. Heat escaping from the planet's interior, leading to crustal melting and volcanism, also produced tectonic forces that compressed the crust, giving rise to faults (rupture zones, along which two blocks of rock move relative to each other) or stretched it to form *grabens* (sunken parts of a planet's crust, bordered by parallel faults).

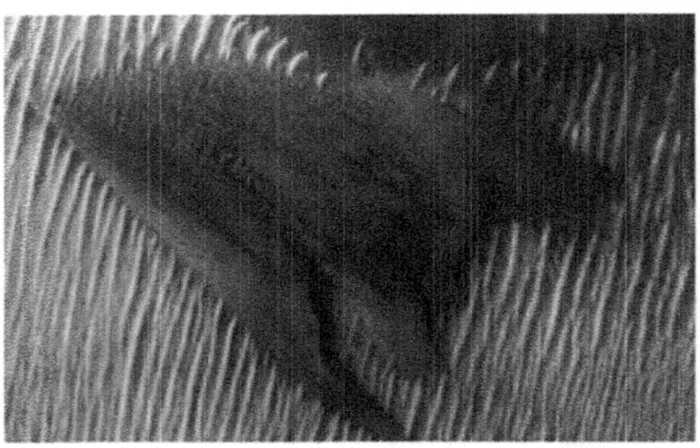

a)

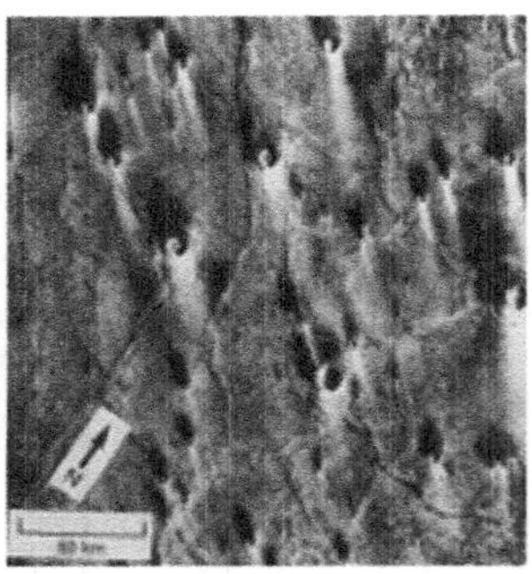

b)

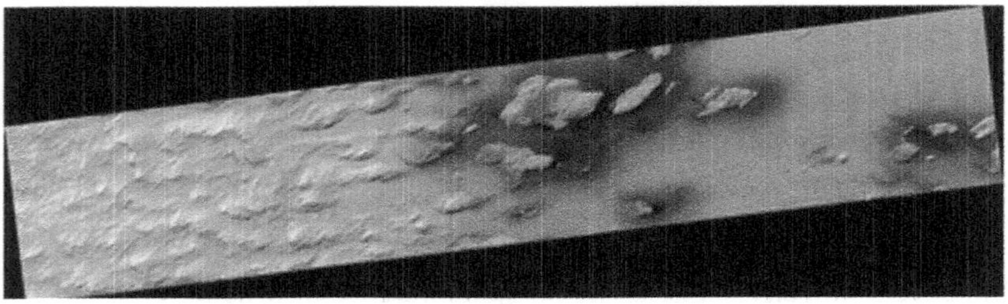

c)

Figure 4.22. *(a) False color image of a dune field inside Lyot crater, on which finer (or differently composed) materials have formed a large complex dune (in blue). The image was taken by the HiRISE camera of the Mars Reconnaissance Orbiter. (b) Dark and bright streaks formed around craters. (c) Yardangs in the southern part of Elysium Planitia (source: NASA). For a color version of this figure, see www.iste.co. uk/encrenaz/solar1.zip*

Aeolian phenomena are present all over the Martian surface, as a direct result of the strong winds that constantly sweep across it. Large areas of sand dunes are visible, especially near the north polar cap and within several craters (see Figure 4.22). Most dunes are, in some way, associated with the local topography. The chances of finding an isolated dune are low, because dunes migrate very quickly until they are trapped by some obstacle, where they can remain for a long time.

Streaks are perhaps the most striking and common wind feature on Mars. Some have undergone changes in a short period of time, indicating that the wind erosion and/or deposition that formed them is currently active. Behind the obstacles, dark trails have been observed, formed by the preferential erosion that lifts the brighter dust. When deposited again, these form bright streaks: these two types of streaks are often found in the same region. Dark streaks associated with sediment deposits, formed by the deposit of dark material displaced by saltation, have also been observed. The directions of the streaks and their temporal variations make it possible to reconstruct the main characteristics of the Martian winds. These are also responsible for permanent wind erosion. There are other significant wind erosion characteristics on Mars, in addition to streaks: yardangs, which are hills sculpted by unidirectional winds. Lastly, dust circulation is the cause of stratified deposits, especially in the polar caps. These deposits serve as tracers of ice accumulation, and can be used as chronometers for Martian paleoclimatology.

4.3.3. *The atmosphere*

We have seen (section 1.3.2.2) that Mercury, the closest planet to the Sun, does not have a stable atmosphere, due to its low escape velocity (4.43 km/s) and its high daytime temperature (700 K). The other three planets – Venus, Earth and Mars – show a remarkable diversity in their surface and atmospheric conditions (see Table 4.1).

Planet	Period of revolution (years)	Rotation period	Obliquity (°)	V_{esc} (km/s)	P_{surf} (Pa)	T_{surf} (K)	Atmospheric composition
Venus	0.723	243.0 d	177.33	10.46	93 10^5	735	CO_2, N_2
Earth	1.000	23.93 h	23.45	11.19	1 10^5	288	N_2, O_2
Mars	1.524	24.62 h	25.19	5.03	6 10^2	215	CO_2, N_2

Table 4.1. *Physical characteristics of telluric planets with an atmosphere*

4.3.3.1. *Atmospheric composition*

The chemical composition of planetary atmospheres depends on the escape velocity at the ground, as well as on their formation scenario. We have seen (section 1.3.2.2) that, under the conditions of thermodynamic equilibrium, carbon and nitrogen are preferentially found in the form of CO and N_2 in the conditions of pressure and temperature, corresponding to those of the telluric planets; CO_2 appears following the reaction of CO with H_2O. Hydrogen and helium escape because the escape velocity of the telluric planets is insufficient to retain them. In telluric planets, one thus expects to find an atmosphere that is mainly made up of carbon dioxide, nitrogen and water; this composition reflects that of the primitive atmospheres of these planets.

Table 4.2 presents the atmospheric composition of Venus, Earth and Mars. It calls for two major remarks. The first is that the atmospheres of Venus and Mars are both dominated by carbon dioxide, with a few percent of molecular nitrogen, whereas on Earth, the major constituents are molecular nitrogen and molecular oxygen, in a ratio close to 4:1. The second remark is that water, very abundant on Earth in liquid form, is virtually absent on Mars and Venus. While the CO_2/N_2 ratio has remained more or less constant throughout Mars and Venus's history, the atmospheric composition of the Earth has evolved considerably since its origin. Initially, the presence of oceans allowed the trapping of carbon dioxide in the form of calcium carbonate; moreover, the appearance of life is at the origin of the formation of oxygen, which has become, with molecular nitrogen, one of the two major constituents of the Earth's atmosphere. It should also be noted that, on Earth, the presence of oceans and the disappearance of almost all the carbon dioxide strongly contributed to limiting the greenhouse effect (see Box 4.1), allowing the Earth to keep its surface temperature relatively constant over time, and thus escape the runaway greenhouse effect observed on Venus.

The greenhouse effect occurs in the case of an atmosphere that is transparent in the visible range, but whose components absorb infrared radiation. The solar radiation (a black body at 5,770 K), at maximum in the visible range, reaches the surface and heats it (see section 1.3.1). The surface (at a temperature of a few hundred Kelvin) mainly radiates in infrared, but this radiation is absorbed by some constituents of the lower atmosphere, whose temperature rises in turn, sending radiation back to the surface, heating it even more and amplifying the phenomenon (see Figure 4.23). The so-called "greenhouse gases" are the molecules that present intense modes of vibration-rotation in the infrared range: this is particularly the case for CO_2, H_2O, CH_4 and NH_3.

In the absence of a counter-reaction, the phenomenon tends to diverge: this is what happened in the case of Venus. Being endowed with a thick CO_2 atmosphere, the greenhouse effect was amplified during the history of the planet: this effect is therefore responsible for the very high surface temperature observed today.

In the case of the Earth, following the trapping of CO_2 from the primitive atmosphere at the bottom of the oceans, the greenhouse effect has remained limited, because the major atmospheric gases, N_2 and O_2, have no spectral signature in the infrared range. However, the continuous increase of terrestrial CO_2, linked to human activities since the beginning of the industrial era, constitutes a major risk for the evolution of the Earth's climate.

Box 4.1. *The greenhouse effect*

Atmospheric constituent	Venus	Earth	Mars
Carbon dioxide (CO_2)	96.5%	400 ppm	95.7%
Argon (Ar)	70 ppmv	0.93%	2.07%
Molecular nitrogen (N_2)	3.5 %	78.1%	2.03%
Molecular oxygen (O_2)	–	20.9%	0.173%
Carbon monoxide (CO)	17 ppmv	0.01 ppmv	0.075%
Water vapor (H_2O)	30 ppmv	0.4–1% (variable)	0.03% (variable)
Molecular hydrogen (H_2)	–	550 ppbv	15 ppmv
Helium (He)	12 ppmv	5.24 ppmv	10 ppmv
Neon (Ne)	7 ppmv	18.2 ppmv	2.5 ppmv
Krypton (Kr)	–	1.14 ppmv	300 ppbv
Xenon (Xe)	–	90 ppbv	80 ppbv
Ozone (O_3)	–	0–70 ppbv	40–200 ppbv (variable)
Hydrogen peroxide (H_2O_2)	–	–	0–40 ppbv (variable)
Nitrous oxide (N_2O)	–	300 ppbv	–
Nitrogen dioxide (NO_2)	–	20 ppbv	–
Methane (CH_4)	–	1.79 ppmv	0–40 ppbv (variable)
Sulfur dioxide (SO_2)	150 ppmv	–	–

Carbonyl sulfide (OCS)	20 ppmv	–	–
Hydrogen sulfide (H_2S)	1–3 ppmv	–	–
Hydrochloric acid (HCl)	100–600 ppbv	–	–
Hydrofluoric acid (HF)	1–5 ppbv	–	–
Average molecular weight	0.0435 kg/mol	0.029 kg/mol	0.0433 kg/mol

Table 4.2. *Atmospheric composition of Venus, Earth and Mars; values refer to the troposphere*

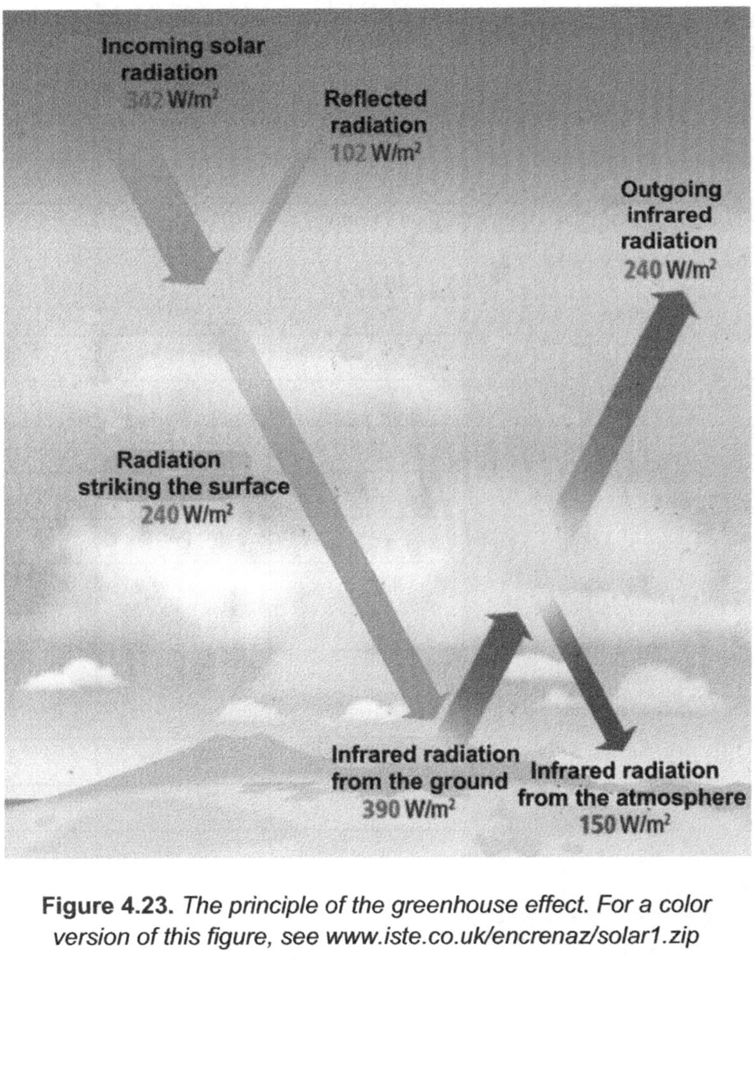

Figure 4.23. *The principle of the greenhouse effect. For a color version of this figure, see www.iste.co.uk/encrenaz/solar1.zip*

COMMENT ON FIGURE 4.23. – *The values shown are averages over the surface of the globe in the case of the Earth. Solar radiation (maximum at visible wavelengths) heats the surface, which emits infrared radiation. This radiation is absorbed by the greenhouse gases in the atmosphere (CO_2, H_2O, CH_4). The temperature of the atmosphere increases, which in turn radiates towards the surface, contributing to further heating, and the mechanism is amplified (source: (Encrenaz* et al. *2019)).*

4.3.3.2. *Thermal structure*

The thermal structure of the atmosphere is defined by the equilibrium between the hydrostatic pressure and the law of perfect gases. As the telluric planets have a surface, the pressure variation with the altitude is written as follows:

$$P(z) = P_0\, e^{-z/H} \quad [4.1]$$

P_0 being the surface pressure and H being the scale height:

$$H = RT/\mu g \quad [4.2]$$

R being the constant of perfect gases, μ being the mean molecular mass, g being the gravity and T the temperature. The value of μ is given in Table 4.2. At the surface, H is equal to about 15 km, 8 km and 11 km in the case of Venus, Earth and Mars, respectively. Since the atmosphere has a negligible thickness compared to the planetary radius, we can assume that g is constant throughout the atmosphere. The variations of temperature being much smaller than those of pressure, we can consider, as a first approximation, that the pressure decreases exponentially as the altitude increases: this is the barometric law. The mean molecular weight is constant with altitude throughout the region in which the atmospheric constituents are uniformly mixed: this is the homosphere, in which the pressure is sufficient for collisions between molecules to ensure the homogeneity of the gas.

In reality, the temperature is not constant with the altitude, and the homosphere is stratified according to the altitude into several regions, depending on the mode of energy transfer that prevails.

1) In the *troposphere*, just above the surface, the temperature decreases as the altitude increases because the surface is heated by solar radiation, which peaks in the visible range. In telluric planets, the pressure in the lower atmosphere is sufficient for the transfer of energy to be convective. In the adiabatic regime (in other words, without heat transfer with the outside), the P/ρ^γ ratio is constant, ρ being the specific mass and γ being the ratio of specific heats at constant pressure and volume, C_P/C_V.

Taking the hydrostatic law and the law of perfect gases into account, we obtain the following relationship:

$$dT/dz = -g\mu/C_P \qquad [4.3]$$

which defines the dry adiabatic gradient, constant throughout the troposphere. The troposphere reaches an altitude of 12 km on Earth, and about 25 km and 60 km on Mars and Venus, respectively. The adiabatic gradients are, respectively, about 8 K/km on Earth, 2 K/km on Mars and 7 K/km on Venus. In the case of the Earth, it is necessary to include the wet adiabatic gradient, which is calculated by taking the latent heat of sublimation of the water molecule into account; the value of the adiabatic gradient then becomes lower than the dry adiabatic gradient. On Earth, the adiabatic gradient goes from about 10 K/km in dry regions to 6 K/km in wet regions.

2) The tropopause marks the upper boundary of the troposphere and corresponds to a temperature minimum. Above this region, energy is transported by radiation. In the whole homosphere, the pressure is sufficient enough to consider the medium to be in local thermodynamic equilibrium: the radiation emitted by an atmospheric layer is thus that of the black body at the temperature of the medium. The temperature $T(z)$ of an atmospheric layer is such that the energy it emits is equal to the sum of contributions received from all other atmospheric layers, integrated over all wavelengths:

$$\sigma T(z)^4 = \iint B(z)\, e^{-\tau(z)} d\lambda dz \qquad [4.4]$$

The optical thickness $\tau(z)$ being defined as follows:

$$\tau(z) = \int_0^{\infty} \alpha(z)\rho(z) dz \qquad [4.5]$$

$\alpha(z)$ being the sum of the absorption coefficients of the various molecules present in the atmosphere, and $\rho(z)$ being the specific mass at altitude z. The absorption coefficients of the various atmospheric constituents (atoms and molecules) are determined, at any wavelength, using spectroscopic databases.

The vertical structure above the tropopause depends on the atmospheric composition and spectroscopic properties of the various constituents:

- in the case of the Earth, the presence of molecular oxygen leads to the formation of an ozone layer by photolysis, at an altitude of about 40 km. This ozone layer absorbs solar ultraviolet radiation, which results in a rise in temperature between 30 and 70 km: this is the stratosphere, which extends from 12 km to about 85 km;

- in the case of Venus and Mars, in the absence of oxygen, the temperature remains approximately constant between the tropopause and an altitude of about 100 km: this is the mesosphere.

3) The region above the mesosphere (or, in the case of the Earth, the stratosphere) is the *thermosphere*, in which the main heating mechanism is the dissociation and ionization of molecules and radicals. This region also coincides with the *heterosphere*, in which the atmospheric constituents are no longer uniformly mixed, the rate of collision becoming insufficient to ensure the mixing. The *ionosphere* begins below the homopause and extends to the border of the thermosphere. In the case of the Earth, the only telluric planet with a strong magnetic field, the interaction of the ionosphere with the magnetosphere manifests itself, at high latitudes, by the formation of polar auroras. Through this mechanism, the planetary atoms and ions, excited by the energetic particles of the solar wind, emit a particularly spectacular fluorescence radiation in the visible range.

Figure 4.24 shows the vertical profiles of Venus, Earth and Mars as a function of altitude.

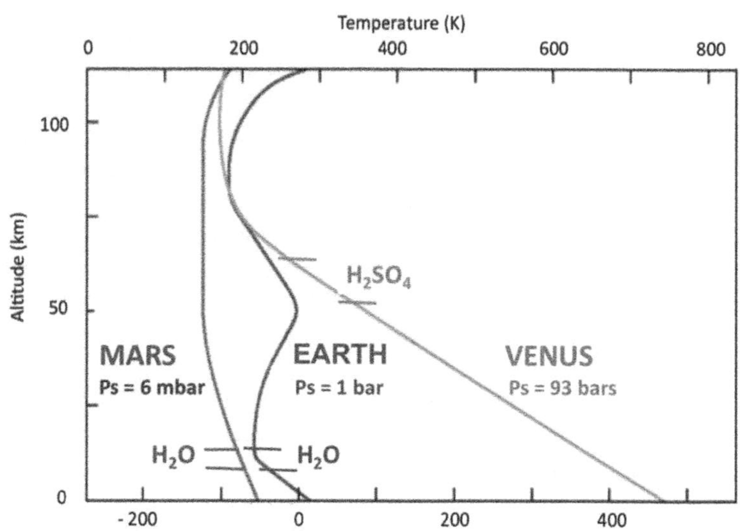

Figure 4.24. *The temperature profile of Venus, Earth and Mars as a function of altitude. It should be noted that the Earth's troposphere is much less extended than that of Mars, and especially Venus. The adiabatic gradient of Mars is significantly weaker than that of the Earth and Venus, because of the reduced gravity field of the planet. For a color version of this figure, see www.iste.co.uk/encrenaz/solar1.zip*

4.3.3.3. *Cloud structure*

Knowing the thermal structure of an atmosphere and its chemical composition, we can determine its cloud structure from the equations of thermodynamic equilibrium, coupled with photochemical equilibrium.

1) In the case of Mars and Earth, clouds are formed directly after the condensation of some minor species (H_2O on Earth, H_2O and CO_2 on Mars).

On Earth, the condensation of water gives rise to the formation of clouds of different forms, depending on altitude. In the lower troposphere, cumulus clouds, with a large vertical extension, are formed by convection following an instability of the air. Stratus clouds are the result of a progressive and stratified lifting of an air mass, and can form very extensive cloudy systems, of uniform texture. Cirrus clouds form in the upper troposphere and are made up of ice crystals. These three main categories are subdivided into multiple names depending on their appearance, their altitude, their composition (liquid or ice) and their evolution. Their distribution is closely linked to the Earth's atmospheric circulation (see section 4.3.3.6).

Mars has the particularity of experiencing phase changes to its main atmospheric constituent, carbon dioxide. Due to the high eccentricity of the planet (0.09) and its obliquity close to that of the Earth (25°), CO_2 condenses alternately at the north and south poles at the winter and summer solstices, and the atmospheric pressure at the ground undergoes seasonal variations of up to 30%. Water, although not abundant, also condenses at the poles in winter. The CO_2 and H_2O cycles are accompanied by seasonal cloud structures. Water clouds form on the sides of volcanoes in summer (see Figure 4.25), when the atmosphere is particularly water-rich, but also around the equator in summer, and around the pole corresponding to the winter solstice. CO_2 cirrus clouds also form at high altitudes in the polar night, where the temperature can be lower than that of the CO_2 condensation.

2) In the case of Venus, the simultaneous presence of SO_2 and H_2O in a troposphere dominated by CO_2 leads to the formation of sulfuric acid (H_2SO_4), following the photodissociation of SO_2.

Sulfuric acid is formed above the clouds, at an altitude of about 70 km, by solar ultraviolet radiation (see section 4.3.3.4). It condenses immediately due to the lower temperature (about 235 K) and the droplets fall by gravity to the base of the cloud zone at about 45 km; the temperature (above 300 K) is then high enough to cause their evaporation. The cloudy layer is divided into several levels with, at around 65 km, an upper layer that is uniformly opaque to visible radiation, an intermediate layer around 50 km, and a thinner and sparser layer around 40 km. This layer has been highlighted by near-infrared measurements made from the ground, and then by space probes. They showed that it was possible, in some places, to probe Venus's lower troposphere down to its surface at some specific wavelengths, between the CO_2 absorption bands (see Figure 4.26).

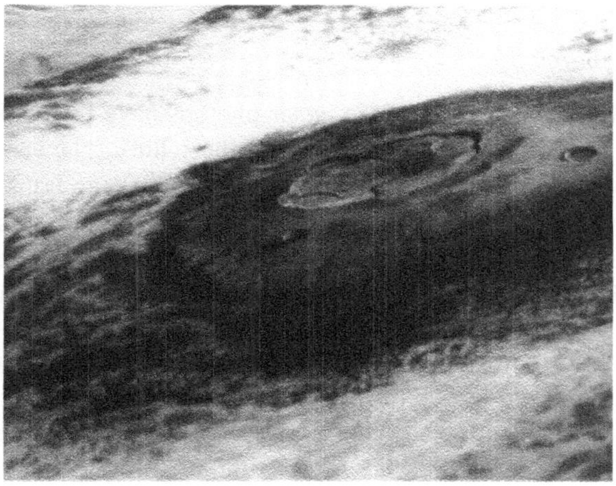

Figure 4.25. *Water clouds observed on Mars in summer around the Olympus Mons volcano, by one of the cameras of the Viking mission (source: NASA). For a color version of this figure, see www.iste.co.uk/encrenaz/solar1.zip*

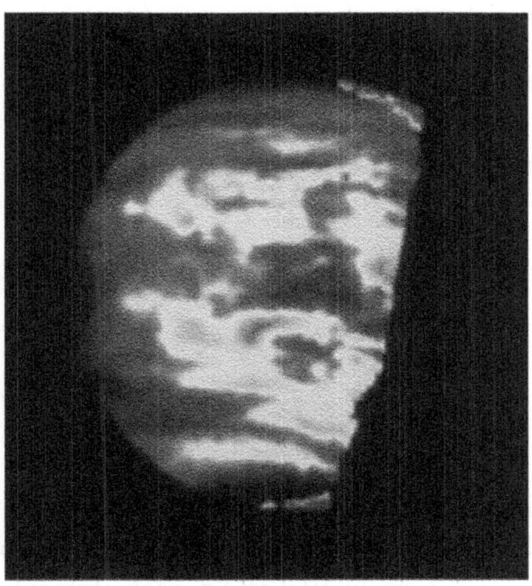

Figure 4.26. *The night side of Venus, observed by the NIMS imaging spectrometer of the Galileo probe during its flyby over Venus in December 1990. For a color version of this figure, see www.iste.co.uk/encrenaz/solar1.zip*

COMMENT ON FIGURE 4.26. – *The image is taken at 2.35 μm, a wavelength located between the CO_2 absorption bands, which makes it possible to probe the deep troposphere. The lower the altitude surveyed, the higher the emitted flux. The image shows that the cloud structure at 40 km is more transparent than at 60 km. The figure is extracted from (Carlson* et al. *1991) (source: NASA).*

4.3.3.4. *Photochemistry*

In the visible and ultraviolet range, solar photons have the ability to dissociate some neutral molecules of the atmosphere of the telluric planets into radicals or atoms. The photons with even more energy ($\lambda < 100$ nm) can transform molecules, radicals and atoms into ions: this is photoionization. Photoionization reactions – as well as the reverse reactions, recombination – can thus modify the atmospheric and cloud composition, and play a major role in the radiative balance of atmospheres.

The photodissociation of atmospheric constituents leads to the formation of an ionosphere, characterized by the presence of free electrons. The electron density is determined by the ionization and recombination rates of the different constituents. On Earth, the electron density reaches a maximum of 10^{12} m^{-3} at an altitude of about 300 km; on Mars and Venus (Krasnopolky 1986), the maximum is reached around 140 km, with respective maximum values of 10^{11} m^{-3} and 4 10^{11} m^{-3}.

1) Earth: photochemistry is governed by the photodissociation of molecular oxygen O_2. This occurs at about 60 km, under the effect of photons with a wavelength lower than 175 nm:

$$O_2 + h\nu \rightarrow O + O$$

Since the direct recombination reaction is very slow, it is carried out via a catalyst M:

$$O + O + M \rightarrow O_2 + M$$

and:

$$O + O_2 + M \rightarrow O_3 + M$$

This last reaction is at the origin of the ozone layer, which peaks at an altitude of 40 km. The vertical distributions of O, O_2 and O_3 result from a balance between these three reactions and their production rate at each atmospheric level.

2) Mars: photochemistry is dominated by the dissociation of CO_2 and H_2O. The dissociation reaction of CO_2 is written as:

$$CO_2 + h\nu \rightarrow CO + O$$

As the reverse reaction is very slow, the recombination of CO_2 is done with the help of a catalyst:

$$CO + O + M \rightarrow CO_2 + M$$

The excess of O_2 (0.17%), compared to CO (0.075%), observed in the atmosphere of Mars suggests that other reactions are involved in the formation of oxygen.

The only possible source seems to be the photodissociation of water:

$$H_2O + h\nu \rightarrow OH + H$$

$$O + H + H$$

$$O + H_2$$

The following reactions are also involved:

$$O + O_2 \rightarrow O_3$$

$$O + OH \rightarrow HO_2$$

$$H + HO_2 \rightarrow H_2O_2$$

leading to the formation of ozone and hydrogen peroxide. However, photochemical models have difficulty in accounting for the measured abundances of the observed photochemical products, and this issue is still subject to discussion.

3) Venus: photochemistry is, as on Mars, dominated by the photodissociation of CO_2 and H_2O, but two other constituents are involved; HCl and SO_2:

$$HCl + h\nu \rightarrow H + Cl$$

$$SO_2 + h\nu \rightarrow SO + O$$

The dissociation of HCl is the source of a flow of hydrogen atoms that escape to the outside. SO_2 combines with oxygen to form SO_3, which in turn reacts with H_2O to form sulfuric acid (H_2SO_4):

$$SO_2 + O + M \rightarrow SO_3 + M$$

$SO_3 + H_2O \rightarrow H_2SO_4$

As we have seen above (section 4.3.3.3), H_2SO_4 condenses to form the cloud cover of Venus. This is at the origin of the particularly high albedo of the planet (0.75).

4.3.3.5. *Atmospheric escape*

A particle can escape from an atmosphere if its kinetic energy is greater than that due to the gravitational field, and if its outward trajectory is not interrupted by a collision. The region of the atmosphere where escape is possible is the *exosphere*, and its lower limit is the *exobase*. The mean free path of a particle of cross-section σ and density N is $1/\sigma N$; the exobase is located at altitude z_{ex} such that $\sigma N(z_{ex})H = 1$, H being the scale height. In an isothermal atmosphere, the exobase is located at the level where the mean free path is equal to the scale height.

The first escape mechanism is thermal escape, or Jeans escape, which applies to gases at thermodynamic equilibrium, with Maxwellian velocities. The Jeans escape provides a lower limit of the escape rate, because non-thermal processes are involved: charge exchange reactions (energy transfer from a fast ion to a neutral one) and sputtering (energy transfer following a collision). Two mechanisms are further added to these processes, which must have been effective at the beginning of the history of the Solar System: hydrodynamic escape and erosion by impact. Hydrodynamic escape occurs when, under the effect of an intense energy source (UV radiation or solar wind), the escape of hydrogen leads to the escape of heavier atoms. Erosion by impact can occur following the fall of a meteorite that is massive enough to generate a plume, causing the escape of surrounding gas.

In the case of telluric planets, the hydrogen flux escaping into space is not limited by the rate of escape at the exobase, but by the flux of hydrogen arriving at the exobase following photochemical reactions. This flux is generated by two mechanisms: turbulent diffusion and molecular diffusion. Turbulent diffusion is a macroscopic process, which has the effect of homogenizing the atmospheric constituents below the homopause, either by convection in adiabatic or super-adiabatic atmospheres, or by gravity waves in sub-adiabatic atmospheres. Turbulent diffusion is measured by the eddy diffusion coefficient K, whose typical value is 10^{10} to 10^{14} m^2s^{-1}. Molecular diffusion tends to equalize the mixing ratios between different species below the homopause, but favors species differentiation above the homopause, with each constituent having its own scale height. The hydrogen flux escaping to the outside cannot be higher than the limit flux, $\Phi = ND/H$ (N being the density of hydrogen, D being the diffusion coefficient and H being the scale height). To estimate the outgoing hydrogen flux, the contributions of the different hydrogen-containing species (H, H_2, H_2O, CH_4, HCl, etc.) must be taken into account. A full discussion of this issue can be found in (Catling and Kasting 2017).

In the case of the Earth, the temperature is about 200 K at the homopause (z = 100 km). The hydrogen sources are H (0.19 ppmv), H_2 (0.55 ppmv), H_2O (3 ppmv) and CH_4 (1.8 ppmv). The scale height H is 6 kilometers. Using the binary scattering parameters of H_2 and H, we obtain $\Phi = 3.5\ 10^8$ atoms cm^{-2} s^{-1}. This estimate is in agreement with measurements made at the homopause. The escape is partially of thermal and suprathermal origin.

On Mars, the temperature at the homopause varies between 150 and 200 K, and H_2 is the main source of hydrogen. The scale height is 9 km. The H_2 abundance is 15 ppmv (see Table 4.2). The result is $\Phi = 3.3\ 10^8$ atoms cm^{-2} s^{-1}, comparable to the terrestrial value.

Lastly, in the case of Venus, the temperature of the exobase is about 300 K. The hydrogen sources are H (a few ppmv), HCl (0.5 ppmv) and H_2O (1 ppmv) (see Table 4.2). The hydrogen escape flux, which is essentially suprathermal, is Φ = $3\ 10^7$ atoms cm^{-2} s^{-1}.

4.3.3.6. *Atmospheric circulation*

We have seen that the solar flux plays a major role in the structure of atmospheres, whether it is their chemical composition or their thermal and cloud structure. However, the solar flux is not constant at all points of the planet; it is at its maximum at the subsolar point. In the case of Venus, whose obliquity is almost zero, and in the case of the Earth and Mars at the equinox, the flux received is at its maximum at the equator. In the troposphere, where the transfer of energy is done by convection, the warm air rises at the equator from the surface to the tropopause and moves to high latitudes where, at equal altitude, the pressure is lower. It then cools and descends to join the equator at low altitude: this is the Hadley circulation.

In the case of the Earth and Mars the situation is complicated by the rotation of the planet. The winds, moving from high pressure regions to low pressure regions, are deflected by the Coriolis force.

1) Earth: three convective cells form; the Hadley cell (at latitudes 0°–30°), the Ferrel cell (at latitudes 30°–70°) and the polar cell. The air rises at the equator, rich in humidity, to form clouds, moving to the higher latitudes and falling back down, dried out, to the tropics; the movement is reversed in the Ferrel cell (see Figure 4.27). In the northern hemisphere, the surface winds blow westward at low latitudes (the trade

winds) and eastward in the Ferrel cell; the situation is reversed in the southern hemisphere.

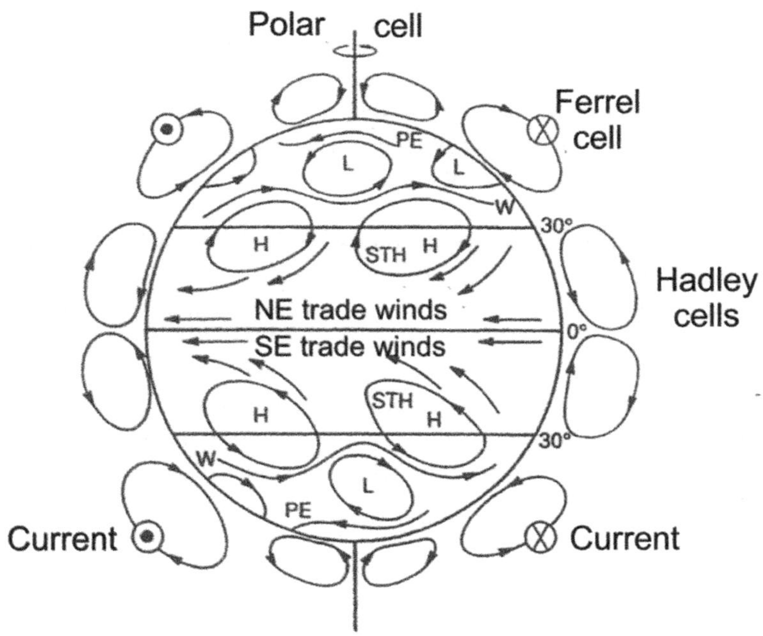

Figure 4.27. *Representation of the Hadley circulation in the case of the Earth; due to the Earth's rotation, three convective cells are formed: the tropical zone, the temperate zone and the polar zone (source: from (Graedel and Crutzen 1993))*

2) Mars: with an obliquity close to that of the Earth, Mars presents a strong seasonal cycle. Moreover, because of the low pressure on the ground, the atmosphere reacts very quickly to variations in solar flux. At the equinox, a Hadley circulation is established, with an upward motion at the equator and a downward motion at higher latitudes; only one cell is found in each hemisphere. At the solstice, the cell moves from the summer hemisphere back down into the winter hemisphere. Ice caps of CO_2 and H_2O form alternately at the poles in winter. The condensation of CO_2, the main atmospheric constituent, causes variations in total pressure of up to 30%, depending on the season.

Moreover, the atmospheric circulation of Mars is modified by the high ellipticity of its orbit, which leads to an asymmetry of temperatures in the north and south at the summer solstice (the southern hemisphere, currently at perihelion in summer, being warmer than the northern hemisphere in summer). The temperature can vary from about 130 K at the pole in winter to more than 300 K near the subsolar point. The strong variations of temperature and pressure lead to the formation of violent winds near the terminators, marking the border between day and night. These winds, generated by thermal tides, can exceed 100 m/s and create dust storms. These can be local tornadoes (dust devils), driven by convection in the presence of a super-adiabatic gradient, or global dust storms, enveloping the entire planet. The global dust storms occur shortly after the autumnal equinox, and peak at the summer solstice in the southern hemisphere; the last one occurred in July–August 2019. Figure 4.28 shows the Martian cycles of dust, water vapor and water ice.

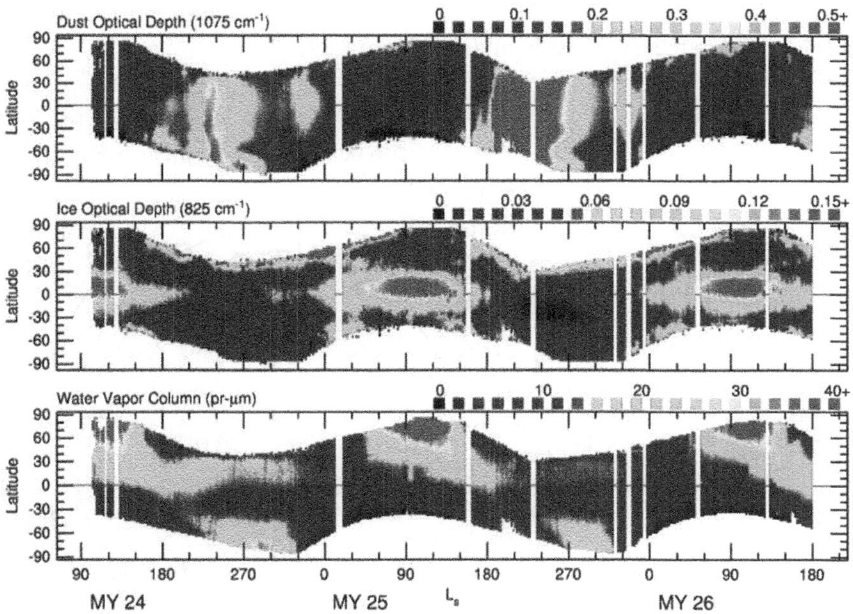

Figure 4.28. *The cycles of dust, water ice clouds and water vapor, observed on Mars by the TES infrared spectrometer aboard Mars Global Surveyor between 1999 and 2003. For a color version of this figure, see www.iste.co.uk/encrenaz/solar1.zip*

COMMENT ON FIGURE 4.28. – *A global storm is present during the Martian year MY 25, during spring in the southern hemisphere. The water vapor maximum appears in early summer in the northern hemisphere, along with water ice clouds at the equator. The local maximum of water vapor at* L_s = *270° is much less* *pronounced than at L_s = 90°; this asymmetry is due to the eccentricity of the Martian orbit (source: from (Smith 2004)).*

3) Venus: as its rotation speed is very low, only one Hadley cell develops in each hemisphere, between the equator and 60° latitude. The convection zone extends to the top of the cloud layer, at 60 km altitude. The polar regions are the seat of vortices, revealed by the Pioneer Venus and very well observed by the Venus Express probe. At the top of the troposphere, the clouds are subjected to a super-rotation, with a period of about four Earth days; the origin of this phenomenon is still poorly understood.

In the mesosphere, another type of circulation takes over. Under the effect of the rise in temperature at the sub-solar point, a global circulation is established, from the sub-solar point to the anti-solar point (see Figure 4.29).

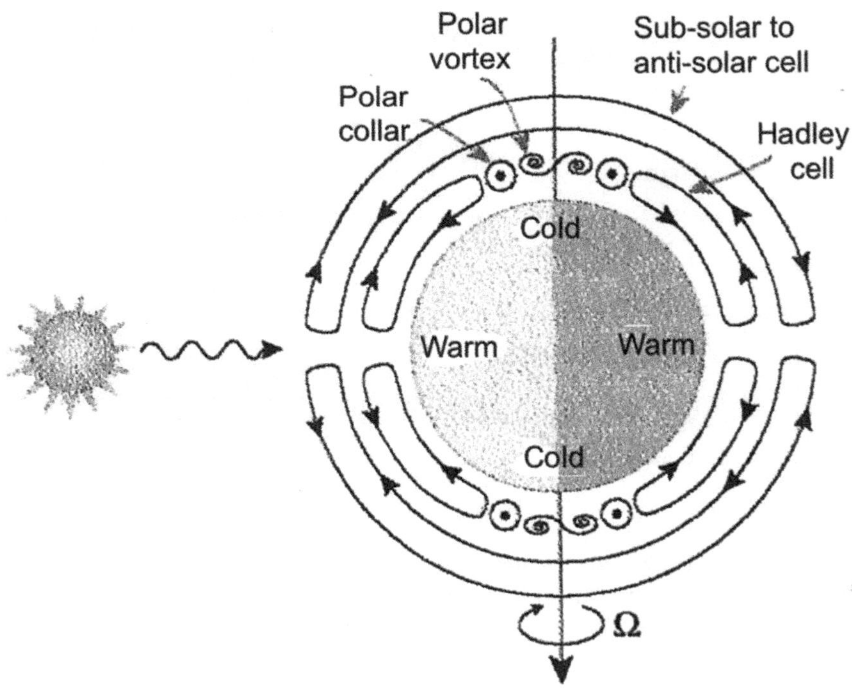

Figure 4.29. *The atmospheric circulation of Venus. In the troposphere, a Hadley circulation extends, in each hemisphere, from the equator to 60° latitude. Above the tropopause, in the mesosphere, a sub-solar → anti-solar circulation encompasses the planet (source: from (Taylor 2014))*

It is difficult for global climate models to understand the sulfur cycle and to account for all the abundance measurements, especially following the Venus Express mission. Indeed, the abundances of SO_2 and H_2O under the clouds are, respectively, about 150 ppmv and 30 ppmv. Above the clouds, they drop to 300 ppbv and 1 ppmv, respectively (see Figure 4.30). In the photochemical cycle described above, the formation of H_2SO_4 consumes almost all of the H_2O, but a high concentration of unused sulfur remains: models suggest that an abundance of 15 ppmv below the clouds is sufficient to account for the observed abundances in the mesosphere. The sulfur is likely to be in the form of sulfur-rich aerosols; these could be responsible for the unidentified UV emission observed around 380 nm. Moreover, submillimeter observations from Earth have shown the presence of SO_2 and SO at an altitude of about 90 km: another source of sulfur, probably in the form of aerosols, must also be present at this level.

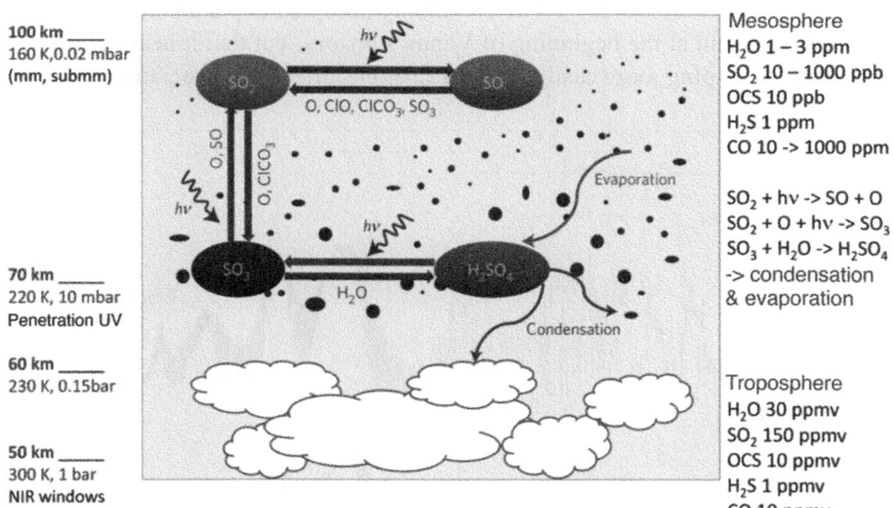

Figure 4.30. *Schematic representation of the sulfur cycle on Venus. For a color version of this figure, see www.iste.co.uk/encrenaz/solar1.zip*

4.3.3.7. *Evolution of atmospheres*

The atmospheres of Venus and Mars are, today, extremely poor in water vapor. However, thermochemical models suggest that water was very abundant in the past. Proof was given to us, for Mars as for Venus, by the measurement of the abundance of monodeuterated heavy water HDO. In the case of Mars, many other indications (networks of branched valleys, presence of water in the subsoil, presence of clays in the ancient terrains, etc.) show that liquid water flowed in abundance in the planet's ancient past.

4.3.3.7.1. Venus

While the cloud layer is opaque to visible radiation, it is partially transparent in some windows of the near-infrared spectrum, between the CO_2 absorption bands. The ground temperature is sufficiently high for the heat flux from the lower stratosphere (or even the surface at certain wavelengths) to be detected on the night side (see Figure 4.26). It was thus possible, by high resolution spectroscopy from Earth, to identify tropospheric molecules (CO, H_2O, HDO, OCS) and to measure their abundances (see Figure 4.31). By comparing HDO and H_2O measurements, it appeared that the D/H ratio on Venus was more than 100 times higher than the value in the Earth atmosphere (Vienna Standard Mean Ocean Water, VSMOW = 1.556 10^{-4}). The commonly accepted explanation is that water was very abundant at the beginning of Venus's history, but that it heavily escaped, the H atoms escaping more easily than the D atoms after the dissociation of water.

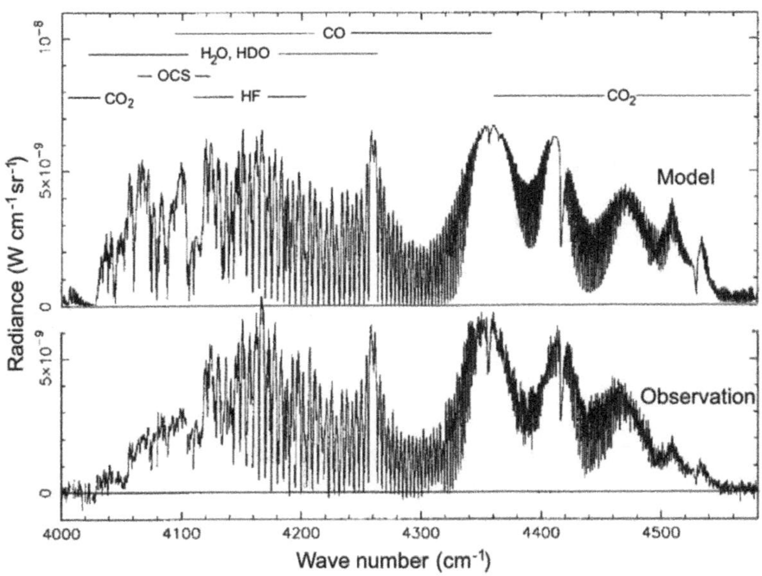

Figure 4.31. *The spectrum of the night side of Venus between 4,000 and 4,600 cm^{-1} (2.17-2.50 cm), observed with the Fourier transform spectrometer of the CFHT telescope at Mauna Kea Observatory. Top: synthetic spectrum; bottom: observed spectrum (source: from (Bézard* et al. *1990))*

We can thus reconstruct the history of the atmosphere of Venus. Originally, according to the models of stellar evolution, the solar flux was 70% of its current value. The equilibrium temperature on the surface was then compatible with the presence of liquid water, and the planet could, at that time, have been covered with oceans. However, as the surface of Venus was remodeled by volcanism over a few hundred million years, all traces of this period have disappeared. The presence of CO_2 and H_2O in large quantities must have generated a significant greenhouse effect, leading to the raise of its surface temperature. The increase of the solar flux led to the dissociation of H_2O, the preferential escape of hydrogen with respect to deuterium, and, finally, the almost total disappearance of water.

The present atmosphere of Venus also presents temporal variations, which are still poorly understood. It has been known since the observations of Pioneer Venus and the Venera probes, in the 1980s, that the amount of SO_2 above the clouds varies strongly with time, up to a factor of 10 (see Figure 4.32). The most plausible hypothesis is the presence of volcanic activity in episodic form, but this could not be proven definitively.

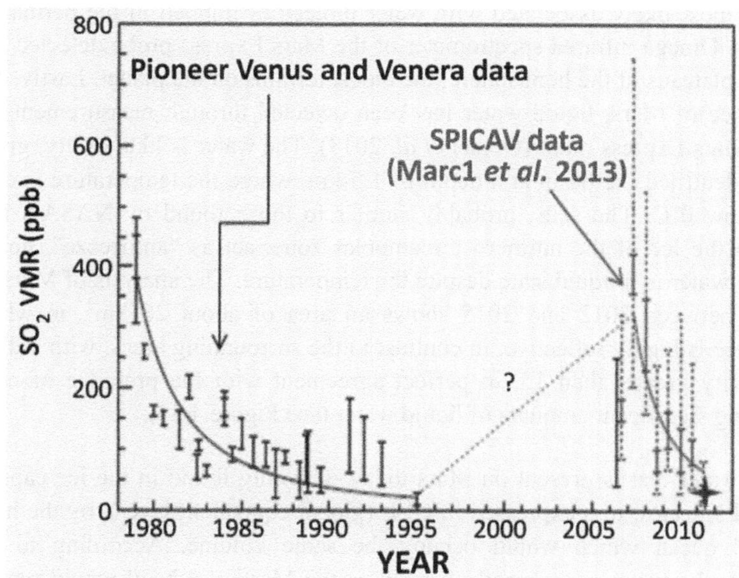

Figure 4.32. *Evolution of the SO_2 mixing ratio above the Venus clouds between 1979 and 2012. SPICAV measurements are from Venus Express. Other measurements recorded in thermal infrared from Earth show a new maximum in 2017, around 600 ppbv (source: from (Marcq* et al. *2013)). For a color version of this figure, see www.iste.co.uk/encrenaz/solar1.zip*

4.3.3.7.2. Mars

There is a whole range of evidence to support the presence of large quantities of water in Mars's early history. The first is the measurement of the D/H ratio. As in the case of Venus, this was first obtained from Earth by high resolution near-infrared spectroscopy, in 1988: a deuterium enrichment by a factor of 6 was measured, confirmed by many subsequent measurements, suggesting the presence of a denser and more humid atmosphere in the past.

As early as the 1970s (see section 4.3.2.3), images from the Viking probes showed valley networks, suggesting that abundant quantities of liquid water were present in the past. In the late 1990s, the Mars Global Surveyor radar altimeter revealed a possible shoreline at the dichotomy line, with constant elevation over very large distances, suggesting the possible presence of a past ocean in the northern hemisphere (see Figure 4.20); this hypothesis was reinforced in 2012 by the measurement of a low dielectric constant under the northern plains, consistent with the presence of hydrated material. In 2002, the γ–ray spectrometer of Mars Odyssey probe detected the presence of hydrogen atoms beneath the surface of the polar regions, most likely associated with water molecules trapped in the permafrost. In 2005, the Omega infrared spectrometer of the Mars Express probe detected clays in the high plateaus of the hemisphere, the oldest terrains on the planet. Lastly, beneath the surface of Mars, liquid water has been detected through measurements by the Marsis/Mars Express radar (Orosei *et al.* 2018). The water is likely salty, given that Marsis identified the basin at a depth of 1.5 km, where the temperature is certainly well below 0°C. The salts, probably similar to those found by NASA's Phoenix probe in the ice of the northern circumpolar zone, act as "antifreeze", helping to keep the water in a liquid state despite the temperature. The analysis of Marsis radar profiles between 2012 and 2015 shows an area of about 20 km^2, in which the subsurface is highly reflective, in contrast to the surrounding areas, with a dielectric permittivity greater than 15, in perfect agreement with the presence of materials containing significant amounts of liquid water (see Figure 4.33).

The water that is present on Mars today is mainly found in the ice caps and in hydrated silicates; it is expressed in GEL (global equivalent layer), by the height of a global ocean which would occupy the same volume. According to various estimates, the water ice currently present in the Martian subsoil would represent a few dozen GEL meters. The estimation of the primitive water reservoir is more difficult, and comes from various sources, including the interpretation of the reliefs and the measurements of the D/H ratio. The results are of the order of a few hundred GEL meters, or about 10 times today's reservoir. The primitive atmosphere of Mars could have had a surface pressure of several hundred hectopascals. Its disappearance could result from two causes: erosion by meteorite impacts (particularly during the

Late Heavy Bombardment, see Chapter 2) and hydrodynamic escape, capable of causing the simultaneous escape of heavy atoms and hydrogen (see section 4.3.3.5).

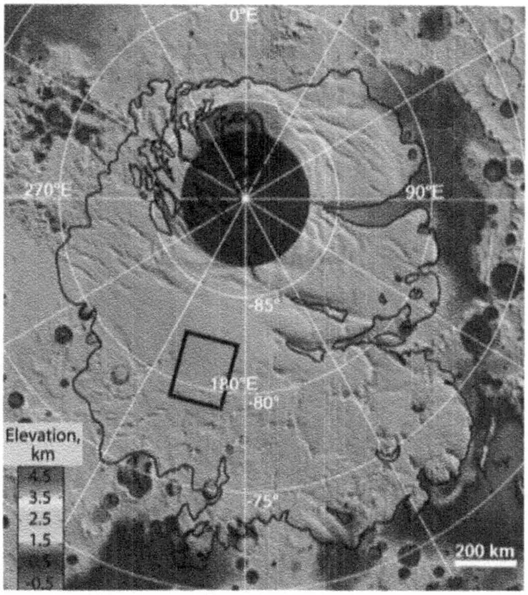

Figure 4.33. *Altimetry map of Mars's south pole; the black rectangle indicates the location of the region below which a subglacial lake was detected by the Mars Express orbiter's Marsis radar (Orosei* et al. *2018) (source: ESA/NASA). For a color version of this figure, see www.iste.co.uk/encrenaz/solar1.zip*

The hypothesis of a denser, more humid and warmer primitive atmosphere than today is however opposed with the "faint young Sun paradox": as mentioned above, the Sun's flux 4 billion years ago was 70% of its present value. The equilibrium temperature at the surface of Mars was then about 200 K, which was not sufficient to allow the presence of liquid water on the surface. How can this contradiction be solved?

Two families of models have been proposed. The first assumes a warm, humid and relatively dense primitive atmosphere, containing sufficient greenhouse gases (CO_2, H_2O, NH_3, CH_4, H_2, SO_2) to raise the temperature to the point of water liquefaction, but calculations show that these contributions are insufficient. The second class of models considers a globally dry and cold climate, with brief temperate episodes following impacts, during which fluvial erosion would have occurred. The issue of the nature of the early Martian atmosphere remains wide open (Catling and Kasting 2017).

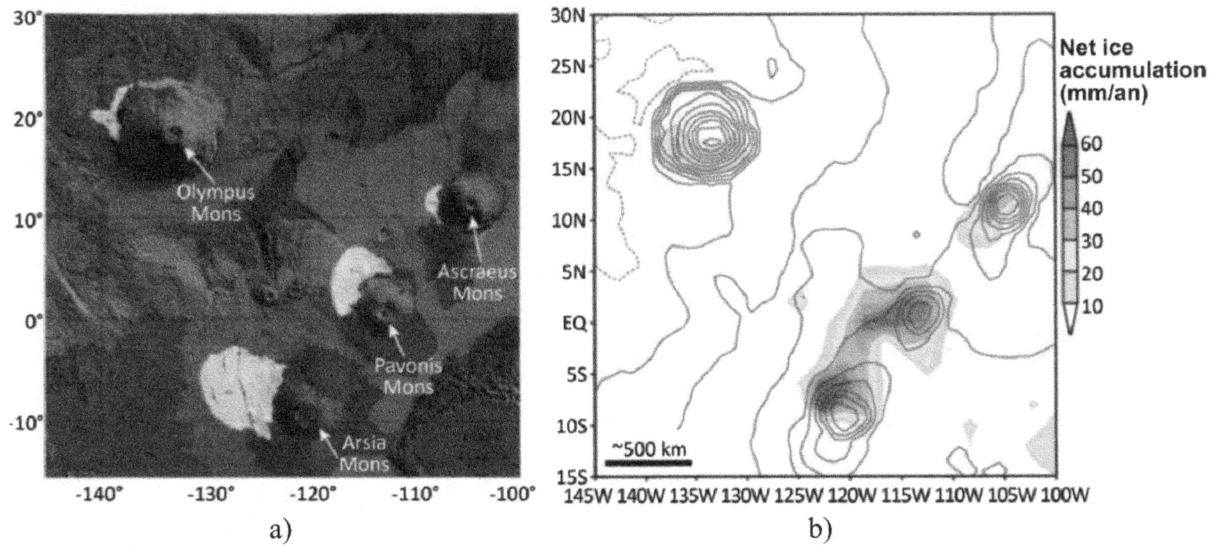

Figure 4.34. *Images of glacial residues (in yellow) on the flanks of the Tharsis volcanoes: (a) observations by the HRMC camera of the Mars Express probe (© ESA); (b) predictions from obliquity variations (Laskar* et al. *2004) and global climate models (Forget* et al. *2006). For a color version of this figure, see www.iste.co.uk/encrenaz/solar1.zip*

In a more recent past, Mars's climate has evolved according to oscillations in its orbital parameters, known as Milankovitch cycles. In particular, the Martian obliquity has changed considerably over time, with a mean value of 40° and maxima up to 80° (Laskar *et al.* 2004). Over the last 10 million years, the obliquity has oscillated with a period of 10^5 years, modulated by an envelope of 1.2 10^6 years. The precession of the equinoxes adds to these oscillations, with a period of 51,000 years. These orbital fluctuations have a great influence on the Martian climate. For an obliquity greater than 54°, the poles receive, on average, more solar flux over a Martian year than the equator, which results in the formation of glaciers at the equator, particularly on the flanks of volcanoes. Remnants of these glaciers have been revealed by the Mars Express camera (see Figure 4.34).

4.3.3.7.3. Earth

Compared to the neighboring planets, Earth has several particularities. First of all, at the very beginning of its history, it experienced a major impact that led to the formation of the Moon. This satellite stabilizes the obliquity of the Earth, thus avoiding the climatic fluctuations observed on Mars. Secondly, its distance from the Sun places it in the "habitable zone", allowing water to remain in liquid form. Lastly, thanks to the presence of oceans or lakes, life has emerged and developed, leading to major climatic changes.

The Earth's early atmosphere was probably, like that of its neighbors, dominated by carbon dioxide, water and molecular nitrogen. Geologists divide the Earth's history into four major periods: the Hadean (4.6–4.0 Gyr), the Archean (4.0–2.4 Gyr), the Proterozoic (2.4 Gyr–540 Myr) and, lastly, the Phanerozoic (540 Myr to the present). The oldest traces of life probably date from 3.7 Myr; they are found in stromatolites, calcareous concretions built by single-celled algae, as well as in micro-organisms close to hydrothermal springs. At the beginning of the Archean, there was enough CO_2 to generate a greenhouse effect, which allowed the Earth to escape a total glaciation, despite the low solar flux. As soon as the oceans appeared, probably at the beginning of the Archean, and in the presence of a silicate soil, the carbon dioxide of the Earth's atmosphere produced limestone at the bottom of the oceans, and the carbon dioxide level decreased. At the same time, the appearance of the first living organisms led to the formation of methane, which, together with carbon dioxide, helped to prevent a total glaciation on Earth (see Figure 4.35).

It seems that the first oxygen-producing bacteria appeared at the end of the Archean. The appearance of oxygen, during the Proterozoic, had the effect of oxidizing the iron present in the seabed, and of causing the decrease of methane by chemical reaction, and thus of the greenhouse effect. This may be the reason for the first global planetary glaciation 2.4 Gyr ago, named "Huronian". Then the CO_2

outgassed from volcanism accumulated, allowing the Earth to escape this episode. Another global glaciation occurred at the end of the Proterozoic, some 600 Myr ago. It was at the beginning of the Phanerozoic that life spread to the continents, following the formation of an ozone layer in the atmosphere, resulting from the photodissociation of molecular oxygen. During the Secondary Era, the temperature increased until it reached about ten degrees above the current value: this was the Carboniferous Era, during which the carbon fossil reserves were built up; during the Mesozoic Era, dinosaurs and birds appeared. The Tertiary Era began 65 million years ago, with the fall of the Chicxulub meteorite, which caused the disappearance of dinosaurs and many animal species. This event promoted the surge of mammals until the appearance of humans.

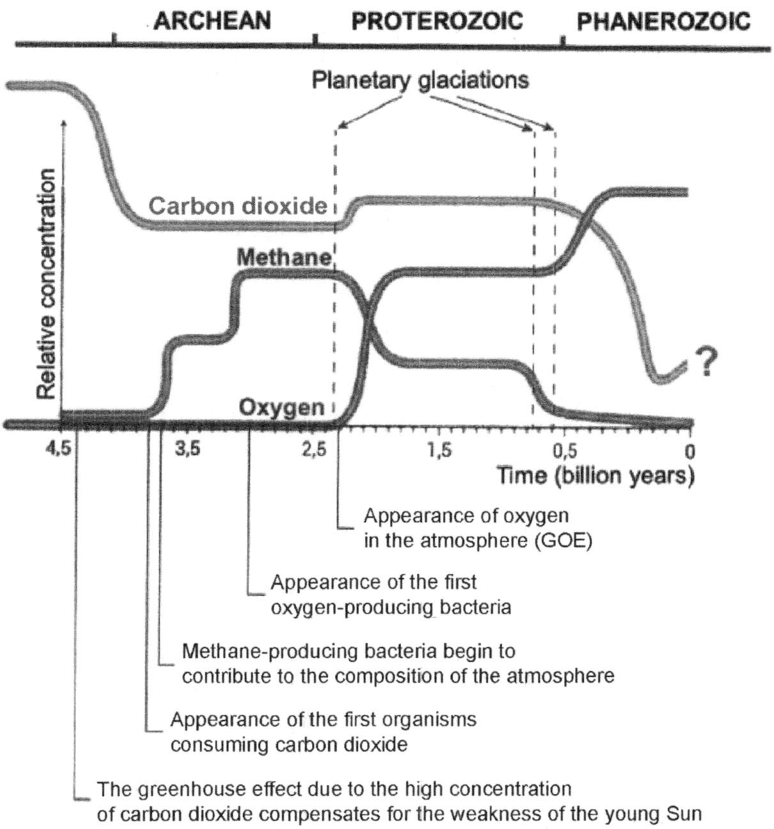

Figure 4.35. *Evolution of the Earth's atmospheric composition according to time (source: from (Ramstein 2017)). For a color version of this figure, see www.iste.co.uk/encrenaz/solar1.zip*

The Quaternary Era, which began 3 Myr ago, is characterized by episodes of glaciation, whose modulations have more or less coincided with variations in solar flux, linked to fluctuations in the Earth's orbital parameters (eccentricity, obliquity, precession of the equinoxes): these are the Milankovitch cycles. As we approach the present time, the situation becomes more complicated. Around 7,000 years ago, the Earth experienced a maximum temperature (the Sahara was covered with vegetation), then a desertification 5,000 years ago. A new medieval climatic optimum occurred between the 10th and 13th centuries, allowing the colonization of Greenland by the Vikings, followed by the Little Ice Age between the 15th and 19th centuries, until the end of the pre-industrial era. The development of human activities then led to a decisive change in the evolution of the climate. The concentration of CO_2 increased from 280 ppmv, at the beginning of the industrial era, to 415 ppmv in 2019, while the temperature increased by 1.2° and the sea level rose by more than 20 centimeters. Given the very long lifetime of CO_2 in the Earth's atmosphere (over 200 years), the phenomenon is irreversible in the short term. We are therefore possibly at risk of a runaway greenhouse effect. Faced with this prospect, global awareness is essential, as well as the implementation of policies to save energy, to search for renewable energy and to safeguard biodiversity.

4.3.4. *The satellites of Mars*

Mars has two small satellites, Phobos and Deimos, discovered in 1877 by Asaph Hall (1829–1907) (Forget *et al.* 2006). They are of irregular shape, their longest axes measuring 27 km and 15 km, respectively. Their orbits, located respectively at 2.7 and 6.9 Martian radii, are circular, in the equatorial plane of Mars, and the two satellites are in synchronous rotation around the planet. In the case of Phobos, tidal effects induce a braking force that slowly decreases the semi-major axis of its orbit: Phobos should crash into Mars in a few tens of millions of years. On the other hand, Deimos is moving away from the planet, like the Earth–Moon couple.

The mean specific mass of the two satellites is close to 2 g/cm^3, and their albedo is very low (about 0.05). Their surface is dominated by opaque carbonaceous minerals, reminiscent of the matrix of carbonaceous chondrites, the most primitive meteorites. The spectra recorded by the Phobos 2 probe in 1989 show a very red slope of the continuum, testifying to the absence of hydration. Measurements by the Mars Global Surveyor probe have confirmed that the surface composition of the two satellites is close to that of D-type asteroids.

The appearance and size of the two objects resemble small asteroids, and their origin has been the subject of much debate. As in the case of the Moon, the hypothesis favored by numerical models involves the protoplanet's collision with a relatively massive body, less than 500 million years after the formation of Mars. The debris, scattered in an equatorial ring, would have assembled in numerous satellites of various sizes. Most of them would have fallen back to Mars, due to tidal forces, and only Phobos and Deimos would have survived.

The Phobos satellite is the subject of particular attention from planetologists because, due to its low gravity field, it is a privileged target for the collection of extraterrestrial samples. Following the Phobos 1 and 2 probes, launched in 1988, the Russian probe Phobos-Grunt, which unfortunately failed at liftoff in 2011, had this objective. The Japanese MMX (Mars Moon Exploration) mission aims to explore Mars's two satellites and collect samples from Phobos. The launch is planned for 2024, and the return of samples to Earth for 2029.

4.4. References

Bézard, B., de Bergh, C., Crisp, D., Maillard, J.-P. (1990). The deep atmosphere of Venus revealed by high-resolution night side spectra. *Nature*, 345, 508–511.

Bézard, B., Russell, C., Satoh, T. , Smrekar, S., Wilson, C. (2020). *Venus III*. Springer-Verlag, Berlin.

Bougher, S.W., Hunten, D.M., Phillips, R.J. (1997). *Venus II*. University of Arizona Press, Tucson.

Canup, R.M. and Righter, K. (2000). *Origin of the Earth and the Moon*. University of Arizona Press, Tucson.

Carlson, R.W., Baines, K.H., Encrenaz, T., Taylor, F.W., Drossart, P., Kamp, L.W., Pollack, J.B., Lellouch, E., Collard, A.D., Calcutt, S.B. *et al.* (1991). Galileo infrared imaging measurements at Venus. *Science*, 253, 1541–1548.

Catling, D.C. and Kasting, J.F. (2017). *Atmospheric Evolution on Inhabited and Lifeless Worlds*. Cambridge University Press, Cambridge.

Drossart, P., Encrenaz, T., Rosenqvist, J., Lellouch, E., Carlson, R.W., Baines, K.H., Weissman, P.R., Smythe, W.D., Kamp, L.W., Taylor, F.W. *et al.* (1993). Earth global mosaic observations with NIMS-Galileo. *Planetary and Space Science*, 41, 551–561.

Encrenaz, T., Bibring, J.-P., Blanc, M., Barucci, M.-A., Roques, F., Zarka, P. (2003). *Le Système solaire*. CNRS Editions/EDP Sciences, Paris.

Encrenaz, T., Bibring, J.-P., Blanc, M., Barucci, M.-A., Roques, F., Zarka, P. (2004). *The Solar System*, Springer-Verlag, Berlin.

Encrenaz, T., Lequeux, J., Casoli, F. (2021). *Planets and Life*. EDP Sciences, Les Ulis.

Forget, F., Haberlé, R.M., Montmessin, F., Levrard, B., Head, J.W. (2006). Formation of glaciers on Mars by atmospheric precipitation at high obliquity. *Science*, 311, 368–371.

Forget, F., Costard, F., Lognonné, F. (2007). *Planet Mars, Story of Another World*. Springer, New York.

Graedel, T.E. and Crutzen, P.J. (1993). *Atmospheric Change, an Earth System Perspective*. Freeman, New York.

Hanel, R., Conrath, B.J., Jennings, D.E., Samuelson, R.E. (2003). *Exploration of the Solar System by Infrared Remote Sensing*, 2nd edition. Cambridge University Press, Cambridge.

Hunten, D., Colin, L., Donahue, T., Moroz, V.I. (1983). *Venus*. University of Arizona Press, Tucson.

Kieffer, H.H., Jakosky, B.M., Snyder, C.W., Matthews, M.S. (1992). *Mars*. University of Arizona Press, Tucson.

Korth, H., Anderson, B.J., Gerschman, D.J., Rained, J.M., Slavin, J.A., Zurbuchen, T.H., Solomon, S.C., McNutt Jr., R.L. (2014). Plasma distribution in Mercury's magnetosphere derived from MESSENGER Magnetometer and Fast Imaging Plasma Spectrometer observations. *Journal of Geophysical Research: Space Physics*, 119, 2917–2932.

Korth, H., Johnson, C., Philpott, L.C., Tsyganenko, N.A., Anderson, B.J. (2017). A dynamical model of Mercury's magnetospheric magnetic field. *Geophysical Research Letters*, 44, 10147–10154.

Krasnopolky, V.A. (1986). *Photochemistry of the Atmospheres of Mars and Venus*. Springer-Verlag, Berlin.

Laskar, J., Correia, A.C.M., Gastineau, M., Joutel, F., Levrard, B., Robutel, P. (2004). Long-term evolution and chaotic diffusion of the insolation quantities of Mars. *Icarus*, 170, 343–364.

Lellouch, E., Encrenaz, T., De Graauw, T., Erard, S., Morris, P., Crovisier, J., Feuchtgruber, H., Girard, T., Burgdorf, M. (2000). The 2.3-45 μm spectrum of Mars observed with the Infrared Space Observatory. *Planetary and Space Science*, 48, 1393–1405.

Lunine, J. (1999). *Earth: Evolution of a Habitable Planet*. Cambridge University Press, Cambridge.

Marcq, E., Bertaux, J.-L., Montmessin, F., Belyaev, D. (2013). Variations of sulphur dioxide at the cloud top of Venus's dynamic atmosphere. *Nature Geosciences*, 6, 25–28.

Moreno, R., Lellouch, E., Forget, F., Encrenaz, T., Guilloteau, S., Millour, E. (2009). Wind measurements in Mars' middle atmosphere: IRAM Plateau de bure interferometric CO observations. *Icarus*, 201, 549–563.

Orosei, R., Lauro, S.E., Pettinelli, E., Cicchetti, A., Coradini, M., Cosciotti, B., Di Paolo, F., Flamini, E., Mattei, E., Pajola, M. *et al.* (2018). Radar evidence of subglacial liquid water on Mars. *Science*, 361, 490–493.

Ramstein, G. (2017). Une brève histoire du climat de la Terre. *Reflets de la Physique*, 55, 6–10.

Slavin, B., Anderson, J., Baker, D.N., Benna, M., Boardsen, S.A., Gloeckler, G., Gold, R.E., Ho, G.C., Korth, H., Krimigis, S.M. *et al.* (2010). MESSENGER observations of extreme loading and unloading of Mercury's magnetic tail. *Science*, 329, 665–668.

Smith, M.D. (2004). Interannual variability in TES atmospheric observations of Mars during 1999–2003. *Icarus*, 167, 148–165.

Taylor, F.W. (2014). *The Scientific Exploration of Venus*. Cambridge University Press, Cambridge.

Vilas, F., Chapman, C.R., Matthews, M.S. (1989). *Mercury*. University of Arizona Press, Tucson.

Wegener, A. (1966). *The Origin of Continents and Oceans*. Dover, New York.

5

Giant Planets

Thérèse ENCRENAZ and Laurent LAMY
LESIA, Paris Observatory, PSL University, Paris, France

5.1. The exploration of giant planets

5.1.1. *From Antiquity to the Space Age*

The four giant planets – Jupiter, Saturn, Uranus and Neptune – fall into two distinct families. With masses 318 and 95 times that of the Earth, the two closest planets to the Sun, Jupiter and Saturn, are essentially made of protosolar gas; they are the gas giants, visible to the naked eye in the night sky and known since Antiquity. The two most distant giants, Uranus and Neptune, which are invisible to the naked eye, have respective masses of 14 and 17 Earth masses, and are essentially made of an ice core; they are called icy giants. Uranus was discovered in 1781 by William Herschel (1738–1822), thanks to the improved quality of telescopes; Neptune was discovered in 1846 by Johann Galle (1812–1910), thanks to the predictive calculations of celestial mechanics, which were independently carried out by Urbain Le Verrier (1811–1877) and John Couch Adam (1819–1892) in order to account for the anomaly observed in the orbit of Uranus.

Our knowledge of the physical nature of Jupiter and Saturn begins with the observations of Galileo Galilei (1564–1642) who, in 1610, discovered the four satellites around Jupiter that now bear his name. In 1655, Christiaan Huygens (1629–1695) discovered Titan, Saturn's largest satellite; in 1659, he solved the enigma of the changing appearance of the disk of Saturn, due to the presence of a ring seen from Earth under a variable inclination. The first good drawings of Jupiter were made in the 1660s by Robert Hooke (1635–1703), as well as Jean-Dominique

The Solar System 1,
coordinated by Thérèse ENCRENAZ and James LEQUEUX. © ISTE Ltd 2021.

Cassini (1625–1712), the first head of the Paris Observatory, who highlighted the Great Red Spot and the structure of the planet in bands and zones; he also discovered the central division of the rings of Saturn, which now bears his name, and four satellites around Saturn (Japet, Rhea, Tethys and Dione). A few years after discovering Uranus, William Herschel discovered its two largest satellites, Titania and Oberon, in 1787, then, two years later, Saturn's Mimas and Enceladus satellites. Shortly after the discovery of Neptune, in 1846, William Lassell (1799–1880) discovered its largest satellite, Triton, followed by Uranus' Ariel and Umbriel satellites in 1851.

Until the beginning of the 20th century, visual observation and drawing were the preferred means to map planetary disks. The photographic plate established itself, particularly at the Lowell and Lick observatories in the United States, and in France's Pic du Midi Observatory with Bernard Lyot's pictures (1897–1952). It then furthered itself in the 1950s, with Gerard Kuiper's pictures (1905–1973) using the American telescopes of Mount Palomar and Mount Wilson. In parallel, the development of infrared astronomy allowed the first measurements of the temperature of Jupiter, Saturn and Uranus by Donald Menzel (1901–1976), in 1926; the first good spectroscopic measurements by Vesto Slipher (1875–1969) led to the detection of methane CH_4 on Jupiter and Saturn, as well as ammonia NH_3 on Jupiter by Rupert Wildt (1905–1976). In 1952, Kuiper suggested that hydrogen was the major constituent of giant planets, but it could only be observed in 1960 by Carl Kiess (1887–1967) and his team.

The 1950s also saw the birth of giant planet radio astronomy. In 1955, Bernard Burke (1928–2010) and Kenneth Franklin (1923–2007) discovered that Jupiter emits an intense and highly variable high frequency signal at 22 MHz; it was the cyclotron maser radiation of weakly relativistic electrons circulating in the polar regions, above the atmosphere. This result is the first direct evidence of the existence of a magnetic field on a planet other than the Earth. A review of our knowledge of Jupiter before the space age can be found in (Gehrels and Matthews 1976). For a general review of the giant planets before the space age, you can refer to (Lewis 1995).

5.1.2. *Space exploration*

Since the 1970s, NASA has been exploring the giant planets from space. The first step consisted of sending two identical spacecrafts, Pioneer 10 and 11, in 1972 and 1973 respectively, which flew over Jupiter in 1974 and 1975; Pioneer 11 then flew over Saturn in 1979. The first objective was to study the interplanetary medium in the asteroid belt in order to assess the potential risks of destruction by impact. The main scientific results were the first detailed images of the disks of Jupiter (see Figure 5.1) and Saturn, and the exploration of their magnetosphere, which, for Saturn, was revealed for the first time.

Figure 5.1. *Jupiter, observed by the Pioneer 10 probe in 1974. The image highlights the complexity of the atmospheric structure in which zones and bands alternate; it also illustrates the variable aspect of the Great Red Spot (on the far right), which appears very contrasted with respect to the South Equatorial Band (source: NASA). For a color version of this figure, see www.iste.co.uk/encrenaz/solar1.zip*

5.1.2.1. *The Voyager mission*

In the 1970–1980 decades, the trajectories of the giant planets presented a particular opportunity to carry out a successive flyby for each of them, using a single probe. Following the success of the Pioneer missions, NASA sent the two Voyager 1 and 2 probes towards the outer Solar System in 1977. Voyager 1 went around Jupiter in 1979, then Saturn and Titan in 1980. Voyager 2 flew over Jupiter in 1979 and Saturn in 1980, then Uranus in 1986 and Neptune in 1989. Then both probes left towards the frontiers of the Solar System. The set of instruments (called payload), which are identical on both probes, included a camera in visible light, ultraviolet, infrared and radio spectrometers, plasma physics instruments, an energetic particle detector and a radio transmitter for occultation measurements. The Voyager mission has gone down in the annals of space research as an extraordinary technological and scientific success (Pyne 2010). In the case of Uranus and Neptune, the Voyager data are still used as a reference, as the icy giants have not been revisited since. Among the great successes of Voyager are:

– the extraordinary complexity of Jupiter's cloud structures (see Figure 5.2); the evidence for active volcanism on Io, the closest Galilean satellite to Jupiter;

– the indications of the presence of a water ocean under the surface of Europa;

– the discovery of prebiotic molecules (hydrocarbons and nitriles) in Titan's atmosphere;

– the discovery of the magnetospheres of Uranus and Neptune, which are very different from one another;

– the presence of geysers on the surface of Triton, Neptune's largest satellite.

The Voyager probes are still active today and continue their journey into the interstellar medium (see Chapter 3).

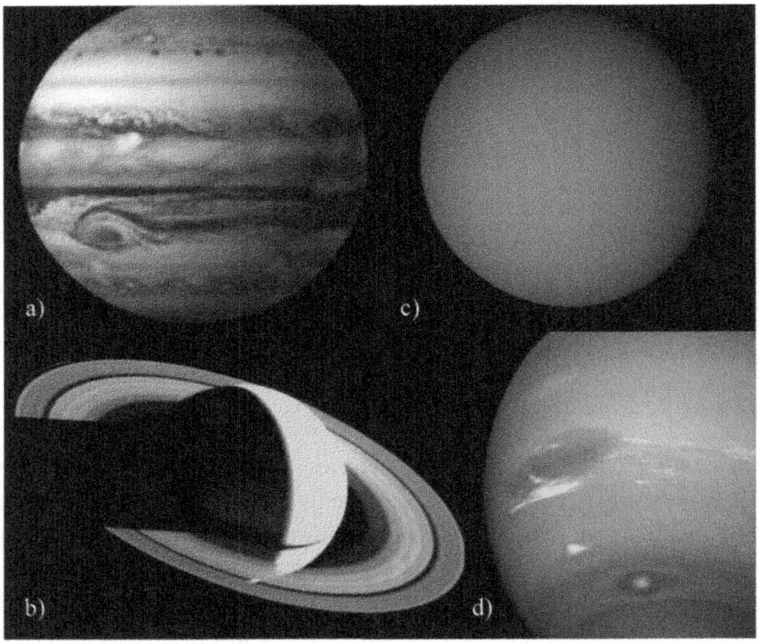

Figure 5.2. *The four giant planets observed by the Voyager mission: (a) Jupiter in 1979; (b) Saturn in 1980; (c) Uranus in 1986; d) Neptune in 1989 (source: NASA)*

Reviews of our knowledge of Uranus and Neptune after the Voyager 2 mission can be found in (Bergstrahl *et al.* 1991; Cruikshank *et al.* 1995), respectively.

5.1.2.2. *The Galileo mission*

Following the discoveries of the Galilean satellites by the Voyager probes, NASA decided to send an ambitious mission to the Jupiter system, including an orbiter designed for several flybys of Europa, Ganymede and Callisto, as well as a descent probe intended to penetrate the Jovian atmosphere. The launch, initially planned at the beginning of the 1980s, but delayed by the Challenger shuttle accident in 1986, finally took place in 1989; the descent probe penetrated the Jovian atmosphere on December 7, 1995, and carried out measurements of elementary composition which were decisive to discriminate among the models of planetary formation. In spite of the failure that made the large telecommunication antenna unusable, the orbiter carried out a continuous monitoring of the planet and the Galilean satellites, and managed to transmit data back to the Earth by means of the small antenna, completing more than 70% of its initial objectives, before plunging into the Jovian atmosphere in September 2003. A description of the Galileo mission can be found in (Harland 2000), a detailed presentation of the morphology of the planet in (Rogers 1995), and a complete review of the state of our knowledge of Jupiter after the Galileo mission in (Bagenal *et al.* 2007).

5.1.2.3. *The Cassini-Huygens mission*

Voyager flybys have highlighted the very particular character of Saturn's satellite Titan, whose atmosphere, rich in nitrogen and prebiotic molecules, could present a certain analogy with the primitive Earth. This idea is at the origin of an ambitious space program jointly led by NASA, ESA and the Italian space agency ASI: the Cassini-Huygens mission. The Cassini orbiter, built by NASA and dedicated to the exploration of the Saturn system, carried the Huygens probe, developed by ESA, intended to land on Titan's soil. Launched in 1997, the mission flew over Jupiter at the end of 2000, arrived near Saturn in 2004 and released the Huygens probe on January 14, 2005. The overall mission, which was the result of an exemplary international collaboration, was a huge scientific and technological success. Some of its most significant results include:

– the exploration of Titan's surface and the discovery of its methane lakes;

– the discovery of a multitude of complex ions in Titan's ionosphere, at the origin of the aerosols which mask the surface;

– the discovery of geysers on the surface of Enceladus, a probable indication of the presence of an ocean under the surface of the satellite and the source of Saturn's E ring;

– the discovery of a complex vortex structure around the two poles of Saturn;

– the monitoring of a giant storm that appeared in December 2010 in Saturn's northern hemisphere (see Figure 5.3);

– the highlighting of the extraordinary complexity of the rings and of interaction phenomena with small neighboring satellites.

The Cassini mission ended on September 15, 2017, with a spectacular dive into Saturn's atmosphere, after a series of passes through the auroral regions and in close proximity to the rings, allowing a very detailed study of these regions. A description of our knowledge of the Saturn system before the Cassini mission, as well as a description of the mission itself, can be found in (Harland 2002). A review of the knowledge gained after the Cassini mission can be found in (Dougherty *et al.* 2009). Irwin (2009) presents the state of our knowledge of the four giant planets following the Voyager, Galileo and Cassini space missions.

Figure 5.3. *Saturn, observed by the Cassini probe in 2011, shortly after the storm started in the northern hemisphere in December 2010. Seasonal storms sometimes develop on Saturn, with variable periodicity (source: NASA)*

5.1.2.4. *The Juno mission*

Launched in 2011 by NASA, the Juno mission, which has been in operation around Jupiter since 2016, is dedicated to the study of Jupiter's atmosphere, its composition and dynamics, its internal structure and its magnetosphere. It is placed on a very elliptical polar orbit that enables it to fly over the poles at low altitudes, and to cross particle acceleration regions and auroral radio sources, making it possible to understand the mechanisms at work. It also aims to measure the abundance of water in the deep troposphere, so as to obtain a value of the O/H ratio that is representative of the whole planet and, thus, better constrain its formation models (see section 5.2.2). Other priority objectives are to better understand the internal structure of the planet, as well as that of its magnetosphere. The results of the different experiments are currently being analyzed; the first analyses show that Jupiter's interior is significantly different from what the models predicted and have revealed that, like the Earth, the Jovian magnetic field presents a secular variation (see below).

5.1.3. *Exploration from Earth and the terrestrial environment*

Fifty years after the beginning of the space exploration of giant planets, astronomical observation using terrestrial telescopes continues to play a major role. Since the 1970s, infrared spectroscopy has allowed the study of a number of minor tropospheric (CO, CH_3D, H_2O, NH_3, PH_3, GeH_4) or stratospheric (C_2H_2, C_2H_6) species, notably using spectral windows around 5 and 10 μm. The stellar occultation experiments carried out in 1977 and 1984 made it possible, for the first time, to discover Uranus' ring system and the arcs around Neptune, which were confirmed during the flyby of the two planets by the Voyager 2 probe in 1986 and 1989.

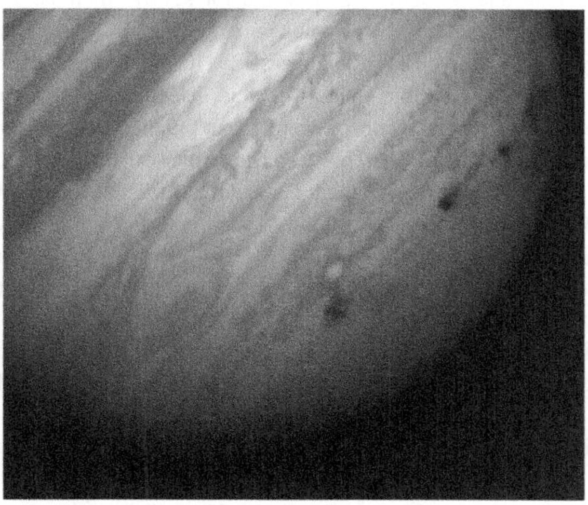

Figure 5.4. *Jupiter, observed by the Hubble Space Telescope in July 1994, when comet SL9 collided with the planet. The impacts occurred successively along the parallel of latitude 44°S, due to Jupiter's rotation. The black spots were caused by the formation of aerosols at each impact site. They remained visible for several months (source: NASA)*

In July 1994, an exceptional phenomenon occurred: comet Shoemaker–Levy 9's collision with Jupiter (see Figure 5.4). This phenomenon made it possible to analyze the response of a planetary atmosphere to a major impact (Noll *et al.* 1996) for the first time, in real time. Box 5.1 provides details on this phenomenon.

On March 24, 1993, a new and very peculiar comet was discovered: comet Shoemaker–Levy 9 (SL9), which appeared as a series of about 20 fragments, moving parallel to each other. Astrometric measurements revealed that this was a comet captured by Jupiter in the 1970s, which came close enough to the planet to be dislocated during its last pass near Jupiter in July 1992. According to predictions, the fragments were to collide with the planet during the next perijove passage in July 1994 (see Figure 2.4, Volume 2).

All the telescopes, on the ground and in space (in particular the Hubble Space Telescope and the Galileo probe on its way to Jupiter), were mobilized to observe the events. Between July 16 and July 23, 1994, the fragments of comet SL9 collided with the planet at a latitude of 44S, producing a series of impacts along this parallel (see Figure 5.4). The same phenomena were observed at each impact, with an amplitude relating to the mass of the impactor fragment. The fall of the object in the atmosphere first produced a fireball with a temperature higher than 20,000 K. The depth of penetration was between 10^4 and 10^5 Pa (0.1 and 1 bar). The material of Jupiter's atmosphere was ejected up to an altitude of 3,000 km, as observed by the Hubble telescope. A quarter of an hour later, the fallout of the ejecta produced the formation of a circular black spot, which was sometimes surrounded by a halo and, in some cases, lasted for several months. The most striking phenomenon was the formation of new molecules created by shock chemistry, at high pressure and high temperature. This was the case of HCN, CO and OCS, which remained present for several years after the impact. It was also the case of water vapor H_2O, detected for the first time in the stratosphere of a giant planet. A few years later, in 1997, the European infrared satellite, ISO (Infrared Space Observatory), detected H_2O in the stratosphere of the four giant planets and Titan. Given the cold trap constituted by the tropopause of these planets, the water couldn't have come from the interior of the planet; it inevitably had an external origin. In the case of Jupiter, this was the remnant of the fall of SL9 on Jupiter, as later confirmed by the Herschel submillimeter satellite.

SL9's collision on Jupiter was a rare phenomenon, but not unique; the Paris Observatory archives revealed drawings by J.-D. Cassini, showing a similar phenomenon in 1690. More recently, in July 2009, an object smaller than SL9 (probably an asteroid) also collided with Jupiter, creating smaller impacts.

Box 5.1. *Comet Shoemaker–Levy 9's collision with Jupiter*

In correlation to telescopic observations from the ground in the visible/near-infrared and low frequency radio ranges, the monitoring of the giant planets from the Earth's orbit or environment has proven to be very fruitful, especially in the case of

Uranus and Neptune, whose *in situ* exploration ended with the Voyager 2 probe. The observations of the HST (Hubble Space Telescope) have been particularly useful for the analysis of the different phases of each explosion on Jupiter, to study the complex aurorae of Jupiter, Saturn and Uranus in the ultraviolet range, and to follow the climatic evolution of Saturn, Uranus and Neptune according to their seasonal cycle.

Astronomical observatories in Earth orbit or in the Earth's environment (in particular at the L2 Lagrange point, see section 1.2.4) make it possible to observe the electromagnetic spectrum of planets in all wavelength ranges. To study cold objects such as giant planets, the infrared and millimeter radio ranges are particularly suitable. The ESA's ISO (Infrared Space Observatory) satellite, in operation between 1995 and 1998, detected water vapor in the stratosphere of giant planets, which was proof of the existence of an interplanetary oxygen source (comets or micrometeorites) in the outer Solar System. These measurements were then refined with the European Herschel satellite, in operation between 2009 and 2013.

5.1.4. *The electromagnetic spectrum of giant planets*

As in the case of telluric planets, the information we have on the physico-chemistry of the atmospheres of the giant planets comes from the observation of their electromagnetic spectrum, from ultraviolet to radio wave ranges. Let us be reminded that the spectral signatures of neutral molecules are mainly in the infrared (vibration-rotation bands) and the millimeter and submillimeter range (rotation lines), while the ultraviolet radiation mainly comes from the upper atmosphere, and gives us information on electronic transitions and dissociation products of molecules (radicals, atoms and ions; see section 4.3.3.4).

As for all Solar System objects, the electromagnetic spectrum of giant planets is characterized by a reflected solar component that peaks at 0.5 μm (black body maximum at 5,770 K, the temperature of the solar photosphere) and a thermal component that dominates beyond a few micrometers (see section 1.3.1 and section 4.1.8). In the reflected component, the spectral signatures of atmospheric constituents are observed in absorption in front of the solar continuum. In the thermal component, the flux $\Phi(\nu)$ emitted at a given frequency ν is given by the following equation:

$$\Phi(\nu) = \int_0^{\infty} \mathrm{B}(\nu)e^{-\tau(z,\nu)}\,\mathrm{d}\tau(\nu) \qquad [5.1]$$

where $\tau(z,\nu)$ is the optical thickness above the altitude level z:

$$\tau(z,\nu) = \int_z^{\infty} \alpha(z_1,\nu)N(z_1)\mathrm{d}z_1 \qquad [5.2]$$

α being the absorption coefficient per molecule at frequency ν and $N(z_1)$ being the density of the molecule at altitude z_1. If several atmospheric absorbers are present, the quantity $\tau(z,\nu)$ is the sum of the contributions from the different molecules. The amount $\alpha(z_1,\nu)$ depends on the properties of the molecules present in the atmosphere and is known from laboratory studies (Hanel *et al.* 2003).

A convenient parameter to measure Φ is the brightness temperature $T_B(\nu)$, which is the temperature of the black body that would emit the same flux at frequency ν. According to the Barbier-Eddington approximation, in the case of a nadir observation, $T_B(\nu)$ is, as a first approximation, the temperature of the atmospheric layer for which the optical thickness $\tau(\nu)$ is equal to 1. In the case where the flux is emitted by the whole planetary disk, $T_B(\nu)$ is, as a first approximation, the temperature of the atmospheric layer for which the optical thickness is equal to 2/3. At the center of a molecular line, the optical thickness is stronger than in the wings; the level $\tau = 1$ (or 2/3) is reached at a higher altitude than at neighboring frequencies. We see that, in the planetary spectrum, the line can appear in emission or absorption, depending on the temperature gradient of the region where the line is formed. In the troposphere (see Chapter 1), the temperature decreases as the altitude increases, so the line appears in absorption. On the other hand, in the stratosphere, the temperature increases with altitude, and the line appears in emission (see Figure 5.5).

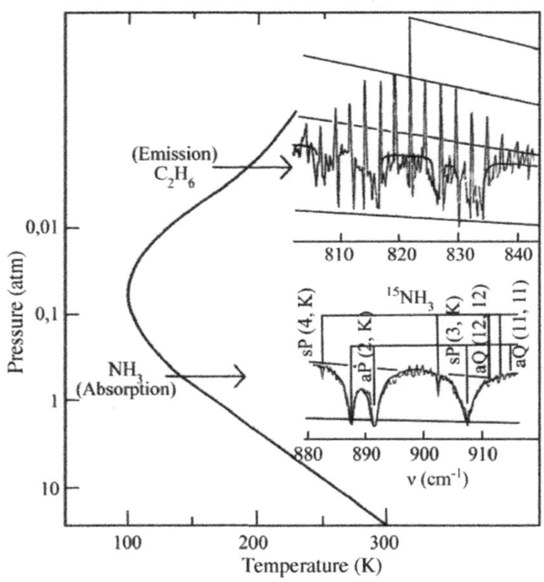

Figure 5.5. *Illustration of the mechanism of formation of atmospheric lines in the thermal domain*

COMMENT ON FIGURE 5.5. – *Left: the thermal profile of Jupiter shows a temperature inversion above the tropopause. Right: the spectrum of Jupiter between 810 and 910 cm^{-1} (11.0–12.3 µm). It shows the emission lines of C_2H_6 (top), formed in the stratosphere, and the absorption lines of NH_3 (bottom), formed in the troposphere (source: from (Encrenaz 1984); data from (Tokunaga* et al. *1979)).*

Figure 5.6 shows examples of spectra of the giant planets, in the solar reflected domain and in the thermal domain. In the near-infrared (reflected solar domain), the spectra of the four giant planets are dominated by the absorption of methane CH_4. In the thermal domain, which dominates beyond 4 µm, the spectra show a succession of molecular lines in emission or absorption, according to the region of the atmosphere where they are formed. A summary of the results, acquired by the remote sensing of planets (both giant and telluric) by infrared spectroscopy, can be found in (Hanel *et al.* 2003).

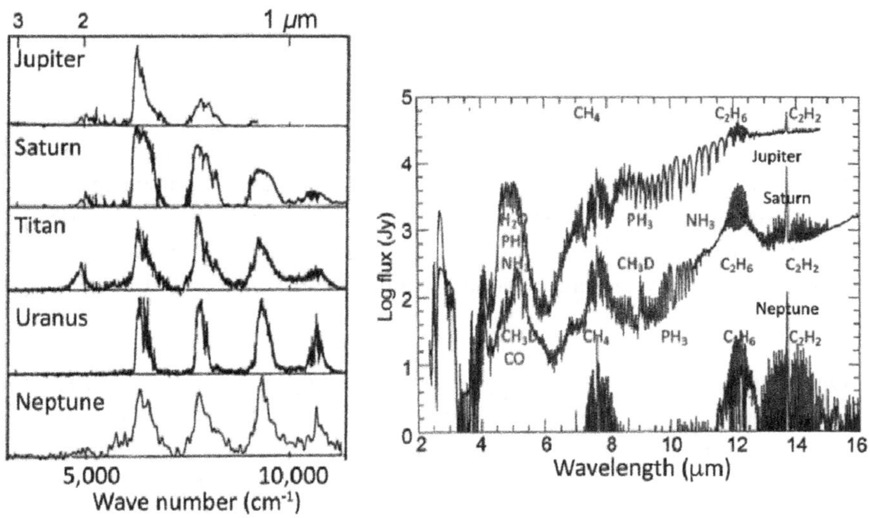

a) b)

Figure 5.6. *Examples of spectra of the giant planets*

COMMENT ON FIGURE 5.6. – *(a) In the near-infrared (1–3 µm), the spectra of the giant planets (as well as that of Titan) are dominated by methane absorption (source: from (Larson 1980)). (b) Spectra of Jupiter, Saturn and Neptune, recorded by the SWS spectrometer of the ISO satellite. The reflected solar component is visible on Jupiter and Saturn below 4 µm. At longer wavelengths, in the thermal domain, the window around 5 µm allows Jupiter and Saturn's deep troposphere to be probed; we can observe PH_3, NH_3, CH_3D, CO, H_2O (indicated in red on the* *figure), as well as other minor species. The CH_4, C_2H_2 and C_2H_6 bands (shown in blue on the figure), formed in the stratosphere, are in emission on Jupiter, Saturn and Neptune. Uranus (not shown in the figure) is only detected at 13.7 μm in C_2H_2 emission, with a flux equal to one third of that of Neptune (source: from (Encrenaz* et al. *1999)).*

5.2. The atmosphere of giant planets

We have seen (in section 1.3.2) that the giant planets are characterized by a high volume and mass, a low density, a large number of satellites and a ring system. Within the giant planets, two subsystems can be distinguished: Jupiter and Saturn have high masses (318 and 95 Earth masses), while Uranus and Neptune, further away, are less massive (14 and 17 Earth masses, respectively). We have seen (Chapter 2) how these differences can be explained, in a first approximation, within the framework of the model of planetary formation by nucleation. According to this model, planets are formed from a solid core of ice and rocks. Located beyond the water condensation line (the ice line), Jupiter and Saturn were able to accrete cores that were sufficiently massive (about ten Earth masses) to capture the surrounding nebula in their gravity field; Jupiter and Saturn, essentially made of protosolar gas, are called gas giants. Further away from the Sun, Uranus and Neptune experienced a slower accretion phase, which was not completed at the time of the dissipation phase of the protosolar gas (the T Tauri phase of the Sun at about 10 million years; see Chapter 2); they are therefore essentially constituted of their initial core; they are called icy giants.

One must also keep in mind the dynamical history of the giant planets, which have undergone a migration during their history; this migration has had important consequences on the evolution of the families of small bodies, in particular certain families of asteroids and the trans-Neptunian objects (see section 1.2.6, Chapter 2 of this book and Chapter 4 of Volume 2).

5.2.1. *Atmospheric composition*

The atmospheric composition of the giant planets can be explained by their formation scenario: formed from protosolar gas, the giant planets are mainly composed (more than 98%) of hydrogen and helium. In this hydrogen-rich medium, the elements C, N and O are mainly in reduced form (CH_4, NH_3, H_2O), in accordance with the rules of thermochemical equilibrium (see section 1.3.2.2). Thus, PH_3, GeH_4 and AsH_3 have also been detected in the tropospheres of Jupiter and Saturn, by spectroscopy in the 4.5 μm window (see section 5.1.4). Besides thermochemical equilibrium, there are other mechanisms responsible for the formation of other minor molecules:

– irradiation by solar ultraviolet radiation leads, in particular, to the formation of various hydrocarbons in the stratosphere of the giant planets (particularly C_2H_2, C_2H_4 and C_2H_6);

– vertical transport motions can bring some species from the deep zones to the regions where they become observable, at pressures lower than about 10^5 Pa (as is the case for PH_3 and CO on Jupiter and Saturn);

– the upper atmospheres of giant planets receive a continuous flow of oxygenated elements, in the form of micro-meteoroids and comets, enriching the stratosphere in H_2O, CO_2 and CO; an example of these impacts was provided by the collision of comet Shoemaker–Levy 9 with Jupiter, in 1994 (see Box 5.1). This flux was revealed in the far-infrared range by the ISO satellite, then in the submillimeter range by the Herschel satellite. Table 5.1 lists the molecules detected in the four giant planets, with their relative abundance.

5.2.2. *Elemental and isotopic abundance ratios*

From the abundance of molecules, it is possible to determine elemental and/or isotopic abundance ratios, which can provide constraints on planetary formation models. The H/He and D/H ratios can be compared to the protosolar values measured by other methods, in particular by the study of the solar wind; the C/H, O/H, N/H, P/H and Ge/H ratios measured in the giant planets, compared to the protosolar values, bring a diagnosis on the models of formation of the giant planets.

5.2.2.1. *The H/He ratio*

The far-infrared spectrum of giant planets depends on the H_2/He ratio through its continuous component, between 30 and 100 μm, induced by H_2-H_2 and H_2-He collisions. Since hydrogen is the main atmospheric constituent, it is possible, by inverting the brightness integral in this spectral range, to simultaneously determine the thermal profile according to pressure and the H_2/He mixing ratio. The results are presented in Table 5.2. We can see that the helium abundance in Uranus and Neptune could be compatible with the protosolar value, while those of Jupiter and Saturn are lower. This helium depletion in Jupiter and Saturn could be due to the changes in the state of the hydrogen and helium inside the giant planets. In the case of Jupiter and Saturn, the internal pressure is high enough for hydrogen to be in metallic form; according to thermochemical models, helium must have condensed there in the form of droplets migrating towards the nucleus, leading to a depletion of the helium vapor phase. In the case of Uranus and Neptune, this phenomenon is not supposed to occur because the internal pressure is insufficient for the hydrogen to reach the metallic phase.

Atmospheric constituent	Jupiter	Saturn	Uranus	Neptune
Hydrogen (H_2)	1.0	1.0	1.0	1.0
Deuterated hydrogen (HD)	1.8 (-5)	2.3 (-5)	5.5 (-5)	6.5 (-5)
Helium (He)	0.157	0.10-0.16	0.18	0.23
Methane (CH_4)-Troposphere	2.1 (-3)	4.4 (-3)	2 (-2)	4 (-2)
Methane (CH_4)-Stratosphere	–	–	3–10 (-5)	7 (-4)
Deuterated methane (CH_3D)-Trop	2.5 (-7)	3.2 (-7)	1 (-5)	2 (-5)
Acetylene (C_2H_2)-Strat	*	3.5 (-6)	2–4 (-7)	1.1 (-7)
Ethylene (C_2H_4)-Strat	7 (-9)	*	–	*
Ethane (C_2H_6)-Strat	4 (-6)	4 (-6)	*	1.3 (-6)
Diacetylene (C_4H_2)	–	9 (-11)	–	–
Methyl-acetylene (CH_3C_2H)-Strat	*	6 (-10)	–	–
Propane (C_3H_8)-Strat	6 (-7)	–	–	–
Benzene (C_6H_6)-Strat	2 (-9)	–	–	–
Methyl (CH_3)-Strat	–	2–10 (-8)	–	–
Ammonia (NH_3)-Trop	2 (-4)	2–4 (-4)	–	–
Phosphine (PH_3)-Trop	6 (-7)	1.7 (-6)	–	–
Germane (GeH4)-Trop	7 (-10)	2 (-9)	–	–
Arsine (AsH_3)	3 (-10)	2 (-9)	–	–
Carbon monoxide (CO)-Trop	1.5 (-9)	2 (-9)	–	–
Carbon monoxide (CO)-Strat	1.5 (-9)	2 (-9)	–	1 (-6)
Carbon dioxide (CO_2)-Strat	3 (-10)	3 (-10)	–	5 (-10)
Water vapor (H_2O)-Trop	1.4 (-5)	2 (-7)	–	–
Water vapor (H_2O)-Strat	1.5 (-9)	2–20 (-9)	5–12 (-9)	1–4 (-9)
Tri-hydrogen ion (H_3^+)	*	*	*	*

Table 5.1. *Atmospheric composition of the giant planets, determined by spectroscopic measurements. The asterisks indicate a detection (source: from (Encrenaz* et al. *2003))*

Planet	Abundance of helium in mass fraction (Y)	Origin
Jupiter	0.234 +/- 0.005	Galileo HDA
	0.18 +/- 0.04	IRIS/Voyager 1–2
Saturn	0.18 – 0.25	Voyager 1–2
Uranus	0.262 +/- 0.048	Voyager 2
Neptune	0.32 +/- 0.05	Voyager 2
Sun	0.24 +/- 0.01	Helioseismology
Protosolar He	0.275 +/- 0.01	Solar wind
Primordial He	0.232 +/- 0.005	Galactic evolution models

Table 5.2. *Abundance of helium in the giant planets (in mass fraction Y) (source: from (Encrenaz* et al. *2003))*

5.2.2.2. *The D/H ratio*

According to the Big Bang model, deuterium was formed in the primordial explosion, along with hydrogen; since then, it has been progressively destroyed by nucleosynthesis within stars, in the reactions transforming hydrogen into helium. The abundance of deuterium in the Universe has therefore been decreasing from the beginning of the Universe until today, the protosolar value being in the middle of these two extremes. As Jupiter and Saturn are essentially formed of protosolar gas, the D/H ratio in these planets can give us an indication of the D/H value in the protosolar nebula.

Moreover, within the Solar System, the D/H ratio can be used as an indicator of the temperature of the medium in which it is measured. Indeed, deuterium is enriched in molecules in the form of ice, following ion-molecule reactions occurring at low temperature; such effects are observed in the laboratory, as well as in the interstellar medium. In the case of Uranus and Neptune, which are essentially constituted by their ice core, we expect the D/H ratio to be enriched in relation to the values measured on Jupiter and Saturn.

In giant planets, the D/H ratio has been measured in two ways:

– in the 1970s, from the CH_3D/CH_4 ratio, by infrared spectroscopy;

– in 1997, from the rotational lines of HD in the far-infrared range, measured by the ISO satellite.

In the case of Jupiter, a direct measurement was obtained by the mass spectrometer of the Galileo probe in 1995. The results are presented in Table 5.3 and Figure 5.7. It can be seen that the D/H values measured in Jupiter and Saturn are compatible with the protosolar value, while those of Uranus and Neptune are enriched (in a lesser proportion, however, than in comets; see section 2.1, Volume 2). The D/H ratio in the giant planets thus provides a diagnosis in favor of the nucleation model of the giant planets, according to which they formed from a core of ice and rock; indeed, if the planets had formed by direct gravitational collapse, the D/H ratio of the four giant planets would have had the protosolar value.

Planet	D/H	Origin
Jupiter	2.6 +/- 0.7 (-5) 2.24 +/- 0.35 (-5)	Galileo GCMS ISO
Saturn	1.70 (+1.70, -0.45) (-5)	ISO
Uranus	5.5 (+3.5, -1.5) (-5)	ISO
Neptune	6.5 (+2.5, -1.5) (-5)	ISO
Protosolar D/H	2.1 +/- 0.5 (-5)	Solar wind
Primordial D/H	5 – 6 (-5)	Galactic evolution models
D/H in the local interstellar medium	1.6 +/- 0.12 (-5)	Spectroscopic measurements in the ultraviolet range

Table 5.3. *The D/H ratio in the giant planets (source: from (Encrenaz* et al. *2003))*

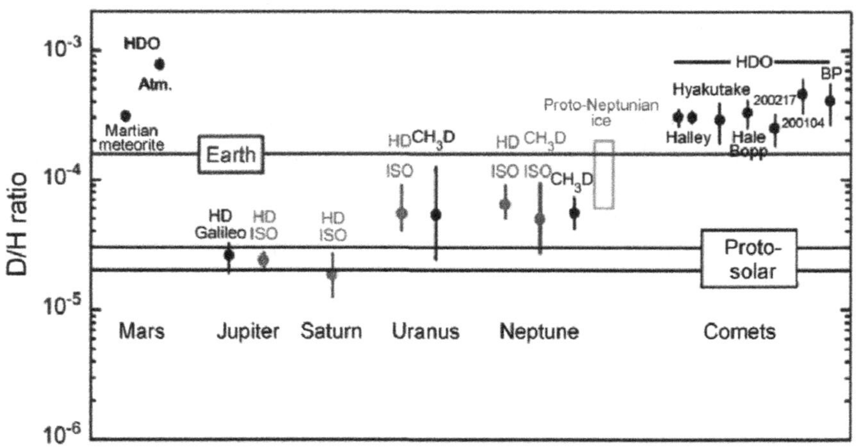

Figure 5.7. *The D/H ratio in the outer Solar System. The D/H value in Jupiter and Saturn is consistent with the protosolar value inferred from measurements in the solar wind, while deuterium in Uranus and Neptune, measured from HDO or CH3D, is enriched due to the presence of an ice core (source: from (Bockelée-Morvan* et al. *2015; Lis* et al. *2019))*

5.2.2.3. *C/H, N/H, O/H, Ge/H and P/H ratios*

Another diagnosis in favor of the giant planet nucleation model is provided by the measurement of the relative abundance (with respect to hydrogen) of "heavy" elements (in other words, heavier than helium) in the planets, compared to the abundance of these elements in the primitive nebula, measured in the Sun (see Chapter 2). The argument is the same as in the D/H ratio case (see section 5.2.2.2): in the scenario of giant planet formation by direct gravitational collapse, the elemental ratios would be unchanged. In the model of formation from an ice core, which is rich in "heavy" elements (C, N, O), the latter must be enriched; this enrichment being all the more prominent as the relative mass of the core, with respect to the total mass of the planet, is bigger. Knowing the relative mass of the "heavy elements" (noted "Z") from the cosmic abundance of the elements (of the order of 1–2%), one can simply assess the expected enrichment according to the following factor: in the case of Jupiter, whose core (estimated at 10 Earth masses) is 3% of the total mass, the expected enrichment is of the order of 3; for Saturn, with a 10% core, the enrichment is 9; for Uranus and Neptune, whose cores are more than 50% of the total mass, the enrichment is 30 to 40 (it would be 50 for an ice core with no atmosphere).

Among the abundance ratios, the best known is C/H, thanks to the very rich spectrum of the CH_4 molecule. Methane is a gas on Jupiter and Saturn, but it condenses on Uranus and Neptune at a pressure level of about 10^5 Pa; in the latter case, it must be measured in the deep troposphere in order for the measurement to be representative of the whole planet, which is possible in some visible and near-infrared spectral windows. Spectroscopic measurements of the giant planets in the near-infrared range show that the CH_4 enrichment is significantly greater on Uranus and Neptune. For all four planets, the results are in agreement with the predictions of the nucleation model, which, together with the D/H ratio, provides additional support for this scenario.

The N/H ratio is more difficult to measure because the NH_3 molecule condenses on the four giant planets (ammonia clouds are responsible for Jupiter and Saturn's white color, see section 5.2.3). Radio measurements, which probe the deep troposphere, indicate an enrichment of 2 to 4 compared to the solar value. In the case of Jupiter, a measurement was provided by the Galileo probe (see section 5.2.2.4) and additional information was provided by the Juno probe (see section 5.2.2.5).

The O/H ratio should *a priori* be measurable on Jupiter from the H_2O lines observed in the deep troposphere, below the water cloud. Measurements, carried out in some regions without clouds (hot spots, see section 5.2.3), surprisingly revealed a depletion with respect to the solar value. The commonly accepted interpretation is that these hot spots are regions of subsidence, devoid of clouds, in which the quantity of H_2O is not representative of the global value of water on the planet; this result was confirmed by the measurements of the Galileo probe in 1995 (see section 5.2.2.4).

As for the germane (GeH_4), it was also detected in the same spectral region, with an abundance well below the solar value. The cause could be the destruction of GeH_4 in the presence of the H_2O cloud, in favor of pure germanium (Ge) or the H_2GeO_3 compound.

The PH_3 molecule, detected on Jupiter and Saturn, presents a special case, because its presence is not expected on the basis of thermochemical equilibrium. It is probably a trace of important vertical movements. Indeed, according to thermochemical models, PH_3 should react with H_2O in the deep troposphere to form P_2O_5. However, the PH_3 molecule is observed on Jupiter and, in even more abundant quantities, on Saturn (see Table 5.1). The commonly accepted explanation is that the molecule is transported from the deep troposphere to the observable regions (10^4–10^5 Pa) by vertical motions, in times shorter than the destruction time of the molecule. Like GeH_4, we cannot use P/H to test the nucleation model of giant planets.

5.2.2.4. *Measurements of the Galileo probe*

In December 1997, the Galileo probe penetrated Jupiter's atmosphere to a depth of 22 10^5 Pa, providing, thanks to its GCMS mass spectrometer, the first *in situ* measurements of the elemental abundances in Jupiter's atmosphere, with the D/H, C/H and S/H ratios, as well as the abundances of the rare gases Ne, Ar, Kr and Xe; the He/H and N/H ratios were measured by other instruments of the probe. The result of all the measurements is presented in Figure 5.8 (Owen *et al.* 1999). Three elements show a notable depletion compared to the solar value. The depletion of Ne, like that of He (see section 5.2.2.1), is attributed to an internal differentiation mechanism within the metallic hydrogen; that of O is attributed to convective circulation, which dries out the regions of subsidence, which are the hot spots; the Galileo probe dived in one of these regions. Apart from these three points, all the elements show an enrichment factor of the order of 3, compatible with the predictions of the nucleation model. Following a re-assessment of the cosmic abundances and of the relative mass of Z (the set of elements other than H and He), this enrichment factor has become closer to 4, both for the measurements and for the predictions of the nucleation model, which means that the agreement with this model remains intact.

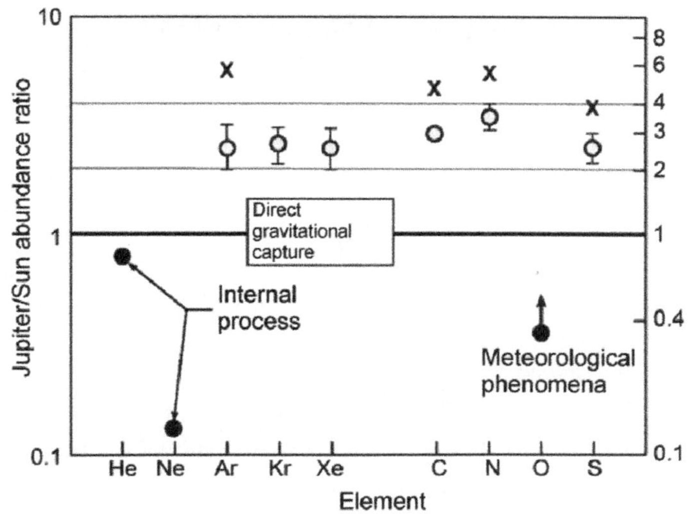

Figure 5.8. *Elemental abundances in Jupiter, measured in relation to hydrogen and normalized to solar values, deduced from Galileo probe measurements (source: from (Owen* et al. *1999))*

COMMENT ON FIGURE 5.8. – *The circles correspond to the values of the initial publication; the crosses correspond to a re-assessment of these ratios, following a new determination of the solar abundances. The average enrichment, initially 3 +/- 1, has been corrected to 4 +/- 2 (source: (Owen and Encrenaz 2006)).*

5.2.2.5. *Measurements of the Juno probe*

With the aim of obtaining a representative measurement of O/H for the whole planet, the American probe Juno, which was launched in 2011, has been operating around Jupiter since 2016, with the MWR (Microwave Radiometer) instrument, a radiometer designed to measure the abundance of H_2O and NH_3 up to pressures of over 10^7 Pa.

Initial results showed that the vertical distribution of NH_3 in the deep troposphere was extremely complex and inhomogeneous (see Figure 5.9). The NH_3 mixing ratio is not at all homogeneous at depth, at pressure levels above 2 10^5 Pa, as previously assumed; it is only constant near the equatorial region, with a value 2.7 times that of the solar value, in agreement with the Galileo probe results. At medium latitudes, the N/H ratio only becomes homogeneous on the disk at pressures above 20 10^5 Pa, where its value remains constant up to pressure levels above 100 10^5 Pa (Bolton *et al.* 2015).

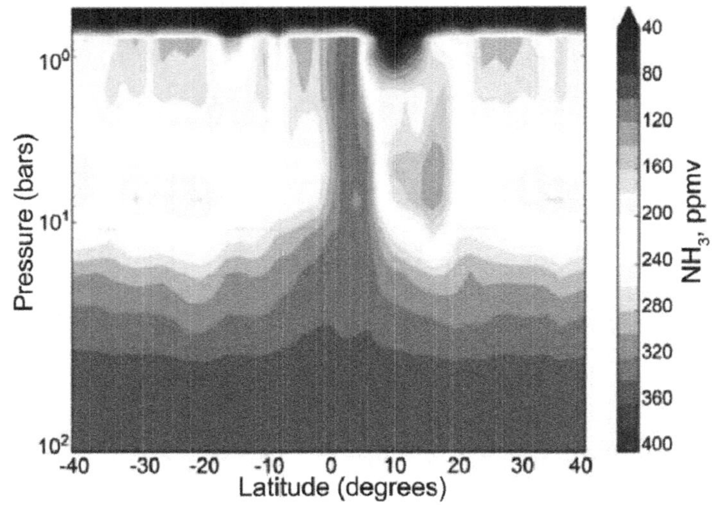

Figure 5.9. *Vertical distribution of NH_3 mixing ratio in Jupiter according to latitude*

COMMENT ON FIGURE 5.9. – *The blue band at the top of the figure represents the region where NH_3 condenses. The high value of this ratio, near the equator, appears to be due to vertical motions that transport ammonia-rich gases upwards from the deepest layers ($P > 60\ 10^5$ Pa) (source: from (Bolton* et al. *2015)).*

The complexity of the vertical and horizontal distribution of NH_3 makes it particularly difficult to measure the O/H ratio, because the thermal spectrum of Jupiter in radio radiation between 0.5 and 22 GHz (λ = 1.4–60 cm) is largely dominated by NH_3. A first estimate made near the equator indicates an O/H ratio value 2.7 times that of the solar value (Li *et al.* 2020), which is in agreement with Jupiter's nucleation model (Owen *et al.* 1999); it remains to be seen whether or not this value is representative of the whole planet.

5.2.3. *Thermal structure*

The thermal structure of giant planets has been determined in several ways:

– from a theoretical point of view, by the radiative transfer models, which make it possible to calculate the temperature at each level by taking into account the flux received from all the other atmospheric layers (see section 4.4.3);

– the inversion of the brightness integral which allows, from the observation of the far-infrared spectrum, to simultaneously determine the thermal profile and the H_2/He ratio by an iterative method (see section 5.2.2.1);

– the stellar occultation method, which consists of observing the passage of a planet in front of a star, and to measure the decrease, followed by the increase of the flux of the latter, when the atmosphere is seen in transmission in front of the star, before and after the occultation.

In the case of Jupiter, a direct measurement of altitude, temperature and pressure was obtained by the sensors of the Galileo probe from the upper atmosphere, up to a pressure of 22 10^5 Pa.

Here above (see section 1.3.2.5), we have presented the laws governing the thermal structure of an atmosphere (law of perfect gases, hydrostatic equilibrium), and we have described (see section 4.4.3) the convective and radiative regimes encountered in the atmospheres of telluric planets. The same laws apply to the giant planets, with the difference that the giant planets do not have a solid surface: in the case of Jupiter and Saturn, their troposphere penetrates to depths of more than 1,000 kilometers, at pressures higher than 10^8 Pa. In the troposphere, the energy transfer regime is convective and the temperature gradient is adiabatic, with a value of about -2 K/km, which varies very slightly with the H_2/He ratio. In the troposphere, the opacity is due to the absorption induced by H_2-H_2 and H_2-He collisions; it is optically thick in the visible and infrared ranges, up to a pressure level of some 10^5 Pa. The spectroscopy in these spectral ranges is thus limited to the high troposphere and the stratosphere; in contrast, the centimetric radio radiation probes the deep troposphere up to some 10^7 Pa (this is what the MWR radiometer of the Juno mission measures).

The tropopause, the limit between the troposphere and the stratosphere, is located at about 10^4 Pa for the four giant planets; it corresponds to a minimum temperature of 110 K for Jupiter, 95 K for Saturn, and 55 K for Uranus and Neptune. The fact that the tropopause is located at the same pressure level for the four giant planets is explained by the fact that their chemical composition, dominated by hydrogen and helium, is very similar, which implies comparable opacities in the infrared range. The temperature at the tropopause naturally decreases as the heliocentric distance increases; the relatively high temperature of Neptune's tropopause is explained by the fact that the planet, unlike Uranus, has (like Jupiter and Saturn) an internal energy source (see section 5.3).

Above the tropopause, the transfer of energy takes place by radiation. The four giant planets have a stratosphere that is characterized by a rise in temperature, due to the absorption of solar radiation in the near-infrared range by methane (CH_4) and various aerosols; these come from the formation of various hydrocarbons, following the photolysis of methane under the effect of solar ultraviolet radiation at a pressure level of about 10 Pa. At higher altitudes, other heating mechanisms are involved, such as gravity waves and energetic particles. Figure 5.10 shows the thermal structure of the four giant planets.

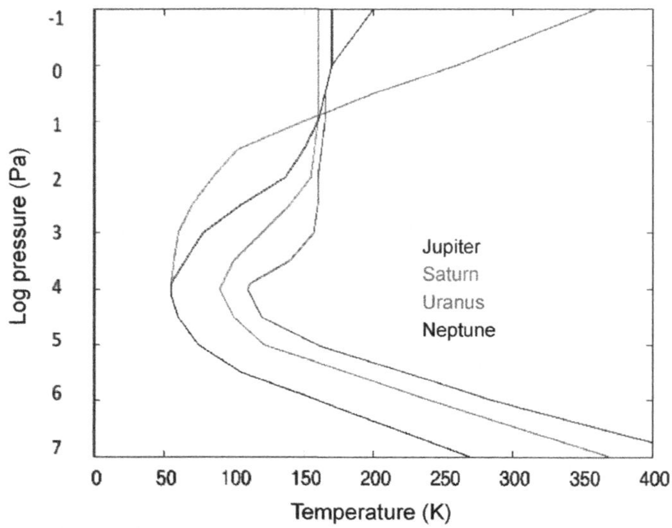

Figure 5.10. *Thermal structure of the giant planets*

5.2.4. *Atmospheric circulation and cloud structure*

While we have a good knowledge of the chemical composition of the gas phase of giant planets, as well as of their thermal structure, there is still considerable uncertainty about the cloud structure and the nature of the condensates present in these atmospheres. The cause of this uncertainty lies in the struggle to interpret the spectral signatures of these condensates, which are often weak and ambiguous, and much more complex than those of gaseous molecules.

5.2.4.1. *Vertical structure of clouds*

From the thermal profile and the relative abundance of the different constituents, thermochemical equilibrium models provide us with initial information on the nature and the altitude of the expected cloud layers. In the case of Jupiter and Saturn, among the spectroscopically detected molecules, only H_2O and NH_3 form observable cloud layers. On Jupiter, the condensation of NH_3 is expected at 5 10^4 Pa, that of ammonium hydrosulfide (NH_4SH) is expected at about 2 10^5 Pa and that of H_2O at about 5 10^5 Pa. A similar structure should be found on Saturn, at slightly higher pressure levels (10^5, 4 10^5 and 10^4 Pa, respectively). In the case of Uranus and Neptune, methane (CH_4) condenses at about 10^5 Pa; the condensation of PH_3 and H_2S (not detected to date) is expected at pressure levels of 10^5 and 3 10^5 Pa, respectively. The condensation of NH_3 and H_2O takes place at pressure levels that are too high to be observed.

Observations on Jupiter and Saturn, by imaging and spectroscopy, confirmed the presence of NH_3 clouds, which are responsible for the yellowish-white color of the zones. They also indicated the presence of deeper cloud layers, due to NH_4SH and H_2O. On the other hand, in the case of Jupiter, the *in situ* measurements of the Galileo probe provided different information: they showed that the "hot spot" in which the probe had descended, in the northern equatorial band, was devoid of clouds; this is a region of subsidence created by convective movements, in which the abundance measurements are not representative of the whole planet (see sections 5.2.2.4 and 5.2.2.5).

5.2.4.2. *Large-scale morphology: zones and bands*

Like the telluric planets, the giant planets have a Hadley circulation, generated by the difference in insolation between the equator and the poles. As a result of the rapid rotation of the planets, this circulation manifests itself, in Jupiter and Saturn's case, by a series of dark bands and light zones alternating in latitude. The latter are more or less symmetrical with respect to the equator (see Figure 5.1 and Figure 5.2), with the direction of the zonal winds reversing between the zones and bands. The direction and amplitude of the zonal winds on Jupiter show a remarkable stability over the years, as shown by the similarity between the observations of Voyager in 1979 and Cassini in 2000. The zones are regions of upward convective motion, in which ammonia condenses as clouds; the bands are drier regions, devoid of ammonia clouds, in which infrared radiation penetrates to pressures of about 10^5 Pa. The width of the zones and bands, inversely proportional to the Coriolis force, decreases as one moves away from the equator.

The models that only consider the latitudinal variations of the solar flux, which results in a Hadley circulation, have difficulty in accounting for the north-south asymmetries observed on Jupiter, as well as the equatorial jet of Saturn, which is four times more intense than that of Jupiter, while the solar flux received is four times weaker. Other models of deep convective circulation have considered the existence of an internal energy source, whose presence has been detected on Jupiter, Saturn and Neptune. In the case of Jupiter, strong winds were recorded by the

Galileo probe, and the MWR radio experiment of the Juno probe revealed a very complex structure of NH_3 abundance up to a pressure over 100 10^5 Pa (see Figure 5.9 and section 5.2.2.5). It is clear that the internal dynamics of giant planets is infinitely more complex than the descriptions made by current models.

5.2.4.3. *Small-scale morphology: cyclones and anticyclones*

In addition to the band and zone structure, the atmospheric circulation of the giant planets results in local phenomena, some of which are stable over long periods of time. The most famous is Jupiter's Great Red Spot, discovered by Jean-Dominique Cassini at the end of the 17th century. Its appearance and size have changed significantly over time, but it has been stable for over three centuries. Thanks to images from the Galileo, Cassini and Juno space probes, we now know that the Great Red Spot is an anticyclonic zone located in a shear zone between the southern equatorial zone and the southern tropical zone, where the zonal winds are in opposite directions. The origin of its red color is still unclear; it could be due to organic compounds, or to phosphorus or sulfur molecules.

Other smaller and temporary anticyclones also appear on the surface of Jupiter, notably in the southern temperate band, and evolve with time. The Juno probe has also revealed a set of cyclones on Jupiter, concentrated around each pole (see Figure 5.11). Lastly, the hot spots, which we have already mentioned above, are subsidence zones, mainly located in the northern equatorial band, and are probably the signature of small-scale convective movements inside this band; the Galileo probe dived into one of these spots in 1995.

In the case of Saturn, the Cassini probe revealed a stable hexagonal structure of cyclones around the north pole (see Figure 5.11). Another remarkable phenomenon is the giant storm that developed in December 2010 in the northern hemisphere of the planet (see Figure 5.3); this would be due to a vertical transport of water vapor followed by its condensation. Such events occur on Saturn every twenty to thirty years, which is probably related to Saturn's period of revolution, with some irregularity that remains to be explained.

The morphology of Uranus and Neptune is much less known than that of Jupiter and Saturn. In addition to the respective flybys of the two planets by Voyager 2 in 1986 and 1989, our knowledge is based on images taken by the Hubble telescope since the 1990s, as well as images taken from the ground (see Figure 5.12). We know that the Great Dark Spot observed on Neptune by Voyager 2 in 1989 (see Figure 5.2) has disappeared with time. However, spots appeared on Uranus as the planet was approaching the equinox in 2007, while the planet appeared devoid of structures during the Voyager 2 flyby in 1986 (see Figure 5.2).

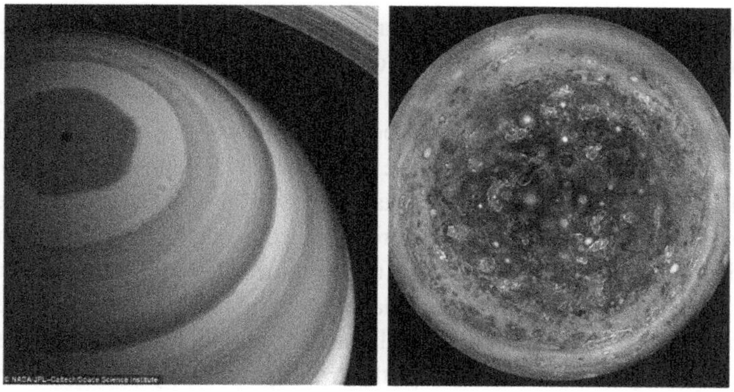

Figure 5.11. *Polar cyclones on (a) Saturn (north pole, seen by Cassini); (b) Jupiter (south pole, see by Juno) (source: NASA)*

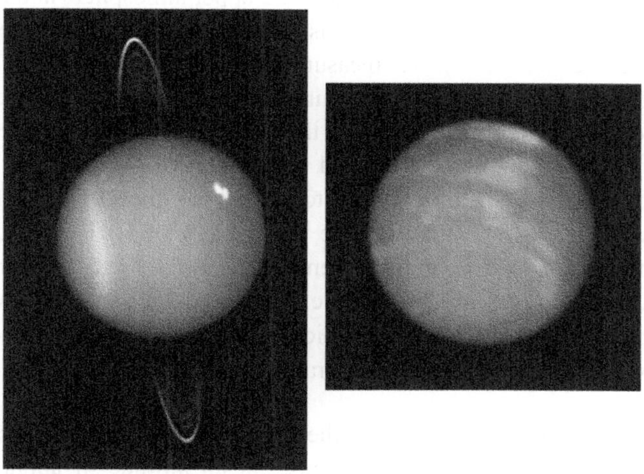

a) b)

Figure 5.12. *Uranus (a) and Neptune (b), imaged by the Hubble Space Telescope in 2005 and 2011, respectively (source: NASA)*

5.2.5. *High atmosphere and photochemistry*

The thermosphere extends above the stratosphere at a pressure lower than 0.1 Pa, in which the photochemistry of methane, dissociated by the solar ultraviolet radiation, is involved. The method of solar and/or stellar occultation on board the Voyager probes has been used in the ultraviolet range, in order to determine the thermal profile in this region, at about a thousand kilometers from the tropopause, corresponding to a pressure of the order of 10^{-4} Pa. In the case of Jupiter and Saturn, the ultraviolet radiation is absorbed by H between 60 and 80 nm, by CH_4 between 90 and 115 nm, and by C_2H_2 and C_2H_6 between 140 and 170 nm; the light curves recorded by the Voyager probes at these different wavelengths allow us to trace the vertical distribution of these constituents. Moreover, in the case of Jupiter, *in situ* measurements of the thermal profile were obtained by the Galileo probe in 1995; at an altitude of 800 km, it measured a temperature of 900 K, higher than model predictions, with strong oscillations that were probably due to the presence of gravity waves. Figure 5.10 shows that, although the gradients of the thermal profiles of the four giant planets are similar in the troposphere, they differ considerably in the stratosphere and the thermosphere, which reflects different heating mechanisms (gravity waves and energetic particles) for each planet.

5.2.5.1. *The turbulent diffusion coefficient*

The distribution of minor constituents in the upper atmosphere and thermosphere depends on vertical transport, which is measured by the turbulent diffusion coefficient (see section 4.4.3). To measure this coefficient, indirect methods are used, such as the Ly α flux measurement, which gives the abundance of atomic hydrogen above the penetration level, and the vertical distribution of methane and its dissociation products. Jupiter's estimated K value is of the order of 10^6 cm^2 s^{-1}, while that of Saturn is of the order of 10^7 to 10^8 cm^2 s^{-1}; the stronger vertical mixing on Saturn results in higher stratospheric abundances of C_2H_2 and C_2H_6 than on Jupiter. In the case of Uranus, the turbulent diffusion coefficient is remarkably low (10^4 cm^2 s^{-1}), while on Neptune, it is of the order of 10^6–10^7 cm^2 s^{-1}. The value of K could be related to the existence of a particularly large internal source of energy on Saturn and Neptune, which is absent on Uranus (see section 5.3.2.1).

In the absence of photochemistry, the atmosphere of giant planets would be homogeneously mixed up to the homopause, at an altitude of about 0.1 Pa (see section 1.3.2.4). However, photochemistry occurs at lower altitudes: this is particularly the case of NH_3, whose photodissociation occurs at the tropopause, by absorption of solar UV photons between 180 and 200 nm.

5.2.5.2. *The external source of oxygen*

We have seen in section 5.2.1 that the upper atmospheres of the giant planets receive a flow of oxygen-rich material, detected by the presence of H_2O and CO_2 molecules in the four giant planets (except for CO_2 on Uranus). Due to the presence of the cold trap constituted by the tropopause, where the temperature is at a minimum, the water present in the stratosphere cannot come from the deep layers, and the origin of these molecules is necessarily external. The H_2O flux is of the order of 10^6 cm^2 s^{-1} for Jupiter, Saturn and Neptune (as well as Titan, see Chapter 1, Volume 2), and a little lower for Uranus. The CO_2 flux could also be of external origin, or result from chemical reactions between H_2O and CH_4.

In addition, carbon monoxide CO has also been detected in the four giant planets, by infrared spectroscopy at 4.7 µm. We have seen (in section 5.2.1) that CO is a non-equilibrium species; according to thermochemical models, its mixing ratio should not exceed 10^{-9}, which is consistent with measurements obtained on Jupiter and Saturn. However, millimeter-range measurements by heterodyne spectroscopy indicated an anomalously high mixing ratio of CO (10^{-6}) in Neptune's stratosphere; in the case of Uranus, CO was detected by fluorescence at 4.7 µm, with a mixing ratio of 3 10^{-8}. In all cases, the existence of an external component is likely. Let us finally add that the HCN molecule was also detected in Neptune's stratosphere, with a mixing ratio of 3 10^{-10}; this probably comes from the reaction of nitrogen atoms with CH_4. The nitrogen atoms could come from Triton's atmosphere, which is rich in molecular nitrogen (N_2), or have an internal origin, N_2 being (according to thermochemical models) present in Neptune's interior at the level of a few percent. The presence of CO and HCN in Neptune's stratosphere, but not in that of Uranus, once again illustrates the difference between these two planets, Neptune being much more dynamically active than Uranus, and endowed with an internal source of energy (see section 5.3.1.2).

5.3. The internal structure of giant planets

The internal structure of giant planets eludes direct observation, and, given the considerable pressures and temperatures involved, cannot be simulated in the laboratory. Our knowledge of the internal structure is therefore acquired through a theoretical modeling of the states of the matter at high pressure, constrained by indirect observable data.

5.3.1. *Experimental data*

5.3.1.1. *Mass, radius, gravitational field*

The mass and radius of the giant planets are known with precision (see Table 1.2). The rotation period of the inner layers establishing the longitude III system (different from that of the atmospheric layers observed in the visible range, establishing the longitude I and II systems) has been historically measured by two complementary methods probing the planetary magnetic field. The remote observation of low frequency radio emissions (produced by the gyration of auroral electrons along the high latitude magnetic field lines, producing a rotating beacon-like modulation) provides accurate radio periods, the relative accuracy being 10^{-6} for Jupiter, which is observed from the ground. The decimetric radio emissions of Jupiter (produced by the synchrotron radiation of relativistic electrons trapped in the radiation belts) provide another way to "see" the oscillation of the magnetic field during the rotation. In the case of Saturn, Uranus and Neptune, radio measurements have been confirmed and completed by *in situ* magnetic measurements made by the Voyager probes, in order to provide the internal rotation periods currently referenced by the IAU. The rotation of the planets results in a deviation from sphericity and a flattening, which make it possible to deduce information from the gravitational field regarding the internal distribution of the masses of the planets. The gravitational potential of a planet, under the assumption of hydrostatic equilibrium, is expressed using Legendre polynomials of even order:

$$\phi(r, \theta) = -\,(\mathrm{G}M/r)\,(1 - \Sigma\,(R_e/r)^{2n}\,J_{2n}\,\mathrm{P}_{2n}(\cos\theta) \qquad [5.3]$$

in which θ is the colatitude, R_e is the equatorial radius, M is the mass of the planet, G is the gravitational constant and $\mathrm{P}_{2n}(\cos\theta)$ is the Legendre polynomial of degree $2n$. The J_{2n} quantities are the gravitational moments. In the case of Jupiter and Saturn, these gravitational moments have been measured, respectively, by the Juno and Cassini probes up to order 10; in the case of Uranus and Neptune, the only measurements available are those of Voyager 2.

5.3.1.2. *Internal energy*

We have already mentioned the existence of an internal energy source in the case of Jupiter, Saturn and Neptune (see section 5.2.3). This was first evidenced in 1969 in the case of Jupiter, by the first measurements of its infrared flux (see equation [1.15], equation [1.16] and Figure 5.13). This result was confirmed in 1979 by the IRIS infrared spectrometer of the Voyager probes, which also detected an internal energy source on Saturn and Neptune, but not on Uranus (see Table 5.4). We have seen (see section 5.2.5.2) that this difference between Uranus and Neptune is accompanied by differences in the vertical circulation, thermal structure and chemical composition in the stratosphere of the two planets; so far, it has not been explained.

The most plausible interpretation of the source of internal energy is the contraction and cooling of the planets, the last stage of their evolution following the collapse of the gas on the initial core. We have seen that, in the case of Jupiter and Saturn, another contribution is the differentiation of helium (and also, in Jupiter's case, of neon) in the metallic phase of hydrogen (see section 5.2.2.1).

Planet	Effective temperature (K)	Φ_{em}/Φ_{sol}
Jupiter	124	1.67
Saturn	95	1.78
Uranus	58	< 1.1
Neptune	58	2.6

Table 5.4. *The internal energy of giant planets. Φem/Φsol is the ratio of the emitted energy to the absorbed solar energy (source: from (Encrenaz et al. 2003))*

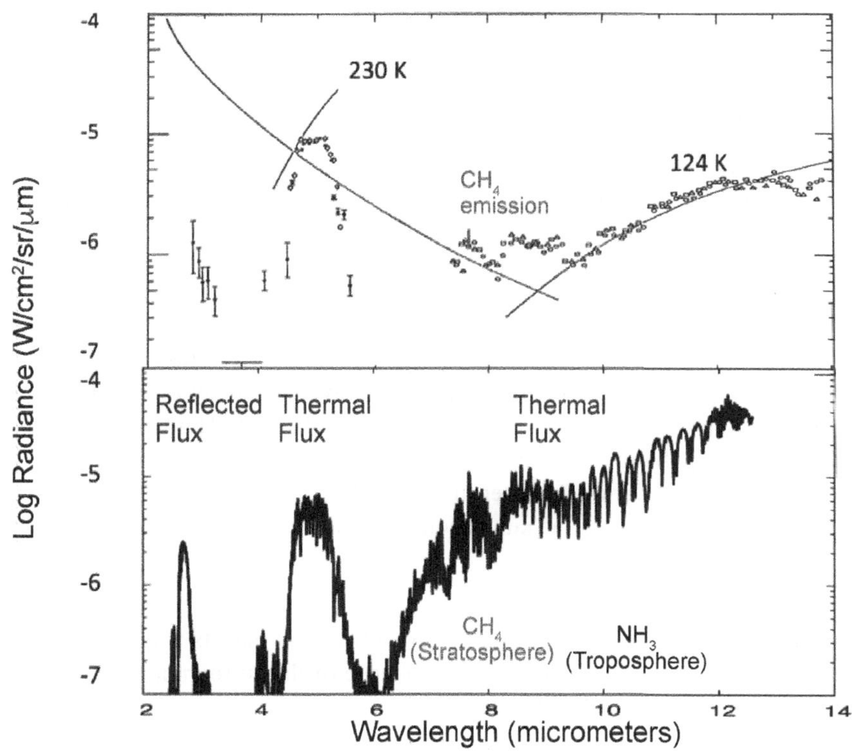

Figure 5.13. *Jupiter's radiation in the near and mid-infrared range*

COMMENT ON FIGURE 5.13. – *(a) First detection of an internal energy source on Jupiter by infrared spectroscopy. The continuum beyond 10 μm corresponds to a black body temperature of 124 K, while the equilibrium temperature of Jupiter is 110 K; at 5 μm, the radiation comes from the deep troposphere at a temperature of 230 K (source: from (Gillett* et al. *1969)). (b) The same domain observed by the ISO* *satellite, with a resolution power of 2,000. One can distinguish the stratospheric emission of CH_4 at 7.7 μm (signature of a temperature inversion) and the absorption of NH_3 in the upper troposphere (source: from (Encrenaz* et al. *1999)).*

5.3.2. *The construction of internal energy models*

5.3.2.1. *Static models*

Static models are based on:

– the equation of hydrostatic equilibrium in the presence of rotation;

– the equations of state of the various components;

– the thermal profile, generally assumed to be adiabatic;

– the relation between the mass distribution and the gravitational moments J_{2n}. The equation of hydrostatic equilibrium in the presence of rotation takes the following form:

$$dp/dr = - GM(r)\rho(r)/r^2 + 2/[3\Omega^2 r\rho(r)] + O(\Omega^4) \quad [5.4]$$

M(r) being the mass inside the radius r and Ω being the rotation velocity. Figure 5.14 shows the mass-radius relation for a self-gravitating body of variable internal composition (pure hydrogen, H-He mixture, ice and rock). The first conclusion is that Jupiter and Saturn are mainly composed of hydrogen, with a helium contribution of the order of a quarter in mass, while Uranus and Neptune are rich in ice; this is a confirmation of the nucleation model described in Chapter 2.

5.3.2.2. *Evolutionary models*

A probable source of the internal energy of the giant planets is the final phase of contraction of the planets, following the collapse of the surrounding protosolar gas onto the core.

Assuming a homogeneous cooling since the formation of the object, we can obtain an estimate of the cooling time of the planet from the excess energy L observed today:

$$L = 4\pi R^2\sigma(T_e^4 - T_0^4) = - d/dt[MC_V T_i] \quad [5.5]$$

R being the radius of the object, T_e the effective temperature, T_0 the effective temperature in the absence of an internal source, T_i a mean internal temperature, and C_V being the mean specific heat. Assuming, as a first approximation, that T_i is equal to

T_e, and assuming a constant R during the contraction phase, we can then obtain the cooling rate τ:

$$\tau = MC_V/[4\pi R^2\sigma]\int dt(T^4 - T_0^4) = MC_V/4L \qquad [5.6]$$

For the model to be plausible, the value of τ found by this equation must be in the range of the age of the Solar System, in other words, 5 10^9 years.

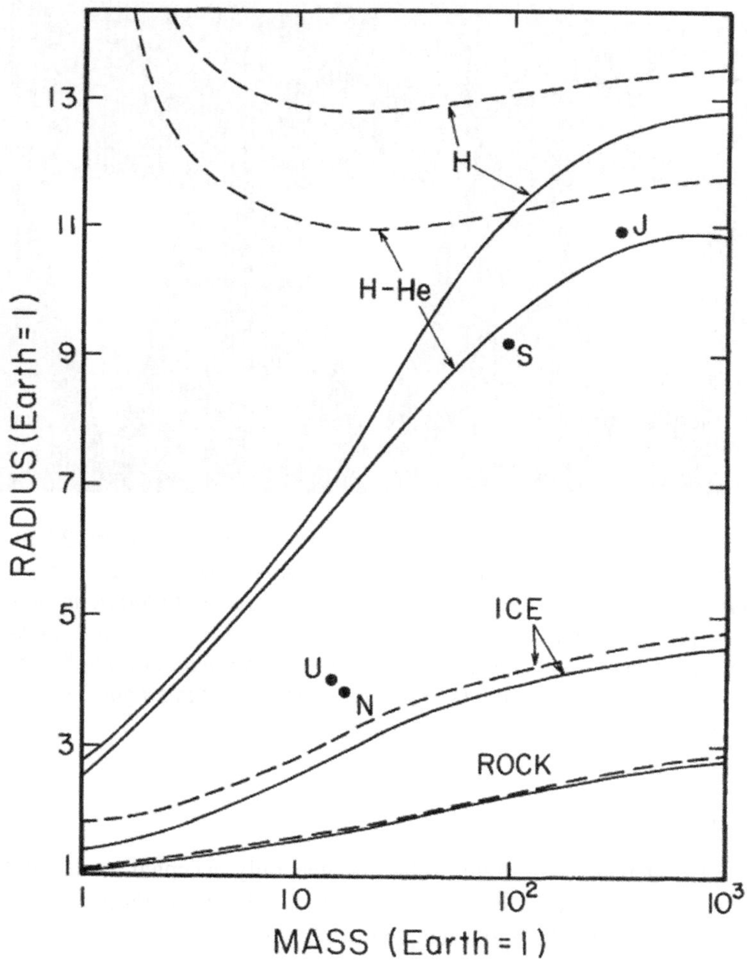

Figure 5.14. *Mass-radius diagram of the giant planets, represented by points; the curves correspond to two assumptions on the temperature (source: from (Stevenson 1982))*

5.3.3. *The results*

Models of the internal structure of the giant planets have been developed by Hubbard, McFarlane, Salpeter, Stevenson, Zharkov and Trubitsyn. In the case of Jupiter and Saturn, recent information has been provided by the Juno and Cassini probes, respectively. A review of the internal structure of giant planets can be found in (Guillot 2005).

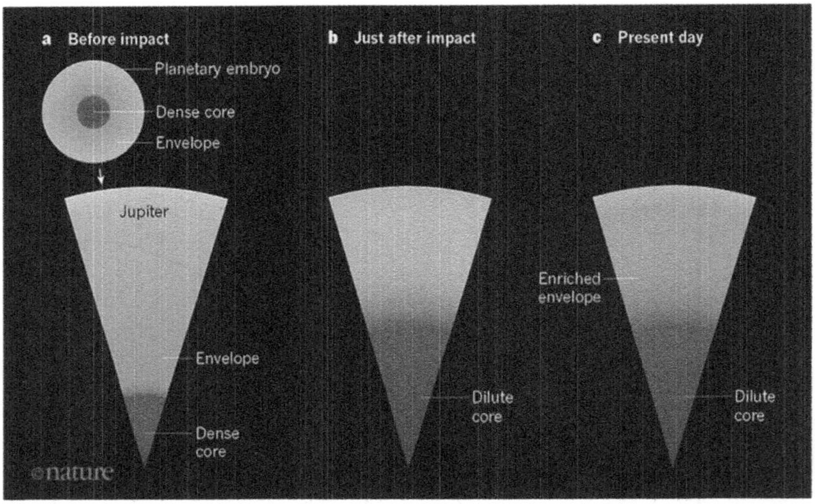

Figure 5.15. *Jupiter's internal structure model, proposed by (Liu* et al. *2019) in order to explain the dilution of the central core observed by the Juno probe gravity measurements. (a) Prior to impact, the dense core is concentrated near the center of the planet. (b) Following impact, the two cores mix and produce a diluted core. (c) The diluted core partially mixes with the surrounding hydrogen shell, producing a shell that is rich in heavy elements. For a color version of this figure, see www.iste.co.uk/encrenaz/solar1.zip*

5.3.3.1. *Jupiter*

Before the arrival of the Juno probe, the generally accepted model of Jupiter's internal structure included a core of ice and rock, about ten Earth masses, with a pressure of about 4 10^{12} Pa and a temperature of 23,000 K at the interface between its core and the outer envelope, made of metallic hydrogen and helium, whose density decreases as the distance from the center increases. It is in this medium that the dynamo responsible for the Jovian magnetic field is generated. According to this model, the interface between the metallic hydrogen and the molecular hydrogen envelope would be located at 0.85 Jovian radii from the center.

Results from the Juno gravimeter suggest that there is no clear separation between the core (which would be about three Earth masses) and the metallic hydrogen layer (see Figure 5.15). The dilution of the core within the hydrogen layer could be the result of a collision between the planet and a protoplanet of about 15 Earth masses (Liu *et al.* 2019).

5.3.3.2. *Saturn*

As in the case of Jupiter, Saturn's models of internal structure involve a core of ice and rock, surmounted by an envelope of metallic hydrogen, with a helium enrichment that is probably more prominent than in Jupiter. The pressure at the interface between metallic hydrogen and molecular hydrogen would be 13 10^{11} Pa and the temperature would be about 12,000 K. The transition would intervene at a distance of 0.6 Saturnian radii from the center.

The Cassini spacecraft's repeated observations of Saturn's gravity field between 2005 and 2017 have provided new constraints to Saturn's internal structure models. However, a major source of uncertainty in the development of these models is related to the absence of a unique rotation period in the internal part, where hydrogen is metallic. While earlier measurements by the Voyager probes indicated a rotation period of 10 hours 39 minutes, new gravimetric measurements by the Cassini probe have led to a shorter period of 10 hours 32 minutes. This difference in the internal structure models is explained by a smaller core. According to current estimates, Saturn's core would have a mass between 5 and 20 Earth masses. Moreover, the mass of the core depends on the level of pressure at which the differentiation of helium takes place. If this pressure is greater than 4 10^{11} Pa, the mass of the core could be less than five Earth masses (Helled and Guillot 2013).

The combination of old astrometric data with Cassini measurements allowed us to determine its Love number k, a parameter measuring the reactivity of the interior with respect to tidal forces ($k = U_i/U_e$, U_i being the induced potential and U_e being the exciting potential, in other words, the centrifugal potential related to the rotation; $k = 0$ in the case of a rigid body). The value found implies significant dissipation that can be explained by viscous friction in a solid core (Lainey *et al.* 2017). Lastly, radio occultation measurements performed during the final orbits of the Cassini probe, between the planet and the closest rings, allowed us to determine the planet's gravitational field and the mass of the rings. The results imply a differential rotation of the atmosphere to a depth of at least 9,000 km. The total mass of the rings was estimated to be 0.4 times that of the Mimas satellite, implying that they would have formed 10 to 100 million years ago (Iess *et al.* 2019).

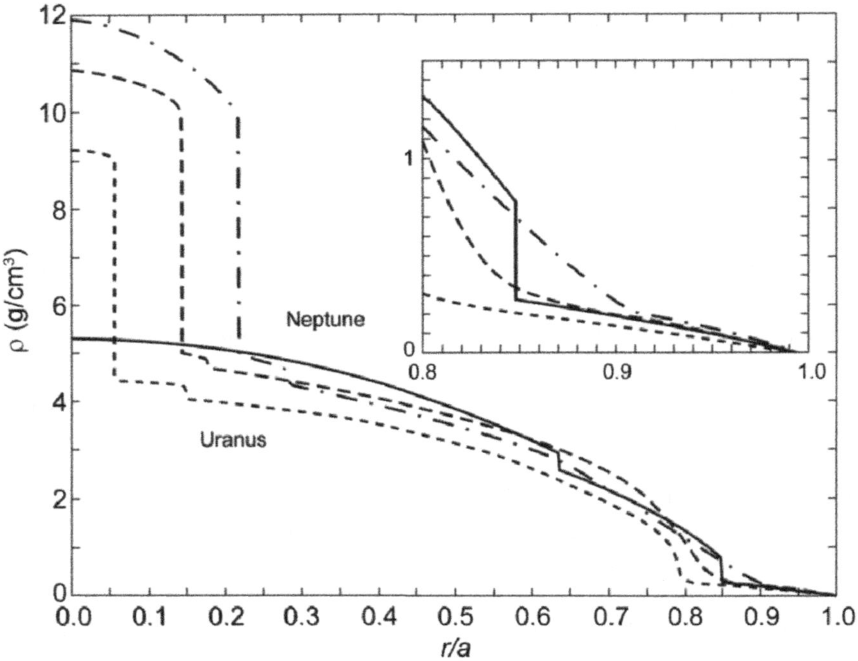

Figure 5.16. *Internal structure models of Uranus and Neptune. The only existing data are those of the Voyager 2 probe. For Neptune, three models are represented, one of them without a central core. The enlarged rectangle shows the transition between the hydrogen-rich region and the ice mantle (source: from (Marley 1999))*

5.3.3.3. *Uranus and Neptune*

Uranus and Neptune have different internal structures, and also present differences in the energy source (absent on Uranus) and dynamic activity (more prominent on Neptune). Gravimetry measurements made by the Voyager 2 probe have shown that, in the case of Uranus, the heavy elements are more concentrated, the central core being smaller than in Neptune. The pressure in the center would be around 6 10^{11} Pa, and the temperature about 3,000 K. The core would be covered by a layer of ice (or a gas-ice mixture) up to a distance of about 0.80–0.85 radii (see Figure 5.16). The temperature and pressure conditions there are such that the hydrogen has to be in molecular, non-metallic form. The magnetic field could originate in the ice-rich central part, in which the molecules, strongly dissociated and ionized, would render the medium conductive.

5.4. The magnetospheres of the giant planets

The magnetospheres of the giant planets (see section 3.3) have been explored *in situ* through the flybys of the Pioneer and Voyager probes in the 1970s and 1980s, and then studied in detail with the Galileo, Juno and Cassini orbital missions for Jupiter and Saturn in the following decades. Additional information has been obtained from distant remote observations of their low frequency radio emissions and optical aurorae.

The source of their magnetic field is attributed, as for the Earth, to a dynamo effect originating in the central, fluid and conductive part of the planet. The cores of Jupiter and Saturn are probably surrounded by a fluid layer of metallic hydrogen, animated by convective movements able to sustain the circulation of electric currents, in turn producing a large-scale magnetic field.

The dynamo effect can also develop in the outer, non-metallic conductive layers. Uranus and Neptune probably host ionic liquid layers. Their strongly non-dipolar magnetic field (their equivalent dipole that is highly offset and inclined with respect to the rotation axis) could result from very inhomogeneous internal convection. The internal magnetic field models of the giant planets are provided in Table 3.1.

The properties of the magnetospheres of the giant planets are controlled by the amplitude of their magnetic field (strong for Jupiter, modest for Saturn, Uranus and Neptune), its inclination (small for Jupiter and Saturn, large for Uranus and Neptune), the incident solar wind ram pressure, the rotation period of their magnetic field (~ 10–11 hours for Jupiter and Saturn, ~ 16–17 hours for Uranus and Neptune) and the existence of internal plasma sources (such as Io for Jupiter, Enceladus for Saturn or Triton for Neptune), which make all of them very specific cases. Our knowledge of these objects is covered exhaustively in books dedicated to Jupiter (Bagenal *et al.* 2007), Saturn (Dougherty *et al.* 2009), Uranus (Miner 1990) and Neptune (Cruikshank *et al.* 1995), and will only be briefly discussed hereafter.

5.4.1. *Jupiter's giant magnetosphere*

5.4.1.1. *The Jovian magnetic field*

Jupiter's magnetic field was discovered in 1955, by the ground-based observation of its auroral radio emission at 22.2 MHz (λ = 13.5 m, named "decametric", or DAM). This radiation, which was elliptically polarized, was early interpreted as being produced by the cyclotron motion of electrons in the Jovian magnetic field (see Figure 5.17), thus proving the existence of the latter and providing an estimate of its surface intensity: about 10 gauss. Thereafter, long-term measurements of the complete DAM spectrum (see Figure 3.39 and Figure 6.5, Volume 2) made it possible to derive a maximum frequency of ~ 40 MHz, corresponding to a maximum surface field of ~ 14 gauss (20 times the maximum value of the Earth's field), and to determine the period of internal rotation of Jupiter, $P = 9$ hours 55 minutes 29 seconds, with excellent accuracy.

As early as 1958, observations at higher frequencies yielded the detection of another, non-thermal, decimetric radiation (DIM, Figure 6.5 and Figure 6.6, Volume 2). Its linear polarization, spatial distribution and spectrum led to identify it as synchrotron emission of very energetic electrons (> 1–10 MeV) trapped in radiation belts similar to the terrestrial Van Allen belts, also discovered in 1958. The DIM polarization distribution indicated that the Jovian magnetic field has a roughly dipolar structure, with an inclination of about 10° with respect to the rotation axis.

The magnetic measurements obtained *in situ* during the flybys of Pioneer 10 and 11, then of Voyager 1 and 2, made it possible to build first models of the Jovian internal field (named D4, O4, O6). It was described by an equivalent magnetic dipole inclined by about 10° with respect to the rotation axis towards the 201° longitude, and shifted by about one tenth of one Jovian radius with respect to the center of the planet, along a direction perpendicular to the rotation axis. This asymmetry explains the different values of the magnetic field at the poles: 14.8 gauss at the north pole and 11.8 gauss at the south pole. At the equator, the surface field reaches about 4.2 gauss. The magnetic north is located in the rotational northern hemisphere, unlike the Earth's case. The Jovian magnetosphere also hosts an equatorial current disk, or *magnetodisk*, which radially stretches the magnetic field lines parallel to the equator. It has been explicitly modeled by a current ring with a thickness of 5 Jovian radii (R_J), extending between 5 and 50 R_J in the magnetic equatorial plane. The azimuthal current density there varies as $1/r$ and the total current through a meridional plane reaches ~ 300 MA. An optimal fit of the data is obtained by placing the magnetodisk along the centrifugal equator, between the rotational equator and the magnetic equator, at 3.2° from the latter.

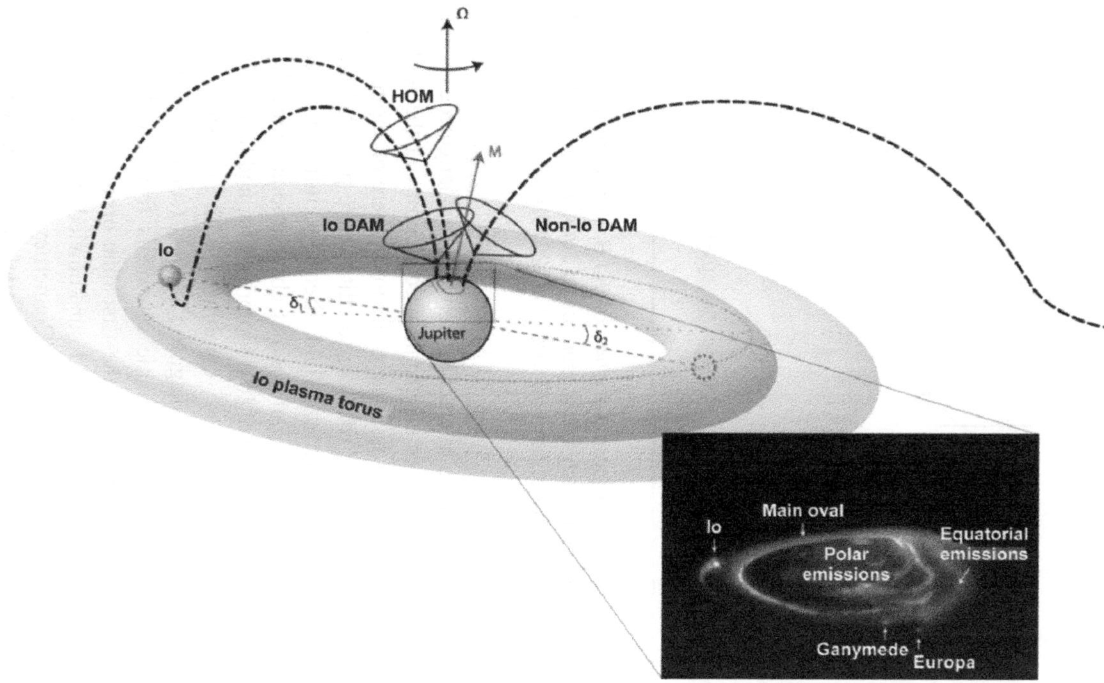

Figure 5.17. *On the left, sketch of Io's plasma torus and of the position of the decametric (DAM) and hectometric (HOM) auroral radio sources along the high-latitude Jovian magnetic field lines. On the right, image of the Jovian UV aurorae, as observed by the HST (source: from (Cecconi* et al. *2012; Clarke* et al. *2002))*

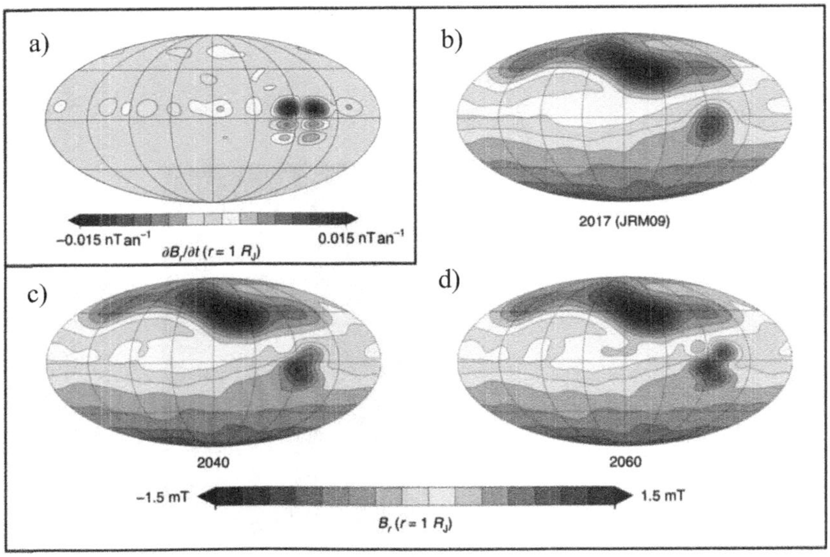

Figure 5.18. *(a) Projection of the temporal variation of the B_r radial component of the Jovian magnetic field and of its intensity at the surface of Jupiter in (b) 2017 (JRM09 model), (c) 2040 and (d) 2060 (source: from (Moore* et al. *2019))*

As we will see, the electrodynamic interaction between Io, Ganymede or Europa and the Jovian magnetic field produces electronic precipitations at the foot of the magnetic flux tubes connecting these satellites to the ionosphere of Jupiter. Optical observations of the auroral “footprints” of these satellites have provided additional constraints on the topology of the magnetic field. Precisely, infrared observations (carried out in the 1990s with IRTF) and ultraviolet observations (since the end of the 1990s with the HST, see Figure 5.17) of Io’s footprint first, and then of Europa’s and Ganymede’s footprints, have been combined with *in situ* magnetic measurements to build better constrained magnetic models (such as VIP4, VIT4, VIPAL and ISaAC). More recently, *in situ* measurements from the Juno probe, which were obtained near the planet during the first nine orbits, covering a wide range of longitudes, have significantly improved our knowledge of the Jovian field and resulted in a model at order 10 (JRM09) (Connerney and Jørgensen 2018). Figure 5.18b plots the B_r radial component of this model on the surface of Jupiter, which reveals the magnetic anomaly in the north. Further Juno measurements at missing longitudes will further improve the modeling of the Jovian field. Already, the comparison of these measurements with those acquired by Pioneer in 1973 has revealed a secular variation of the magnetic field (Moore *et al.* 2019), which indicates a temporal evolution of the structure of the inner conducting layers that generate it. The temporal variation of B_r at the surface is shown in Figure 5.18a, and its current and future intensity in Figures 5.18b to 5.18d. This was the first time that a secular variation of a planetary magnetic field was detected elsewhere than at Earth.

5.4.1.2. *Magnetosphere configuration and plasma sources*

The discovery of Jupiter's magnetic field before its *in situ* exploration led scientists to hypothesize the existence of a magnetosphere, produced by the equilibrium between the pressures of the planetary magnetic field and of the solar wind. In this regard, it was the first magnetosphere to have been discovered, and the term "magnetosphere" was formulated at this time by T. Gold, in 1959. Our knowledge of the Jovian magnetosphere is based on numerous observations: flybys by Pioneer and Voyager, but also by the Ulysses (high latitude), Cassini or New Horizons probes, as well as orbital exploration by Galileo from 1995 to 2002 (equatorial orbits), and then by Juno since 2016 (polar orbits). After the first characterization of the structure of the Jovian magnetosphere by Pioneer and Voyager, Ulysses determined the position of auroral radio sources and highlighted the strong compressibility of the magnetosphere. Galileo studied its dynamics and temporal variations. Cassini and New Horizons carried out the first exploration of its "evening" sector in local time and of the distant magnetotail. Lastly, Juno explored the near and polar regions in order to study the magnetic field and auroral processes in detail. These space-based observations were complemented by ground-based observations, using more sophisticated instrumentation and allowing higher data rates.

The *Jovian magnetosphere*, shown in Figure 5.19a, is structured into four main regions (Khurana *et al.* 2004). The *ionosphere* overlies the stratosphere, and is produced by the ionization of the upper atmosphere by solar UV and X radiations and the precipitation of energetic particles from the magnetosphere (see section 3.2.1.2). It extends to more than 3,000 km above the clouds. Analysis of Voyager 2 occultations indicated a maximal electron density of 3.5 10^5 cm^{-3} (in other words, an electron plasma frequency f_{pe} of ~ 5.3 MHz) at 1,600 km above the 1 bar pressure level, and an exponential decay at higher altitudes with a scale height of 900 km. Below the main peak, Pioneer measurements revealed low layers of high electron density (up to 10^5 cm^{-3}), possibly of meteoric origin and strongly absorbing in radio, preventing, for example, the radio detection of atmospheric lightnings observed in the visible range. *In situ* Juno measurements have identified the presence of cold (a few eV) ionospheric protons, up to 0.5 R_J altitude, along flux tubes whose apex ranges from 10–15 R_J (auroral oval) to 5.9 R_J (Io), where magnetospheric ions precipitating into the atmosphere coexist. This suggests that the latter contribute to the heating of the ionosphere and to the increase of its scale height at mid-latitudes (Valek *et al.* 2020).

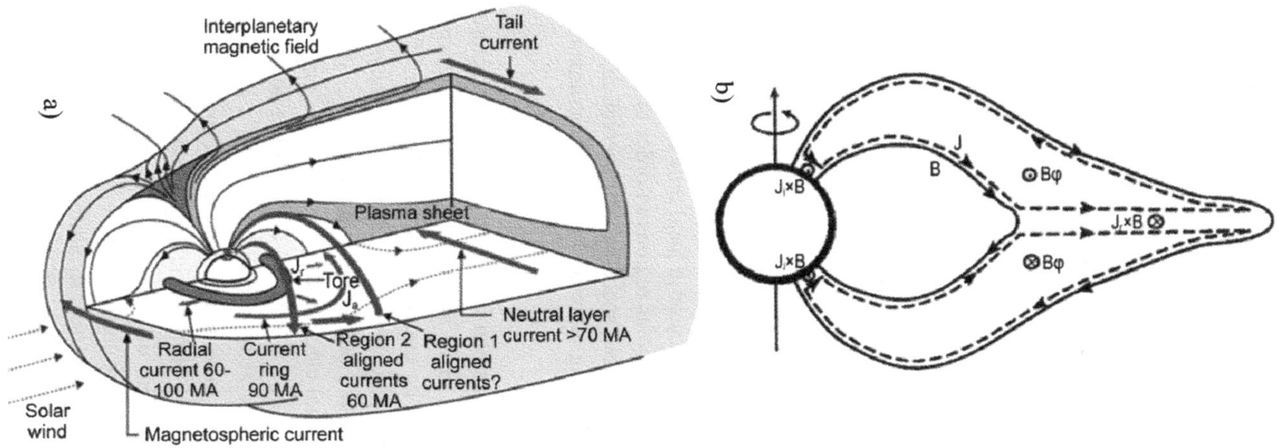

Figure 5.19. *(a) Sketch of Jupiter's magnetosphere and of the main electric currents. (b) Sketch of field-aligned current systems, ensuring the transfer of angular momentum between the ionosphere and the magnetosphere (dotted lines). The radial equatorial current J_r is associated with a Laplace force $J_r \times B$, which drives the equatorial plasma into corotation. It partially flows along the field lines (solid lines) associated with the auroral oval and then closes in the ionosphere, shown as Ji, where the $J_i \times B$ force opposes to corotation. The magnetic disturbance $B\varphi$ produced by this current warps the field lines out of the meridian plane (source: from (Khurana* et al. *2004; Cowley and Bunce 2001))*

The *inner magnetosphere* corresponds to apex flux tubes smaller than ~ 5 R_J. It is characterized by a quasi-dipolar field, an absence of electric currents and the presence of highly energetic particles belts (> 1–10 MeV) that are trapped in the magnetic field. These radiation belts are the source of the synchrotron DIM radiation, whose mapping directly probes the topology of the field and the populations of trapped particles.

The *middle magnetosphere* lies between ~ 5 and ~ 50 R_J and includes the Galilean satellites. The field there is strongly distorted by the azimuthal current J_a of the disk (see Figure 5.19a), produced by the drift motion of electrons and magnetospheric ions in opposite directions, which is necessary to balance the centrifugal acceleration of the plasma due to corotation by the $\boldsymbol{J_a} \times \boldsymbol{B}$ Laplace force. This equatorial current disk produces a radial magnetic field above and below it of a few tens of nT in amplitude, varying in $r^{-3/2}$. Its amplitude exceeds that of the internal field beyond ~ 15 R_J, and stretches the field lines radially. This disk is quasi-equatorial up to ~ 30 R_J, and gradually aligns with the solar wind flow beyond. Lastly, it undulates in the azimuthal direction, like the "ballerina skirt" of the solar wind (described in section 3.1.3), as the magnetic field perturbation, induced by the periodic rotation of the centrifugal equator with the planet, propagates at the finite Alfvén velocity: $V_A \sim 750$ km s^{-1} ~ 40 R_J h^{-1}.

Beyond ~ 50 R_J, the *outer magnetosphere* extends to the daytime magnetopause. It has a southward-facing, disordered, variable field of ~ 5–15 nT, affected by edge-on effects from the current disk of the middle magnetosphere. This region hosts bubbles of cooler, denser plasma, which probably come from the magnetodisk. Should the pressure of the Jovian internal field alone balance that of the solar wind (P_{SW}), the subsolar point of the magnetopause would stand at ~ 60 R_J from the planet and its position would vary in $P_{SW}^{-1/6}$. Instead, its real position lies at ~ 90 R_J and varies as $P_{SW}^{-1/4}$ or $P_{SW}^{-1/5}$, which shows a strong compressibility. On the night side, the magnetotail, with a diameter of 300–400 R_J, is formed by two lobes (north and south) of intense magnetic field, separated by a layer of equatorial current, which forms the extension of the magnetodisk, in which plasma bubbles (plasmoids) propagate in the antisolar direction. Its length reaches several astronomical units. Voyager 2 has shown *in situ* that it can even extend up to the orbit of Saturn, 5 au farther. Jupiter's magnetosphere thus stands as the largest astrophysical object in the sky.

A particularity of the Jovian magnetosphere is the presence of large internal plasma sources (Bolton *et al.* 2015), dominated by Io. The satellite's volcanism produces ≥ 1 ton of material per second (mostly SO_2, as well as dust and Na atoms) that, after partial ionization by solar UV radiation and collisions with energetic particles, produces ~ 3 10^{28} ions s^{-1} (including S^+, O^+, and as many electrons), or ~ 250 kg s^{-1}. Europa also produces ~ 10^{27} ions s^{-1}, that is to say ~ 25 kg s^{-1}, by sputtering of its surface water vapor by energetic magnetospheric particles and by volcanism. The magnetospheric light ions are sustained, from the interior, by the ionosphere (H^+, H_2^+, H_3^+), and, from the exterior, by the solar wind (H^+ and He^{++}) at a rate of ~ 10^{28} ions s^{-1}, that is to say, a few tens of kg s^{-1}. The matter ejected by Io at $R = 5.9$ R_J has a Keplerian orbital velocity V_K = 18 km s^{-1}. As soon as they are created, the S^+ and O^+ ions are "frozen in" with the magnetic field and are carried away by the rotation at $|\boldsymbol{\Omega} \times \boldsymbol{R}|$ = 74 km s^{-1} (the so-called pick-up phenomenon). They thus sustain a plasma torus about 1 R_J thick around Io's orbit. The gyration energy of the newly injected ions around the local magnetic field corresponds to a gain of δV_{Io} = 56 km s^{-1}, or 250 to 500 eV. This gain is only ~ 0.02 eV for the electrons. These will be further energized by the Coulomb collisions which will transfer part of the energy of the ions to the electrons. Europa also supplies a plasma torus at R = 9,4 R_J of thickness similar to that of Io, although it is much less dense.

5.4.1.3. *Magnetospheric dynamics and plasma sinks*

The result of the density of neutrals being very low in the Jovian magnetosphere is that the plasma is non-collisional and therefore very conductive. It is therefore permanently "frozen in" with the planetary magnetic field lines. The only existing electric field comes from the motion of the plasma and is expressed by the equation [3.9]: $\boldsymbol{E} = -\boldsymbol{V} \times \boldsymbol{B}$, which implies that the magnetic field lines are also electric equipotentials with $\boldsymbol{E}.\boldsymbol{B} = 0$.

5.4.1.3.1. Corotation of the plasma

Section 3.3.3.3 has already described how the circulation of magnetospheric plasma is governed by the combination of solar wind induced convection and the rotation of the planetary magnetic field. The convective electric field is written as:

$$\boldsymbol{E}_{\mathbf{conv}} = -\eta \boldsymbol{V}_{SW} \times \boldsymbol{B}_{SW} \quad [5.7]$$

η being the reconnection efficiency at the magnetopause (~ 0.1 for Earth) and B_{SW} being the interplanetary magnetic field strength. With B_{SW} ~ 1 nT at 5 au for Jupiter, we obtain E_{conv} ~ 4 10^{-4} V m^{-1} and the potential difference across the magnetopause is $\Delta V_{conv} = E_{conv} \times 2R_{MP}$ ~ 5 MV with R_{MP} ~ 90 R_J. The corotation electric field, radially oriented, is written as:

$$E_{corot} = |-(\boldsymbol{\Omega} \times \boldsymbol{R}) \times \boldsymbol{B}| = \Omega R B_0/r^3 = \Omega B_0 R_J/r^2 \quad [5.8]$$

where B_0 is the equatorial Jovian field strength at 1 R_J (equation [3.27]), Ω is the angular frequency of Jupiter's rotation and $R = rR_J$. The potential difference across the magnetopause due to corotation is thus: $\Delta V_{corot} = \int E_{corot} dR = \int_{1 \to \infty} (\Omega B_0 R_J / r^2)\, R_J dr = \Omega B_0 R_J^2 \sim 400$ MV. The boundary radius r_{lim} separating convection-dominated ($r > r_{lim}$) or corotation-dominated ($r < r_{lim}$) regions can be estimated by solving the equality $E_{conv} = E_{corot}$. Using the $B_{SW} \sim B_0/r_{MP}^3$ approximation for the interplanetary magnetic field, we obtain:

$$r_{lim} / r_{MP} = (\Omega R_J r_{MP} / \eta V_{SW})^{1/2} \qquad [5.9]$$

The region dominated by corotation grows with r_{MP} (therefore, with the planetary field strength) and with the rotation frequency Ω. For Jupiter, which has a strong field and fast rotation, we find $r_{lim}/r_{MP} > 1$. The magnetosphere is very largely dominated by corotation, except for the outermost regions and the magnetotail.

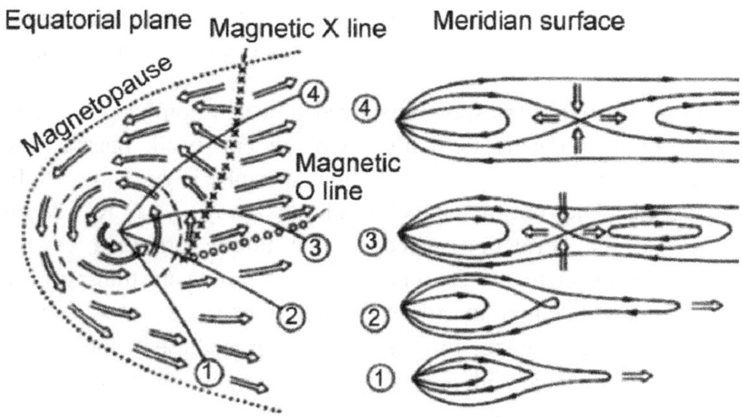

Figure 5.20. *Circulation of magnetospheric plasma (double arrows) and magnetic field lines (solid lines) in the Vasyliunas model. The neutral X line delineates the inner region, where the plasma is driven in corotation, from the outer region, where it is evacuated in the tail in the form of plasmoids (source: from (Vasyliunas 1983))*

The azimuthal component B_φ of the Jovian field created by the plasma corotation results from the appearance of a radial outward current J_r, weak with respect to J_a, which closes into the ionosphere along the magnetic field lines (Figure 5.19a and b). Near the equator, the Laplace force $\boldsymbol{J}_r \times \boldsymbol{B}$, with $\boldsymbol{B}$ facing south accelerates the magnetospheric plasma toward corotation. In the ionosphere, the $\boldsymbol{J}_i \times \boldsymbol{B}_i = - \boldsymbol{J}_r \times \boldsymbol{B}$ force slows down the ionospheric plasma, which continues to be dragged by friction with the neutral atmosphere, where field and plasma are not frozen in anymore because of collisions. There is thus a transfer of momentum from the rotating planetary ionosphere to the magnetospheric plasma. The transport of Jovian magnetospheric plasma is described by the Vasyliunas model (Vasyliunas 1983) (depicted in Figure 5.20), which differs from the Dungey model for the Earth. Here, the rapid rotation governs the accumulation of plasma in the outer magnetosphere and its episodic release into the tail in the form of plasmoids, evacuated on the night side from a midnight-morning neutral line. Inside this line, the reconnected field lines are re-dipolarized and driven towards the morning side.

5.4.1.3.2. Radial transport and deviation from corotation

For Jupiter, the centrifugal force becomes predominant over the gravitational force beyond ~ 2.3 R_J. The plasma of the Io torus therefore undergoes a fast radial transport towards the exterior, up to the internal part of the magnetodisk, via the exchange of flux tubes (interchange instability). The "heavy" tubes, which are dense in cold plasma, "fall" towards the exterior and are replaced by "light" tubes populated by tenuous hot plasma, coming from the external magnetosphere. A fraction of the plasma is also driven towards the exterior by charge exchange.

Rapid outward plasma transport is slowed down by ion-neutral friction at the ionospheric foot of magnetic field lines. On the one hand, plasma injection at Keplerian speed at 5.9 R_J slows down the magnetospheric flow to ~5% below the corotation speed. The plasma travels radially away ~ 2 R_J, before being re-accelerated to the corotation speed by the torque provided by ion-neutral collisions in the ionosphere. On the other hand, radial outward transport requires an increase in plasma momentum in order to remain in corotation at $\boldsymbol{V}_\perp = \boldsymbol{\Omega} \times \boldsymbol{R}$. This drive is provided by the Laplace force $\boldsymbol{J}_r \times \boldsymbol{B}$, *via* an increase of the current up to a limit value $J_{r\,max}$, associated with $J_{i\,max}$, beyond which the ionosphere can no longer provide the momentum required to accelerate the magnetospheric plasma to the corotation speed, causing the magnetic field to warp beyond 20 R_J. These "field-aligned" currents have been early proposed as the origin of energetic particles (particularly electrons) precipitating along auroral flux tubes down to the main oval (Cowley and Bunce 2001; Hill 2001).

5.4.1.3.3. Radiation belts and other plasma sinks

Let us return to the inner magnetosphere and the synchrotron DIM emission of its highly energetic particle belts. The theory of synchrotron radiation is well known: it consists of the emission of electrons of energy E that are in a gyration motion in a magnetic field B, exciting all harmonics of the cyclotron frequency. Dispersion effects in energy and in angle of attack $\theta = (\boldsymbol{v},\boldsymbol{B})$ broaden these lines, which produce a continuous spectrum, whose peak frequency is proportional to $E^2B\sin\theta$. This is a non-thermal, but incoherent emission (unlike DAM), whose total power, proportional to $E^2B^2\sin\theta$, is simply the product of the sum of intensities emitted by each electron, resulting in brightness temperatures of "only" 10^3 to 10^6 K (see

Figure 6.5, Volume 2). This radiation has been observed regularly since its discovery in 1958, with 2D and 3D maps of the spatial distribution of the radiation between 20 and 240 cm wavelength (corresponding to E = 1-30 MeV), from ground-based radio interferometers, the Cassini probe telemetry antenna, and then the Juno probe radiometer down up to 1 cm (E > 40 MeV). The source of the emission is a torus encircling the planet, whose mean plane is 10° to the equator and whose large diameter is ~ 4 R_J (see Figure 6.6, Volume 2). The emitted polarization is almost completely linear, the electric field vector being in the instantaneous gyration plane of the electron: at the equator, it is parallel to the central plane of the torus. The emission is essentially constant with time, apart from an oscillation linked to the rotation of the planet, and the intensification, of magnetospheric origin, which followed the collision of comet SL9 with Jupiter. Thanks to numerical codes modeling the synchrotron emission by taking into account all the physical processes affecting the Jovian radiation belts, it has been possible to compare simulated and observed radio emission maps. These results showed that the main source of energetic particles in the belts is the radial diffusion of energetic plasma from the current disk, while loss processes include absorption by satellites and rings, and angular diffusion that can lead to particle precipitation in the mid-latitude atmosphere, as well as synchrotron losses themselves. Satellites thus form both sources and sinks of magnetospheric plasma. Initial comparisons between models and Juno data show that the observed amplitude of the DIM radiation still differs significantly from simulations, which is attributed to the unconstrained physical conditions prevailing in the inner magnetosphere (Santos-Costa *et al.* 2017).

5.4.1.4. *Plasma energization process*

5.4.1.4.1. Energy plasma and acceleration mechanisms

In addition to the cold and energetic plasma populations discussed above, the Jovian magnetosphere contains a rarefied population (N ~ 10^{-2}–10^{-3} cm^{-3}) of much more energetic H^+, S^{n+} and O^{n+} ions and electrons. Their characteristic energy increases from ~ 10 keV in the outer magnetosphere to more than 2 MeV at ~ 7 R_J, while their density reaches ~ 1 cm^{-3} in the Io torus. Density and energy then rapidly decrease towards Jupiter. This energetic plasma forms the external part of the current disk. Its composition indicates a source internal to the magnetosphere, and its distribution reveals a radial transport towards the interior. The origin of this slow radial diffusion is to be found in the random fluctuations of the ambient electric field, linked to turbulent motions and winds in the thermosphere, to a pressure gradient towards the interior, or to a large-scale convection caused by compressions of the magnetosphere under the effect of the solar wind ("magnetospheric storms"). The energy density and pressure of this energetic plasma dominate those of the rest of the magnetospheric plasma, except in the inner regions of the torus. This pressure is comparable to that of the magnetic field in the middle magnetosphere, and explains the subsolar distance of the magnetopause at 60–100 R_J, as well as its strong compressibility, which we described above.

The heating and acceleration mechanisms of the plasma to energies far above that of corotation (~ 1 keV at maximum) remain poorly understood. During the inward transport, the plasma encounters an increasing magnetic field and the flux tubes compress. Considering this slow transport as adiabatic, it causes a (1) Betatron and (2) Fermi type of plasma heating:

1) the former comes from the conservation of the first adiabatic invariant $\mu = mv_\perp^2/2B$ (magnetic moment), which implies that $v_\perp^2$ grows with $B \alpha R^{-3}$;

2) the latter comes from the conservation of the second adiabatic invariant $J = \int mv_{//}dl$ (bounce integral) along a field line, between the mirror points of the charged particle. The increase in $v_{//}$ is inversely proportional to the length of the field line l: $v_{//}^2 \alpha l^{-2} \alpha R^{-2}$.

Perpendicular acceleration dominates, and if a diffusion angle-of-attack of the particles redistributes the heating isotropically, then the total energy gain reaches $E \alpha v^2 \alpha R^{-8/3}$. These adiabatic mechanisms allow energies of a few MeV per particle to be reached in a few tens of hours. These processes are probably at the origin of the particles populating the radiation belts of the inner magnetosphere.

Much faster, non-adiabatic processes are required to accelerate the particles of a few keV to a few tens of keV that precipitate in the auroral zones. The field-aligned current system associated with the main auroral oval, briefly discussed above (corotation breakup from 20 R_J), is an example of this. Another one concerns the interaction between Jupiter and its Galilean satellites.

5.4.1.4.2. Interaction between Jupiter and the Galilean moons

The electrodynamic interaction between the Io satellite and Jupiter's magnetic field was discovered in 1964, by the observation of recurrent radio DAM bursts corresponding to certain preferential phases of the satellite with respect to the Jupiter-observer direction. This interaction forms a dynamo, capable of accelerating particles along the magnetic flux tube linked to Io. Two models of this interaction have been proposed (Saur *et al.* 2004).

The first model, termed "the unipolar inductor", considers that the corotating electric field in Io's frame of reference: $\boldsymbol{E}_{corot} \sim \delta \boldsymbol{V}_{Io} \times \boldsymbol{B} \sim 0.1$ V m^{-1} applied to the conductive ionosphere of the satellite induces a current which quasi-instantaneously closes into the planetary ionosphere via field-aligned currents. The second model, that of "Alfvén wings", considers Io's ionosphere as an almost impenetrable obstacle for the incident Jovian field lines. These are therefore stacked upstream, while the foot of these field lines, anchored in the ionosphere, remain in corotation. The magnetic field is stretched and distorted, then bypasses the obstacle to catch up with the corotation. This flow, being subsonic and sub-Alfvénic, causes no shock upstream of the obstacle. The perturbation of the Jovian field induced by Io then propagates towards Jupiter along the magnetic field, in the form of Alfvén waves at the following velocity: $V_A = B/(\rho\mu_o)^{1/2}$. In the frame of reference related to Io, the field lines catch up with the satellite with a relative speed of 56 km s^{-1} at the equator, and the magnetic perturbation propagates along two northern and southern "Alfvén wings" tilted downstream, along the direction of the magnetospheric plasma flow, as shown in Figure 5.21.

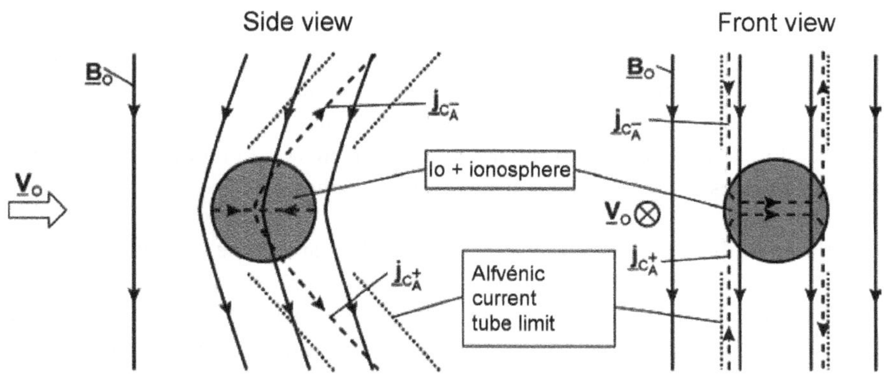

Figure 5.21. *"Alfvén wings" model of the Io-Jupiter interaction*

COMMENT ON FIGURE 5.21. – *The magnetic field and current lines are displayed by solid and dashed lines, respectively. The currents flowing in Io's ionosphere propagate along the "Alfvén wings", whose inclination, with respect to the direction perpendicular to the Jovian field, depends on the Alfvén velocity. The Alfvén current tubes thus have a component that is not parallel to the field lines, and close away from Io through the magnetospheric plasma or Jupiter's ionosphere. Furthermore, the Alfvén waves are associated with electric fields that are parallel to the magnetic field, capable of accelerating electrons up to a few tens of keV (source: from (Saur* et al. *2004)).*

The first model corresponds to an infinite conductivity of the obstacle, the second to a finite conductivity. The complex reality (very high, but finite conductivity) probably corresponds to a hybrid interaction between these two ideal models. Based on Voyager magnetic measurements, the current flowing between Io and Jupiter has been estimated to 1–3 MA. Combined with the total potential difference across Io: $\Delta V = 2R_{Io}E_{corot} \sim 400$ kV, it involves a power dissipation along the current circuit of ~ 10^{12} W per hemisphere, which allows for electron acceleration and the production of intense auroral emissions (Hess 2011).

These emissions have been detected as UV and IR spots (see Figures 5.17 and 3.37) in the vicinity of the foot of the magnetic field tube disturbed by Io, at ~ 65° north and south latitude. They correspond to emitted powers of ~ 10^{10} W, and require precipitation of some 10^{11} W. These intense precipitations are concentrated on a surface of ~ 5 10^4 km^2, that is to say, ~ 10 W/m^2 (20% of the incident solar flux), and produce violent local atmospheric perturbations. Above these spots, the precipitating electrons produce DAM radiation, whose emitted power reaches ~ $10^{9\text{-}10}$ W. The very particular morphology of these radio emissions, in the form of arcs in the time-frequency plane (see Figure 3.39), results from the combination of the observation geometry with the anisotropy of the radio emission. The UV and IR spots, as well as the radio arcs induced by Io, are sometimes multiple and extended by a wide trail of several tens of degrees in longitude. Those phenomena are attributed to the dense plasma wake downstream of Io, and to multiple reflections of the Alfvén waves in the torus and on Jupiter's ionosphere.

Although the Io-Jupiter couple forms the archetype of planet-satellite interactions, HST observations in the 2000s also revealed UV auroral footprints associated with Europa and Ganymede (see Figure 5.17), while the analysis of Voyager, Cassini and Nançay Decameter Array observations in the 2010s confirmed the existence of associated decametric emissions. Europa and Ganymede orbit at larger distances (9.6 and 15 R_J), in a weaker magnetic field and in a more tenuous plasma than Io, corresponding to lower E_{corot} values and dissipated power. The case of Ganymede is particular. Galileo indeed discovered an internal magnetic field, which generates a magnetosphere inside the Jovian magnetosphere (Kivelson *et al.* 1997). Given that the incident Jovian flow is sub-Alfvénic and subsonic, there is no shock or tail formation, and its magnetosphere differs from the other planetary magnetospheres described in section 3.3. Ganymede's field lines at the equator are permanently oriented in the opposite direction to those of the ambient Jovian field, and the magnetic reconnection upstream of the satellite governs the dynamics of its magnetosphere. The detailed exploration of the environments of Europa and Ganymede, as well as their interaction with Jupiter, is the subject of the NASA Europa Clipper and ESA Juice missions that are currently under development.

5.4.1.5. *Auroral electromagnetic emissions*

Generally speaking, Jupiter's upper atmosphere forms a screen that reveals a wide variety of magnetospheric acceleration processes, through the auroral electromagnetic emissions that they drive (see section 3.3.4) (Badman *et al.* 2015;

Grodent 2015). In addition to those related to the moons, auroral emissions are observed at high latitudes ≥ 70°, mainly along and within a circumpolar oval. They highlight electron (and ionic) precipitation from a fraction of a keV to several hundred keV, observed *in situ* by Galileo, and especially by Juno, in both hemispheres between ~ 7 and ~ 40 R_J. The total precipitated power is estimated to be ~ $10^{13\text{-}14}$ W, or 0.1–1 W/m^2 in the polar atmosphere. Jovian auroral emissions account for >20% of the incident power and cover the entire spectrum, from X-rays to low frequency radio waves. In the atmosphere, the UV aurorae have been observed by IUE, HST, Hisaki, and their on-board UV imaging spectrometers, in the Lyman-α line of atomic hydrogen, and the Lyman and Werner bands of H_2, mostly between 120 and 170 nm (see Figure 5.17). The visible emissions of atomic hydrogen were detected with Galileo (see Figure 3.34). The IR emissions, observed with ground-based telescopes and Juno's imaging spectrometer (see Figure 3.37), are due to the heating of the upper atmosphere by currents (Joule effect), as well as to certain transitions of hydrocarbons (CH_4, C_2H_6, C_2H_2) and especially of the H_3^+ ion. Lastly, the soft (Jupiter-specific) and hard X-ray aurorae, respectively produced by energetic electron braking radiation and by charge exchange (see Figure 3.38), have been observed with Chandra and XMM-Newton.

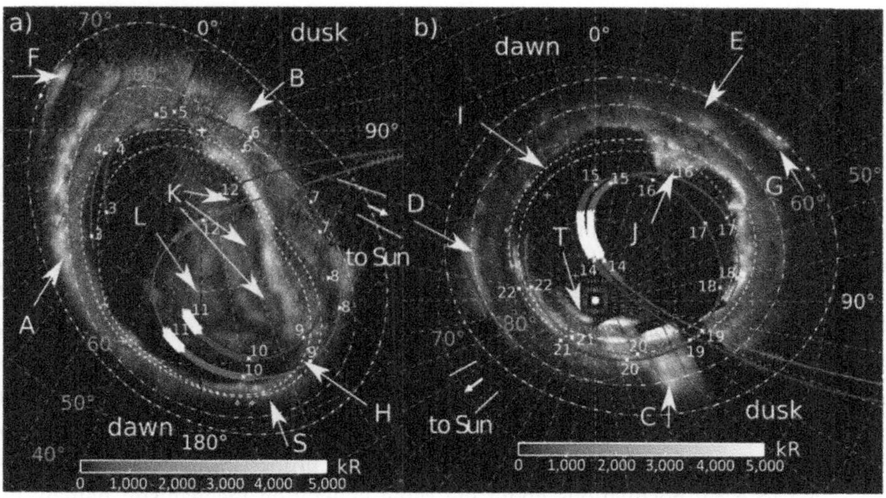

a) b)

Figure 5.22. *Jupiter's UV aurorae (a) in the north and (b) in the south, as observed by Juno in the vicinity of its first perijove, reveal a wide variety of intense and diffuse emissions. Juno's magnetic footprint, plotted in red or green with different magnetic field models, crosses several auroral components (source: from (Bonfond* et al. *2017)). For a color version of this figure, see www.iste.co.uk/encrenaz/solar1.zip*

The auroral oval is the main atmospheric emission region. It forms a quasi-continuous arc at higher (and, exceptionally, lower) latitude than the Ganymede footprint, which is weakly sensitive to the solar wind, with a curved shape toward the north due to the northern magnetic anomaly (Pallier and Prangé 2001). It has been observed at all wavelengths and is associated with the upward region of the current system introduced above, accelerating electrons towards Jupiter, a region observed to be located between ~ 15–20 and ~ 50 R_J. The UV and IR emissions correspond to a total emitted power of ~ 10^{12-13} W, while the visible and X-ray aurorae are much weaker. The polar emissions, inside the main oval, come from the polar cap. They have been observed as pulsating UV, IR and soft X-ray emissions (0.1–1 keV), produced by charge exchange of heavy ions such as $O^{6+,7+}$ (see Figure 3.38). The origin of these precipitations, associated with the solar wind or with a return current directed towards Jupiter, remains poorly understood. A variety of other auroral emissions, such as injections from the middle magnetosphere at lower latitudes than the auroral oval, have been successively identified with the improvement of instrumental sensitivity and resolution. Early Juno observations of the aurorae (see Figure 5.22 and Figure 3.37), and associated particles, have revealed a degree of complexity far beyond the general ideas given here (Bonfond *et al.* 2017).

Jovian auroral radio emissions, produced above the atmosphere, cover the kilometric to decametric wavelength range (see Figure 3.39). They are directly produced by 1–10 keV electrons, precipitating at high latitudes along field lines connected to optical aurorae along the main oval, or associated with satellite footprints (as mentioned above). The instantaneous sources appear to be small-sized and fluctuating, as in the case of the Auroral Kilometric Radiation (AKR) at Earth. The Jovian auroral radio emissions, like those of the other magnetized planets, are also strongly anisotropic, elliptically polarized and sporadic (bursts of variable duration, from <1 s to a few minutes). Their total emitted power exceeds 10^{10} W. Burst episodes last from a few minutes to a few hours, and are synchronous with the rotation of the planet. The precise rotation period of the Jovian magnetic field (or inner rotation rate) was measured from long-term DAM observations, making it possible to define a system of fixed longitudes with respect to the internal layers of the planet (system III, 1965.0). These macroscopic properties, similar to those of AKR, led to early postulations about the same generation mechanism: namely the cyclotron maser instability (CMI), which allows very efficient amplification of radio waves from non-equilibrium electron distributions. Juno's first *in situ* observations in the Jovian radio sources have validated the CMI and the conditions under which it operates in Jupiter (Louarn *et al.* 2017), with implications for the search for radio exoplanets (Zarka 2007). Jupiter produces other low-frequency radio emissions, caused by local instabilities, such as the ~ 40-minute quasi-periodic bursts (correlated to UV/X aurorae) that probe the interaction with the solar wind at the magnetopause, or the narrow-band kilometric emissions produced at the boundaries of the plasma torus.

Numerous questions about the mechanisms driving transfer and dissipation of energy and momentum in the polar regions remain. For example, the solar wind influences most of the auroral emissions, whose intensity variations appear to be fairly well correlated with those of the interplanetary dynamic pressure. However, theory predicts that the brightness of the main auroral oval should be quite anticorrelated to the dynamic pressure of the solar wind, because a decrease of this pressure yields a larger magnetosphere, producing an increased radial outward transport, and consequently, an intensification of the field-aligned currents at the edge of the corotation zone. The ongoing analysis of Juno's unique observations in the auroral regions should lead to significant progress in understanding the microphysics of these auroral processes.

5.4.2. *Saturn's symmetrical magnetosphere*

5.4.2.1. *Kronian magnetic field*

Unlike Jupiter, Saturn's magnetosphere was discovered *in situ* by the Pioneer 11 magnetometer in 1979 (the probe did not carry a radio instrument), and then explored by Voyager 1 and 2 (flybys in 1980 and 1981), and particularly by the Cassini orbital mission (from 2004 to 2017). These data have been completed by remote observations of Saturn's auroral radio emissions from the Ulysses spacecraft between 1990–2000, and of its optical aurorae from Earth-based telescopes.

In situ measurements obtained during the flybys revealed a magnetic field that was quasi-aligned with the rotation axis, equivalent to a very slightly eccentric dipole, ~ 4% north, giving rise to the first axisymmetric models corresponding to a centered dipole of 0.215 $G.R_S^3$ (named P11-84, Z3, SPV). *In situ* Cassini measurements, obtained during the final near-polar orbits in 2017, allowed the construction of an internal field model of order 11 aligned with its rotation axis at less than 0.06° (Cassini 11 model) (Dougherty *et al.* 2018). This unique Solar System characteristic should result in a weak rotational modulation of the magnetosphere, but, as discussed below, observations of Saturn's Kilometric Radiation (SKR) by the Voyager probes revealed a strong modulation, which was used to determine Saturn's internal rotation period, P = 10 hours 39.4 minutes, and to construct an associated longitude III system.

Like Jupiter, Saturn is surrounded by an equatorial current ring with a thickness of ~ 5 Kronian radii (R_S), extending from ~ 8 to ~ 16 R_S. The azimuthal current density there varies by 2.9 $10^6/r$ (A/R_S), and the total integrated current crossing the meridian plane is ~ 10 MA, corresponding to a ~ 10 nT magnetic contribution radially stretching the magnetic field near the equator.

5.4.2.2. *Characteristics of the Kronian magnetosphere*

Saturn's magnetosphere displays similarities with both magnetospheres of the Earth and of Jupiter. It shares a modest equatorial surface field (B_e = 0.215 G) with the Earth, as well as a comparable relative subsolar distance from the magnetopause (16–22 R_s), a moderate role of the plasma pressure (the magnetopause would be at ~ 14 R_s as a result of the magnetic field pressure alone) and a strong correlation of magnetospheric processes with the solar wind pressure. Like Jupiter, Saturn has a fast rotation inducing, in principle, a prominent corotation which determines the general structure of the magnetosphere.

Saturn's magnetosphere is also unique in several ways. In addition to its axisymmetric character, it has extremely diverse internal plasma sources and an associated neutral/plasma density ratio far greater than at Earth and Jupiter (Blanc *et al.* 2015). The main source of magnetospheric plasma is formed by the geysers of Enceladus, discovered by the Cassini mission, which produce ~ 10^{28} H_2O molecules every second by cryovolcanism. The ionization of these molecules by impact sustains the magnetosphere with plasma at a ~ 100 kg s^{-1} mass loading rate.

Secondary sources of plasma include Titan (H^+ and N^+ ions injected at 20 R_S) and the torus of neutral hydrogen and nitrogen (probably escaped from Titan) that dominates the outer magnetosphere; the icy satellites, whose surface properties are modified by sputtering by magnetospheric energetic particles; the rings (water ions); the ionosphere (H^+ and H_3^+ ions), probed *in situ* by Cassini, with a variable maximum electron density in latitude (>10^4 cm^{-3} at the equator at low altitudes, and <10^2 cm^{-3} at higher latitudes and altitudes) and in local time (with greater density, 10^3 cm^{-3}, on the daytime side), and a complex structure with several height scales and a layer interacting with the rings between 2,500 and 4,500 km altitude (Persoon *et al.* 2019); and, of course, the solar wind, whose precise contribution remains poorly understood. Saturn also has radiation belts, the innermost part of which, an energetic proton belt (> 25 MeV) inside the rings, was also discovered *in situ* by Cassini. These sources, which are much weaker than Io in the Jovian magnetosphere, are, for the most part, sinks (absorption, diffusion, etc.) for the magnetospheric plasma.

The Kronian magnetospheric structure, shown in Figure 5.23, is quite similar to that of Jupiter. As Enceladus is located beyond the synchronous orbit, the inner plasma torus that the moon generates flares out radially to sustain a current disk in the middle and outer magnetosphere.

The interaction between periodic injections of energetic plasma from the outer magnetosphere and the neutral tori has been highlighted with the observation of energetic neutral atoms by Cassini (an example of which is given by the bottom insert of Figure 5.23). Between 12 and 15 R_S, slightly below the limit between open and closed field lines, there is a system of field-aligned currents giving rise to intense aurorae (top insert of Figure 5.23). This system of currents is partly produced by the solar wind/magnetosphere interaction, and partly produced by corotation (see below). Lastly, in the outer magnetosphere, like for Jupiter, plasmoids detached from the current disk propagate along the magnetotail, which has a diameter of 80–100 R_S and probably extends over several au.

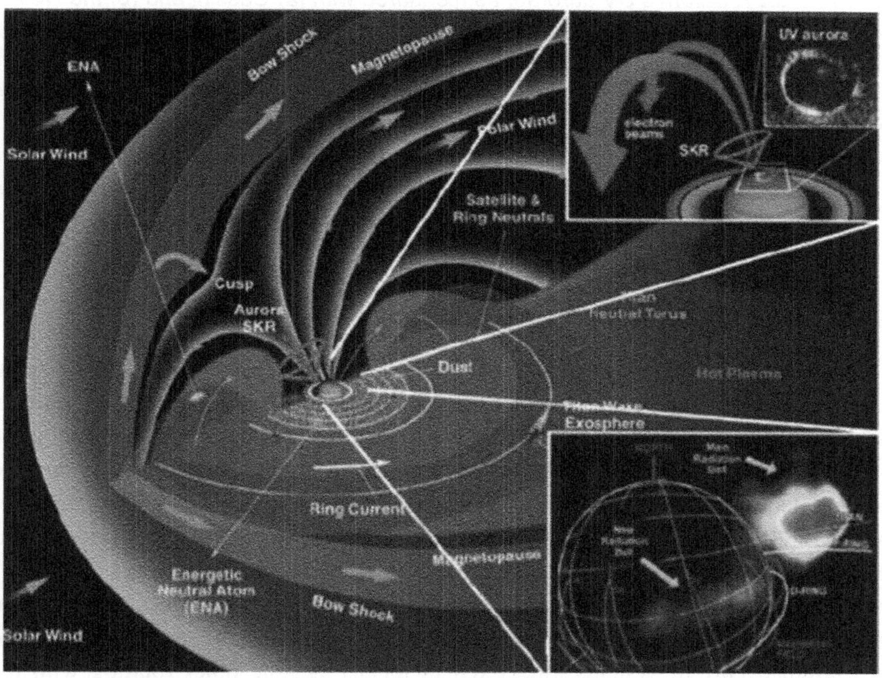

Figure 5.23. *Schematic diagram of Saturn's magnetosphere; the inserts illustrate the UV and radio auroral emissions and the equatorial neutral torus (source: NASA, ESA). For a color version of this figure, see www.iste.co.uk/encrenaz/solar1.zip*

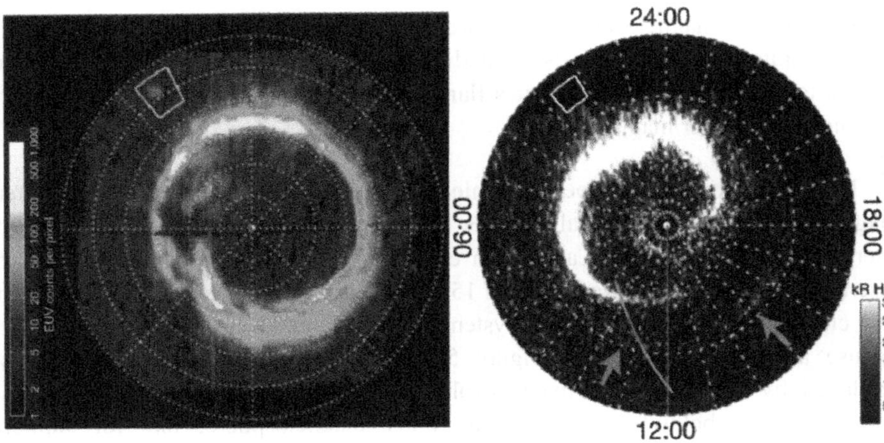

a) b)

Figure 5.24. *Saturn's UV aurorae, as measured with (a) Cassini and (b) the HST*

COMMENT ON FIGURE 5.24. – *The Cassini image shows the main oval and one (out of three) case of the Enceladus auroral footprint (within the white box). The HST image shows an intense, widened, auroral oval following an interplanetary shock and a weaker secondary oval, at lower latitude (green arrows), on field lines that have been crossed by the Cassini probe (magnetic footprint in gray/red) (source: from (Pryor* et al. *2011; Lamy* et al. *2018a)).*

5.4.2.3. *Auroral electromagnetic emissions*

The Kronian auroral emissions have a comparable, although simpler, morphology than their Jovian counterpart (Badman *et al.* 2015). The UV aurorae, detected by Voyager and then observed regularly by HST and Cassini (see Figure 5.23), are highly variable and can radiate up to ~100 GW, corresponding to precipitated power 10 times larger. They are roughly distributed into three categories (Grodent 2015). A circumpolar oval, at ~ 70° latitude and a few degrees wide, forms the dominant emission. It is associated with electron precipitation from 1 to 20 keV. It is generally more intense on the morning side, and is subject to rotational modulation. These two Kronian specificities are explained by two superimposed field-aligned current systems: one fixed in local time, the other rotating with the planet. These were measured *in situ* during the Cassini high latitude passes. Following sudden magnetospheric compressions induced by the solar wind, the oval intensifies on the night/morning side at very high latitude, before degenerating into a spiral pattern to return to its standard quasi-circular state in a few rotations (see Figure 5.24b). On the noon side, along the oval or at higher latitudes, Earth-like polar cusp emissions appear sporadically. They can radiate ~ 10 GW and are signatures of dayside intermittent magnetic reconnection between the planetary and interplanetary magnetic fields. At lower latitudes, between 60° and 70°, stands a diffuse and weaker secondary oval, attributed to the precipitation into the atmosphere of ambient hot plasma of the middle magnetosphere (see Figure 5.24b). Lastly, Cassini revealed the existence of an Enceladus-Saturn electrodynamic interaction, comparable to the Io-Jupiter case, with a UV auroral footprint at the foot of Enceladus' flux tube (see Figure 5.24a). The low intensity of the footprint reveals a weaker coupling than for Io, which is in agreement with predictions (Hess 2011), while the observation of only 3 UV signatures so far over 13 years of observations reveals a weak and very intermittent interaction. More sporadic observations of the visible Kronian aurorae (see Figure 3.34) have allowed us to study the fine spatial distribution of electron precipitations. In the IR, ground-based observations from the late 1990s, followed by Cassini observations, revealed auroral emissions of the H_3^+ molecule, associated with hot polar regions (> 400 K) radiating several 10^{11} W.

At radio wavelengths, the SKR (already described) was detected by the Voyager probes (at a distance of more than 1 au, see Figure 3.39), and then observed on long-term timescales by Ulysses and Cassini. This radiation, very similar to AKR at Earth, has a maximum frequency of 1.2 MHz (limited by the maximum gyrofrequency reached near the planet), and also has small-sized fluctuating sources that radiate a total of ~ $10^{8\text{-}9}$ W. Its intensity increases by several orders of magnitude during magnetospheric compressions, giving an excellent proxy of auroral storms. The solar wind is the main engine of the acceleration of electrons giving rise to the radiation. A spectacular illustration of this was provided by Voyager 2, which measured episodes of sudden SKR extinctions when Saturn passed through the tail of Jupiter's magnetosphere. The SKR is also subject to an intriguing rotational modulation, discussed below. The radiation was first imaged by Cassini (see Figure 3.40), whose unique *in situ* observations from its emission region first validated the CMI mechanism for a planet other than Earth. Triggered by 6–12 keV electrons, SKR sources are co-localized with the UV oval as long as the f_{pe}/f_{ce} ratio << 1 in the source (Lamy *et al.* 2018b). Like Jupiter, Saturn produces a variety of other low-frequency radio emissions such as lightning storms, whose low cutoff frequency (~ 1 MHz) probes the maximum plasma density of the ionosphere, or narrow-band emissions (< 100 kHz) produced by local instabilities in the inner regions of the magnetosphere.

5.4.2.4. *Dynamics of the magnetosphere*

Saturn's magnetosphere is, like that of the Earth, very sensitive to the solar wind. The interplanetary dynamic pressure fronts cause violent magnetospheric compressions, which trigger the ejection of plasmoids in the tail, allowing the plasma that has been injected into the internal magnetosphere to evacuate. Saturn mainly differs from the Earth by its fast rotation of ~ 10.7 hours, which implies that the field lines which have been re-dipolarized after episodes of magnetic reconnection in the tail are quickly driven toward the morning side, which they supply with energetic plasma. These reconnection episodes are not only induced by the solar wind, but also by the rotation, the plasma transport cycle being a composition of the terrestrial Dungey and Jovian Vasyliunas cycles. The injected/evacuated mass balance produced from *in situ* Cassini measurements have shown that the number of plasmoids must be much higher than estimated.

Another important effect of planetary rotation is the modulation of the whole magnetosphere (auroral emissions, magnetic field, particle populations, shock and magnetopause position, etc.). This rotational modulation was initially detected distantly by Voyager SKR remote observations, and then *in situ* from Voyager magnetic data which revealed magnetic oscillations. These measurements made it possible, like for the other giant planets, to accurately determine a period, which was initially interpreted as being the rotation rate of the planetary internal core producing the magnetic field. However, with Saturn's magnetic field being aligned with the rotation axis, the existence of a modulation itself remained ununderstood. The long-term observations of SKR by the Ulysses spacecraft later revealed that the radio period actually varied by ~ 1 %, on timescales of a few years (see Figure 5.25a), and could therefore not directly probe the internal rotation period (Galopeau and Lecacheux 2000). Cassini radio, optical, magnetic and plasma measurements then revealed the existence of two periods of ~ 10.6 and ~ 10.8 hours, one per hemisphere, each varying with time in anti-correlation (Gurnett *et al.* 2009). This global, dual, modulation of the magnetosphere was attributed to the existence of two field-aligned current systems, each rotating in one hemisphere and mainly closing near the equator (Figure 5.24b).

The origin of these current systems, and the reason why both periods vary with time, remain open questions beyond the Cassini mission. One scenario, which appears to be supported by MHD models, suggests that these currents are produced by vortices in the thermosphere of each hemisphere (Jia and Kivelson 2012). This region of the atmosphere is difficult to probe and this hypothesis has yet to be validated observationally.

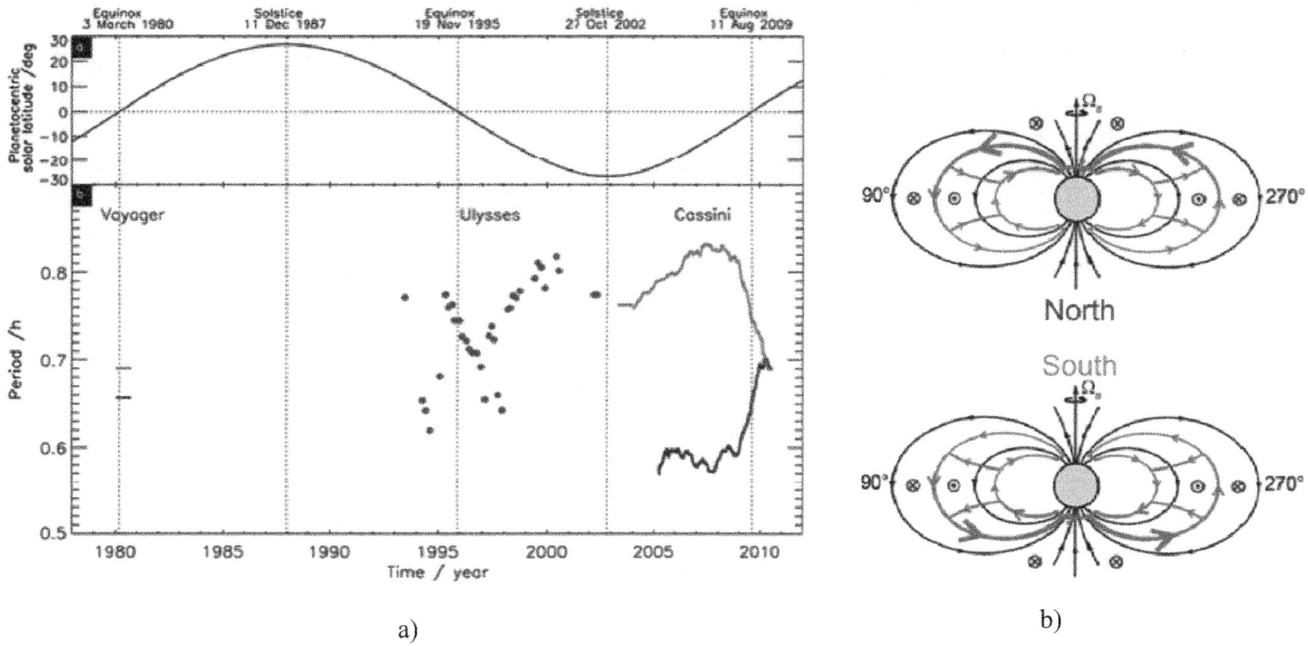

Figure 5.25. *(a) Temporal variation of radio/magnetic rotation periods, as measured by Voyager, Ulysses and Cassini. (b) Schematic diagram of the rotating hemispherical field-aligned current systems responsible for auroral emissions and magnetic oscillations (source: from (Provan* et al. *2014)). For a color version of this figure, see www.iste.co.uk/encrenaz/solar1.zip*

5.4.3. *The asymmetric magnetospheres of Uranus and Neptune*

5.4.3.1. *Complex internal magnetic fields*

The magnetospheres of the icy giant planets are much less known than those of the gas giant planets. They have, in fact, only been explored by a single flyby of the Voyager 2 probe, on January 24, 1986, for Uranus (at solstice) and on August 25, 1989, for Neptune. It was suspected that both planets had a magnetic field, but for different reasons. For Uranus, variable radio and UV emissions, potentially of auroral origin, had been detected from Earth in the 1970–1980s by the IMP6 probe and the IUE UV telescope, respectively. For Neptune, the excess of internal energy suggested the existence of an internal dynamo instead.

Voyager 2 remote radio measurements confirmed the existence of kilometric auroral radio emissions (termed UKR and NKR by analogy with AKR and SKR, see Figure 3.39), just one week before each flyby. These observations provided the first evidence of the existence of a planetary magnetic field, which was confirmed *in situ* by magnetic measurements a few days later. The latter revealed magnetic fields of intensity comparable to those of the Earth or Saturn, but of unexpected configurations. Their equivalent dipoles were very inclined with respect to the rotation axis and offset with respect to the planetary center, respectively by -60° and 0.31 R_U for Uranus and by -47° and 0.55 R_N for Neptune (see Table 3.1). These characteristics reflect important multipolar terms and a B_{max}/B_{min} ratio of ~ 10 at the surface of these planets (versus 3 to 4 for the Earth, Jupiter or Saturn). It is not known yet if the internal dynamo, which is at the origin of these fields, is of similar nature or radically different to that prevailing at the other planets. Magnetic measurements obtained along a single flyby trajectory only weakly constrain the high-degree terms. The best fit of *in situ* magnetic data yielded a quadrupole model (Q3) for Uranus, and an octupole model (O8) for Neptune (Connerney and Acuna 1991). In the case of Uranus, Voyager 2 observations of UV aurorae were additionally used to symmetrize the foot of the north/south magnetic field lines, where electron precipitation occurs, resulting in a better constrained model (AH5) (Herbert 2009).

5.4.3.2. *Complex and dynamic magnetospheres*

The asymmetric magnetospheres resulting from these magnetic configurations are extremely complex and dynamical, due to the rapid planetary rotation. Their general characteristics and global plasma circulation have been briefly discussed in section 3.3.3. They share very low magnetospheric plasma densities, a weak current ring, auroral zones located at middle planetographic latitudes, and a wide variety of auroral radio emissions, because the rarefaction of plasma implies that the cyclotron emission condition f_{pe}/f_{ce}<<1 is satisfied in many regions of those magnetospheres, both at high latitudes but also near the magnetic equator.

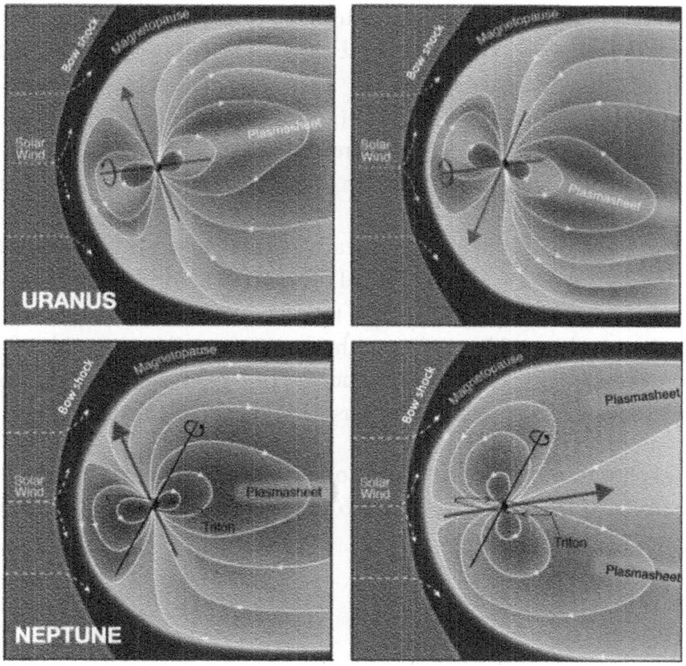

Figure 5.26. *Configuration of the magnetospheres of Uranus at solstice and Neptune, during their flyby by Voyager 2, with half a rotation period difference between the left- and right-handed panels. The large and time-variable inclination of the magnetic field with respect to the incident solar wind produces very dynamical and complex magnetospheric processes (source: C. Arridge, F. Bagenal and S. Bartlett). For a color version of this figure, see www.iste.co.uk/encrenaz/solar1.zip*

At solstice, Uranus' magnetosphere (Figure 5.26a) resembles Earth's magnetosphere in precession around the planet-Sun axis at a rate of one revolution in 17.24 hours. Therefore, the absorption of plasma and magnetospheric energetic particles by the satellites and rings is achieved throughout a very complex topology, and the magnetosphere undergoes very large seasonal variations during the 84-year Uranian revolution period around the Sun. Despite the lack of plasma reservoirs, Uranus possesses intense radiation belts whose origin is not understood. The Voyager UV spectrometer has identified that the source of the UV emission previously observed by IUE was not aurorae, but the emission of a neutral hydrogen halo that extends up to ~ 2 R_U in altitude on the daytime side, resulting from a high exospheric temperature. Recent detection from Earth of Uranus' UV aurorae in the 2010s, shortly after the equinox and during episodes of active solar wind, revealed transient localized spots (see Figure 3.35), different from the night-side signatures observed by Voyager 2 at solstice, and attributed to polar cusp emissions induced by magnetospheric compressions. These observations have thus provided the very first information on this magnetosphere, in an unknown geometry, more than two decades after the Voyager 2 flyby (Lamy *et al.* 2017). At equinox, the magnetosphere oscillates between a "terrestrial" configuration and a configuration where the pole is directed towards the Sun, in half a planetary rotation period.

The same situation permanently prevails for Neptune's magnetosphere, with a slightly faster period of 16.1 hours. The Triton satellite, orbiting at 14.6 R_N, injects ~ 200 g s^{-1} of nitrogen and hydrogen ions into the magnetosphere, through a toroidal neutral cloud around its orbit, similar to the one surrounding the orbit of Titan. This plasma then radially diffuses towards the interior over a few days, and this fast transport prevents the accumulation of plasma in the magnetosphere.

The re-exploration of one and/or the other of these magnetospheres will be one of the objectives of the future missions to explore the Uranus and Neptune systems (Fletcher *et al.* 2020).

5.5. References

Badman, S., Branduardi-Raymont, G., Galand, M. *et al.* (2015). Auroral processes at the giant planets: Energy deposition, emission mechanisms, morphology and spectra. *Space Science Reviews*, 187, 99–179.

Bagenal, F., Dowling, D.E., McKinnon, W.B. *et al.* (2007). *Jupiter: The Planet, Satellites and Magnetosphere*. Cambridge University Press, Cambridge.

Bergstrahl, J.T., Miner, E.D., Matthews, M.T. (1991). *Uranus*. University of Arizona Press, Tucson.

Blanc, M., Andrews, D.J., Coates, A.J. (2015). Saturn plasma sources and associated transport processes. *Space Science Reviews*, 192, 237–283.

Bockelée-Morvan, D., Calmonte, U., Charnley, S. *et al.* (2015). Cometary isotopic measurements. *Space Science Reviews*, 197, 47–83.

Bolton, S.J., Bagenal, F., Blanc, M. *et al.* (2015). Jupiter's magnetosphere: Plasma sources and transport. *Space Science Reviews*, 192, 209–236.

Bonfond, B., Gladstone, R., Grodent, D. *et al.* (2017). Morphology of the UV aurorae Jupiter during Juno's first perijove observations. *Geophysical Research Letters*, 44, 4463–4471.

Cecconi, B., Hess, S., Hérique, A. *et al.* (2012). Natural radio emission of Jupiter as interferences for radar investigations of the icy satellites of Jupiter. *Planetary & Space Science*, 61, 32–45.

Clarke, J.T., Ajello, J., Ballester, G. *et al.* (2002). Ultraviolet emissions from the magnetic footprints of Io, Ganymede and Europa on Jupiter. *Nature*, 415, 997–1000.

Connerney, J.E.P. and Acuna, M.H. (1991). The magnetic field of Neptune. *Journal of Geophysical Research*, 96, 19023–19042.

Connerney, J.E.P. and Jørgensen, J. (2018). A new model of Jupiter's magnetic field from Juno's first nine orbits. *Geophysical Research Letters*, 45, 2590–2596.

Cowley, S.W.H. and Bunce, E.J. (2001). Jupiter: Magnetosphere. In *Encyclopedia of Astronomy and Astrophysics*. IOP Publishing, Bristol.

Cruikshank, D.P., Matthews, M.S., Schumann, A.M. (1995). *Neptune and Triton*. University of Arizona Press, Tucson.

Dougherty, M.K., Khurana, K.K., Neubauer, F.M. *et al.* (2006). Identification of a dynamic atmosphere at Enceladus with the Cassini magnetometer. *Science*, 311, 1406–1408.

Dougherty, M., Esposito, L., Krimigis, S. (2009). *Saturn from Cassini-Huygens*. University of Arizona Press, Tucson.

Dougherty, M.K., Cao, H., Khurana, K.K. *et al.* (2018). Saturn's magnetic field revealed by the Cassini Grand Finale. *Science*, 362(6410).

Encrenaz, T. (1984). Primordial matter in the outer Solar System – A study of its chemical composition from remote sensing analysis. *Space Science Reviews*, 38, 35–87.

Encrenaz, T., Drossart, P., Feuchtgruber, H. (1999). The atmospheric composition and structure of Jupiter and Saturn from ISO observations: A preliminary review. *Planetary & Space Science*, 47, 1225–1242.

Encrenaz, T., Bibring, J.-P., Blanc, M. *et al.* (2003). *Le Système solaire*. CNRS Editions/EDP Sciences, Paris.

Encrenaz, T., Bibring, J.-P., Blanc, M., Barucci, M.-A., Roques, F., Zarka, P. (2004). *The Solar System*, Springer-Verlag, Berlin.

Fletcher, L.N., Helled, R., Roussos, E. *et al.* (2020). Ice giant systems: The scientific potential of missions to Uranus and Neptune. *Planetary & Space Science*. Available at: doi:10.1016/j.pss.2020.105030.

Fortney, J.J. and Nettelmann, N. (2010). The interior, structure, composition and evolution of giant planets. *Space Science Reviews*, 152, 423–447.

Galopeau, P. and Lecacheux, A. (2000). Variations of Saturn's radio rotation period measured at kilometer wavelengths. *Journal of Geophysical Research*, 105(A6), 13089–13101.

Gehrels, T. and Matthews, M.S. (1976). *Jupiter.* University of Arizona Press, Tucson.

Gehrels, T. and Matthews, M.S. (1984). *Saturn.* University of Arizona Press, Tucson.

Gillett, F.C., Low, F.J., Stein, W.A. (1969). The 2.8–14-micron spectrum of Jupiter. *Astrophysical Journal*, 157, 925–934.

Grodent, D. (2015). A brief review of ultraviolet auroral emissions on giant planets. *Space Science Reviews*, 187(1–4), 23–50.

Guillot, T. (2005). The interiors of giant planets: Models and outstanding questions. *Annual Reviews of Earth and Planetary Science*, 33, 493–530.

Guillot, T. (2019). Signs that Jupiter was mixed by a giant impact. *Nature*, 572, 315–317.

Gurnett, D., Lecacheux, A., Kurth, W.S. *et al.* (2009). Discovery of a north-south asymmetry in Saturn's radio rotation period. *Geophysical Research Letters*, 36(L16102).

Hanel, R.A., Conrath, B.J., Jennings, D.E. *et al.* (2003). *Exploration of the Solar System by Infrared Remote Sensing*, 2nd edition. Cambridge University Press, Cambridge.

Harland, D.M. (2000). *Jupiter Odyssey: The Story of NASA's Galileo Mission.* Springer/Praxis, Berlin.

Harland, D.M. (2002). *Mission to Saturn: Cassini and the Huygens Probe.* Springer/Praxis, Berlin.

Helled, R. and Guillot, T. (2013). Interior models of Saturn, including uncertainties in shape and rotation. *Astrophysical Journal*, 767(2).

Herbert, F. (2009). Aurora and magnetic field of Uranus. *Journal of Geophysical Research*, 119(A11206).

Hess, S.L.G. (2011). Comparative study of the power transferred from satellite-magnetosphere interactions to auroral emissions. *Journal of Geophysical Research*, 116(A01202).

Hill, T.W. (2001). The Jovian auroral oval. *Journal of Geophysical Research*, 106(A5), 8001–8017.

Iess, L., Militzer, B., Kaspi, Y. *et al.* (2019). Measurements and implication on Saturn's gravity field and ring mass. *Science*, 364(6445).

Irwin, P.W. (2009). *Giant Planets of Our Solar System: Atmospheres, Composition and Structure.* University of Arizona Press, Tucson.

Jia, X. and Kivelson, M.G. (2012). Driving Saturn's magnetospheric periodicities from the upper atmosphere/ionosphere: Magnetotail response to dual sources. *Journal of Geophysical Research*, 117(A11219).

Khurana, K.K., Kivelson, M.G., Vasyliunas, V.M. *et al.* (2004). The configuration of Jupiter's magnetosphere. In *Jupiter*, Bagenal, F., Dowling, T.E., McKinnon, W.B. (eds). Cambridge University Press, Cambridge.

Kivelson, M.G., Khurana, K.K., Coroniti, F.V. *et al.* (1997). The magnetic field and magnetosphere of Ganymede. *Geophysical Research Letters*, 24(17), 2155–21588.

Lainey, V., Jacobson, R.A., Tajeddine, R. *et al.* (2017). New constraints on Saturn's interior from Cassini astrometric data. *Icarus*, 281, 286–296.

Lamy, L., Prangé, R., Hansen, K.C. *et al.* (2017). The aurorae of Uranus past equinox. *Journal of Geophysical Research*, 122(4), 3997–4008.

Lamy, L., Prangé, R., Tao, C. *et al.* (2018a). Saturn's northern aurorae at solstice from HST observations coordinated with Cassini's grand finale. *Geophysical Research Letters*, 45(18), 9353–9362.

Lamy, L., Zarka, P., Cecconi, B. *et al.* (2018b). The low frequency source of Saturn's kilometric radiation. *Science*, 362(6410).

Larson, H.P. (1980). Infrared spectroscopic observations of the outer planets, their satellites, and the asteroids. *Annual Review of Astronomy & Astrophysics*, 18, 43–75.

Lewis, J.S. (1995). *Physics and Chemistry of the Solar System*. Academic Press, Cambridge.

Li, C., Ingersoll, A., Bolton, S. *et al.* (2020). The water abundance in Jupiter's equatorial zone. *Nature Astronomy*, 4, 609–616.

Lis, D.C., Bockelée-Morvan, D., Güsten, R. *et al.* (2019). Terrestrial deuterium-to-hydrogen ratio in water in hyperactive comets. *Astronomy & Astrophysics*, 625(L5).

Liu, S.F., Hori, Y., Müller, G. *et al.* (2019). The formation of Jupiter's core by a giant impact. *Nature*, 572, 355–357.

Louarn, P., Allegrini, F., McComas, D.J. *et al.* (2017). Generation of the Jovian hectometric radiation: First lessons from Juno. *Geophysical Research Letters*, 44, 4439–4446.

Marley, M.S. (1999). *Encyclopedia of the Solar System*. Academic Press, Cambridge.

Miner, E.D. (1990). *Uranus: The Planet, Rings and Satellites*. Ellis Horwood Ltd, New York.

Moore, K.M., Cao, H., Bloxman, J. *et al.* (2019). Time variation of Jupiter's internal magnetic field consistent with zonal wind advection. *Nature Astronomy*, 3, 730–735.

Noll, K.S., Weaver, H.A., Feldman, P.D. (1996). *The Collision of Comet Shoemaker–Levy 9 with Jupiter*. Cambridge University Press, Cambridge.

Owen, T.C. and Encrenaz, T. (2006). Compositional constraints on giant planet formation. *Planetary & Space Science*, 54, 1188–1196.

Owen, T.C., Mahaffy, P., Niemann, H.B. *et al.* (1999). A low temperature origin for the planetesimals that formed Jupiter. *Nature*, 402, 269–270.

Pallier, L. and Prangé, R. (2001). More about the structure of the high latitude Jovian aurorae. *Planetary and Space Sciences*, 49, 1159–1173.

Persoon, A.M., Kurth, G.B., Gurnett, W.S. *et al.* (2019). Electron density distributions in Saturn's ionosphere. *Geophysical Research Letters*, 46(6), 3061–3068.

Provan, G., Cowley, S.W.H., Lamy, L. *et al.* (2014). Planetary period oscillations in Saturn's magnetosphere: Comparison of magnetic oscillations and SKR modulations in the postequinox interval. *Journal of Geophysical Research*, 119, 7381–7401.

Pryor, W.R., Rymer, A.M., Mitchell, D.G. *et al.* (2011). The auroral footprint of Enceladus on Saturn. *Nature*, 472, 331–333.

Pyne, S.J. (2010). *Voyager: Exploration, Space, and the Third Great Age of Discovery*. Penguin Books, London.

Rogers, J.H. (1995). *The Giant Planet Jupiter*. Cambridge University Press, Cambridge.

Santos-Costa, D., Adumitroaie, V., Ingersoll, A. *et al.* (2017). First look at Jupiter's synchrotron emission from Juno's perspective. *Geophysical Research Letters*, 44, 8676–8684.

Saur, J., Neubauer, F.M., Connerney, J.E.P., Zarka, P., Kivelson, M.G. (2004). Plasma interaction of Io with its plasma torus. In *Jupiter*, Bagenal, F., Dowling, T., McKinnon, W. (eds). Cambridge University Press, Cambridge.

Stevenson, R. (1982). Interiors of the giant planets. *Annual Reviews of Earth and Planetary Science*, 10, 257–295.

Tokunaga, A.T., Ridgway, S.T., Wallace, L., Knacke, R.F. (1979). High-resolution spectra of Jupiter in the 744-980 inverse centimeter spectral range. *Astrophysical Journal*, 232, 603–615.

Valek, P.W., Allegrini, F., Bagenal, S.J. *et al.* (2020). Jovian high-latitude ionospheric ions: Juno *in situ* observations. *Geophysical Research Letters*, 46, 8663–8670.

Vasyliunas, V.M. (1983). Plasma distribution and flow. In *Physics of the Jovian Magnetosphere*, Dessler, A.J. (ed.). Cambridge University Press, Cambridge.

Zarka, P. (2007). Plasma interactions of exoplanets with their parent star and associated radio emissions. *Planetary and Space Science*, 55, 598–617.

Appendix

Web links

A list of Internet links that are particularly used in planetology can be found here:

– list of physical and orbital parameters of Solar System objects (with associated tools for trajectory calculation and observability): http://ssd.jpl.nasa.gov/?; https://www.imcce.fr/services/ephemerides/;

– list of orbital parameters of small Solar System bodies, IAU Minor Planet Center: https://minorplanetcenter.net/;

– catalog of exoplanets (with associated tools, histograms, etc.): http://exoplanet.eu/;

– catalog of all stellar objects with positions, motions and physical characteristics (Strasbourg Stellar Data Center): https://cdsweb.u-strasbg.fr/index-fr.gml;

– *Science for Exoplanets and Planetary Systems* digital book (L3 level): http://sesp.esep.pro/fr/index.html;

– list of past and current planetary space missions: https://en.wikipedia.org/wiki/List_of_Solar_System_probes;

– list of the largest ground-based optical telescopes: https://en.wikipedia.org/wiki/List_of_largest_optical_reflecting_telescopes;

– list of the largest radiotelescopes (not completely updated): https://en.wikipedia.org/wiki/List_of_radio_telescopes;

– Martian climate model (with physical atmospheric parameters for any solar longitude, planetary latitude and longitude): http://www-mars.lmd.jussieu.fr/mars/access.html;

– spectroscopic database (provides the molecular spectroscopic data needed for radiative transfer calculations in planetary atmospheres): https://geisa.aeris-data.fr/; https://hitran.org;

– transitions in the submillimeter, millimeter and radio ranges: https://spec.nasa.gov; https://splatalogue.online;

– database for radiative transfer in exoplanets and brown dwarfs: https://exomol.com;

– Planetary Spectrum Generator (calculation of planetary and exoplanetary spectra): https://psg.gsfc.nasa.gov/;

– NASA planetary database: https://pds.nasa.gov;

– ESA planetary database: https://archives.esac.esa.int/psa/;

– Nançay Radio Observatory: http://obs-nancay.fr;

– Planetary auroral spectro-imaging database: https://apis.obspm.fr;

– OV access portal to planetary data: http://vespa.obspm.fr/.

Glossary

Accretion: capture of surrounding matter by the gravitational attraction of an object, for example, a star in formation.

Achondrite: stony meteorite containing less than 35% metal, which comes from the surface of a large asteroid.

Adaptive optics: an observational technique that allows real-time elimination of distortions in the image of a celestial object, due to atmospheric turbulence, by correcting wave surface defects using a thin deformable mirror.

Aerosol: a set of fine particles, solid or liquid, suspended in an atmosphere.

Albedo: fraction of the incident light reflected by a body at a given wavelength.

Ambipolar diffusion: diffusion of charged particles in a quasi-neutral plasma, in other words, in which the charge density is zero at any point, but with microscopic gradients leading to the presence of an electric field.

Antitail: type of cometary tail due to dust, sometimes observed in the direction of the Sun when the Earth crosses the plane of the cometary orbit.

Aphelion: the furthest point from the Sun in a heliocentric orbit.

Asteroid: small solid body in orbit around the Sun. Most are located in the main asteroid belt between the orbits of Mars and Jupiter.

Asthenosphere: ductile part of the Earth's upper mantle, located between the lithosphere and the Earth's lower mantle at a depth of 700 km.

Astrometry: measurement of the position, distance and motion of celestial bodies (also called positional astronomy).

Astronomical unit (AU): unit of distance equal to the average Sun–Earth distance, or about 150 million km. Its exact value is 149.6 10^9 m, or 4.848 10^{-6} pc, or 15.812 10^{-6} ly.

Aurora: luminous phenomenon caused by the precipitation of energetic particles from the planetary environment in the upper atmosphere, generally located near the magnetic poles of a planet.

Brown dwarf: "aborted" star, with a mass between 0.01 and 0.08 solar mass (in other words, between 13 and 80 masses of Jupiter), insufficient for the fusion of hydrogen into helium to be triggered in its core.

Centaur: asteroid orbiting the Sun between the orbits of Jupiter and Neptune.

Chondrite: stony meteorite (less than 35% metal) containing chondrules.

Chondrule: elements of undifferentiated meteorites or chondrites. These small submillimeter-sized beads, generally composed of silicates, are inserted into a matrix.

Clathrate: compound formed of gas molecules trapped in a network of host molecules, most often H_2O. The clathrate most often encountered in planetology is methane clathrate (CH_4, $6H_2O$).

Coma: nebulosity surrounding the nucleus of a comet, created by the sublimation of ice and the ejection of dust that accompanies it.

Comet: small body of the Solar System with an often very eccentric orbit, possibly disturbed by the presence of the most massive planets. Far from the Sun, it is made up of a core of ice and rocks. When approaching the Sun, the sublimation of ice leads to the appearance of a coma, a dust tail and a plasma tail.

Conjunction: the moment when the position of a planet, seen from the Earth, is closest to the Sun. Mercury and Venus have lower conjunctions (when the planet is in front of the Sun) and higher conjunctions (when the planet is behind the Sun).

Continental drift: see *plate tectonics*.

Corona: circular volcanic structure bordered by numerous faults. It is found, in particular, on Venus and on Miranda.

Coronagraphy: astronomical technique consisting of masking the central object (generally the Sun or a star) to reveal its immediate environment (the solar corona or possibly a circumstellar disk or exoplanets). This technique is widely used for direct imaging of exoplanets and circumstellar disks.

Coronal hole: dark zone on the surface of the Sun from which the solar plasma escapes in the form of solar wind.

Cosmic abundance: abundance of elements in the Universe. This abundance is commonly related to the quantity of hydrogen, the most abundant element. The other elements are the results of nucleosynthesis. They could also have been produced during the Big Bang, like helium, or later, essentially by stars. For light elements, solar abundances are used; for heavy elements, measured with respect to silicon, the most primitive meteorites are used.

Cryovolcanism: a process similar to volcanism occurring on some icy satellites (Enceladus, Titan, Triton, etc.). The material ejected is water, ammonia or methane under pressure, which immediately recondenses at the surface.

Cyclotron (emission): low-frequency radio electromagnetic radiation produced in the auroral regions of the Earth and giant planets, by energetic electrons (weakly relativistic) circulating along magnetic field lines, at a frequency close to the electron cyclotron frequency.

Dating: method of determining the age of a body in the Solar System. A first method consists of counting the craters present on the surface: the most cratered bodies are the oldest. A second method consists of using samples (lunar or meteoritic) by measuring what remains of long-period radioisotopes of certain elements (uranium, thorium, etc.).

Daughter molecule: dissociation product (molecule, radical, atom or ion) of a cometary parent molecule.

Doppler-Fizeau (effect): variation $\Delta\lambda$ of the wavelength received from a moving source compared to the wavelength emitted λ_0. In the non-relativistic case, $\Delta\lambda/\lambda_0 = v/c$, v being the speed of the source and c the speed of light.

Dwarf planet: according to the definition adopted by the International Astronomical Union in 2006, it is a body of the Solar System that meets the following four criteria: (1) it is in orbit around the Sun, (2) it has sufficient mass to ensure hydrostatic equilibrium (and thus have a quasi-spherical shape), (3) it does not produce a vacuum in its orbital vicinity, and (4) it is not a satellite. There are currently five dwarf planets recognized by the IAU: one asteroid (Ceres) and four TNOs (Pluto, Haumea, Makemake and Eris).

Eclipse: apparent and temporary disappearance, or decrease in brightness, of a celestial object seen from the Earth when another body passes in front of it. Eclipses of the Sun by the Moon are the most common phenomena. This phenomenon is at the origin of the transit method for the detection of exoplanets.

Ecliptic: plane of the Earth's orbit. In the sky, it is the apparent trajectory of the Sun among the stars during the year.

Ecliptic coordinates: coordinates used to locate a point in the sky relative to the plane of the ecliptic, in which the Sun moves according to the seasons. The ecliptic coordinates are the ecliptic longitude λ and ecliptic latitude β.

Edgeworth-Kuiper (belt): see Kuiper Belt.

Equatorial coordinates: coordinates used to locate a point in the sky relative to the plane of the Earth's equator. The equatorial coordinates are the right ascension α and declination δ.

Equinox: moment of the year when the Sun crosses the equatorial plane of the planet.

Escape velocity: the minimum velocity that a body must reach to escape the gravitational attraction of the object (planet, satellite, etc.) to which it belongs and move away from it indefinitely.

Exobiology: field of research aimed at studying the physical, chemical and biological processes that may be at the origin of the appearance of life in the Universe (in the Solar System or on exoplanets).

Exoplanet: planet in orbit around a star other than the Sun.

Gegenschein: faint glow along the ecliptic in the opposite direction to the Sun, hence its name (German: opposite glow). It is caused by the scattering of sunlight by interplanetary dust, mostly located near the ecliptic.

Gravitational lens: phenomenon occurring when a massive object is located between a distant object and the observer. The deviation of the light rays of the distant object by the massive object concentrates them on the observer, which produces an amplification of the signal. If the distant object is a star and if the intermediate object is a star or a planet, it is called a gravitational microlens.

Great Red Spot: a very large anticyclone (about 15,000 km longitude extension) located in the southern hemisphere of Jupiter, at a latitude of 22°S. It has been stable since its discovery by Cassini in 1665.

Greenhouse effect: process leading to the heating of a planetary atmosphere when its constituents are transparent to visible solar radiation but absorb infrared radiation. The most efficient greenhouse gases are CO_2, H_2O and CH_4. The solar radiation (maximum in the visible range) heats the surface, which, as it heats up, reflects infrared radiation that the atmosphere absorbs, accelerating the heating process, and the mechanism can run away; the best example is that of the planet Venus.

Habitability: defines the capacity of a star to accommodate life. The criteria of habitability include the presence of liquid water, a source of energy and elements involved in the synthesis of prebiotic molecules (C, H, N, O, P, S).

Habitability zone: theoretical sphere surrounding a star where the temperature on the surface of a planet orbiting this star would be compatible with the presence of liquid water. The more massive the star, the brighter it is, and the further away the sphere is from it; the smaller the star is, the closer it is to it.

Hadley (circulation): planetary convective circulation induced by the variation of insolation as a function of latitude, to which is added another effect (Coriolis force) if the planet is rotating.

Haser (model): model describing the distribution of cometary molecules (parents and daughters) in the coma of a comet as a function of their distance from the nucleus, under the effect of photodissociation due to solar radiation.

Heliosphere: elongated envelope generated by the solar wind. It extends to the heliopause, which defines the boundary between the zone of influence of the solar wind and the interstellar medium. Its elongated shape is due to the motion of the Sun with respect to the surrounding interstellar matter.

Hertzprung-Russell (diagram): diagram representing the luminosity of stars as a function of their effective temperature. It makes it possible to distinguish the different types of stars and to understand their evolution.

Heterosphere: region of the upper atmosphere in which the constituents are no longer mixed, each constituent decreasing as the altitude increases with its own scale height.

Homosphere: region of the atmosphere in which the constituents are uniformly mixed. In the case of the Earth, this region extends from the surface to an altitude of 90 to 100 km.

Horizontal coordinates (also called local or alt-azimuth): coordinates used to locate a point in the sky relative to the horizontal plane where the observer is located. The coordinates are the height h (or its complement, the zenithal distance z) and the azimuth A.

Ice line: in a protoplanetary disk, it is the distance at which simple volatile molecules are in the form of ice. It depends on temperature and is specific to each molecule (H_2O, CH_4, CO). In general, the expression refers to the condensation of water.

Interferometer: instrument made up of several telescopes or radio telescopes that are interconnected, allowing high spatial resolution and mapping astronomical objects. In spectroscopy, the interferometer is a spectrometer that makes it possible to obtain a high spectral resolution (Michelson interferometer or Fourier transform interferometer).

Ionosphere: region of the upper atmosphere in which constituents are ionized under the effect of ultraviolet and solar X-rays.

Kepler's laws: empirical laws stated by Johannes Kepler to describe the motion of the planets around the Sun. They were explained by Newton within the framework of the law of gravity.

Kirkwood gaps: regions located in the main asteroid belt, at certain distances from the Sun, in which there are very few asteroids. They are due to orbital resonances related to the proximity of Jupiter.

Kuiper Belt: ring-shaped zone located beyond Neptune's orbit, considered as the reservoir of periodic comets of low inclination. It is also the reservoir of the Centaurs and certain types of trans-Neptunian objects.

Lagrange points: points located on the orbit of a planet around the Sun or perpendicular to it, or of a satellite around a planet, where the forces of gravity induced by the two bodies balance each other. Three points (L1, L2, L3), located on the axis of the two bodies, are positions of unstable equilibrium; the other two (L4 and L5), located on the orbit of the small body, at 60° upstream or downstream of it, are positions of stable equilibrium.

Lambert's law: law according to which the luminous intensity of an isotropic source is proportional to the cosine of the angle between the emission angle and the surface normal: $I(\alpha) = I_0 \cos \alpha$, I_0 being the emission along the normal axis to the surface.

Late Heavy Bombardment (LHB): episode that occurred at the beginning of the Solar System's history (between 4.1 and 3.9 billion years ago), characterized by a high rate of meteorite impacts on the surfaces of bodies in the Solar System.

Libration: phenomenon of oscillation of a satellite as observed from the planet around which it is rotating. It is due to the eccentricity of the satellite's orbit and the angle that this orbit makes with the equatorial plane of the planet.

Light-year (ly): distance travelled by light in one year. It is equal to $9.461\ 10^{12}$ km.

Lithosphere: rigid envelope surrounding the Earth, comprising the Earth's crust and part of the upper mantle.

Magma: fully or partially molten rock, formed by partial melting of the Earth's crust or mantle. By extension, the term can be applied to similar material in other planets or exoplanets.

Magnetic storm: episode of sudden and intense increase in the Earth's magnetic field.

Magnetopause: the boundary between a planet's magnetosphere, governed by its internal magnetic field, and the interplanetary medium.

Magnetosheath: region of the magnetosphere between the magnetopause and the shockwave.

Magnetosphere: plasma envelope surrounding a star, planet or satellite, in which physical phenomena are governed by the object's intrinsic magnetic field.

Magnitude: logarithmic scale allowing the measurement of the brightness of a star. The magnitude is higher the lower the brightness of the star. If m is the magnitude and B the brightness of the star, we have $m = -2.5 \log B + \text{Cst}$.

Main asteroid belt: main reservoir of asteroids, located between the orbits of Mars and Jupiter.

Mantle: in a solid body, the intermediate layer between the crust and the core.

Mesosphere: the atmospheric layer below the thermosphere. In the case of the Earth, it surmounts the stratosphere; in the case of Mars and Venus, in the absence of a stratosphere, it directly surmounts the troposphere.

Meteor: a luminous trail left in the sky by the fall of an extraterrestrial body that vaporizes by friction in the atmosphere. It is also called a "shooting star".

Meteor shower: a meteor shower (also called shooting stars) is caused by the passage of the Earth through the orbit of a comet, resulting in the fall of many of its cometary dusts into the Earth's atmosphere.

Meteorite: a solid object of extraterrestrial origin that has passed through the Earth's atmosphere without being completely volatilized during impact.

NEO (Near Earth Object): small body in the Solar System whose orbit is close to that of the Earth, with a perihelion at less than 1.3 AU. This class of objects includes about 20,000 asteroids (NEA) and about a hundred short-period comets (NEC).

Nucleosynthesis: formation of the nuclei of different chemical elements by nuclear reactions. The lightest elements (H, He, Li, Be, B) were formed just after the Big Bang (primordial nucleosynthesis). The heavier elements were synthetized in stars.

Nutation: periodic oscillation of the axis of rotation of an object around its mean position, in addition to precession.

Occultation: phenomenon by which a celestial body (star, planet, asteroid, etc.) is entirely or partially masked by an object. Stellar occultations are used to determine the properties of objects in the Solar System that occult them.

Olivine: silicate frequently observed on the surface of the bodies of the Solar System. The most common form is forsterite (Mg_2SiO_4).

Oort cloud: large spherical shell located between 30,000 and 100,000 au (one to two ly from the Sun), supposed to be the reservoir of new comets, formed closer to the Sun like other small bodies, but ejected in this region following gravitational interactions with giant planets.

Opposition: the moment when a planet, seen from the Earth, is exactly in the opposite direction to the Sun. The distance from the Earth to the planet is then minimal. Only Mars and the giant planets, whose orbit is outside that of the Earth, can be seen in opposition.

Orbital resonance: in the three-body problem, orbital resonance occurs when the ratio of the periods of two bodies orbiting a third is a rational fraction. This is the case, in particular, of the Io, Europa and Ganymede satellites around Jupiter.

Parent molecule: molecule sublimated from the ices of the cometary nucleus and released directly into the coma.

Parsec (pc): unit of length used to measure the distance of stars in our Galaxy. It is the distance of a star whose parallax is equal to one arc second. 1 pc = 3.26 ly = 3.08 10^{16} meters.

Perihelion: the point closest to the Sun on a heliocentric orbit.

Permafrost: constituent of the subsoil of a planet or a small body that has the characteristic of being permanently frozen.

Photodissociation: chemical reaction undergone by a molecule under the effect of visible, ultraviolet or higher energy radiation. This radiation can ionize the molecule or dissociate it into several other molecules or atoms, called dissociation products.

Planetary migration: motion of a planet within a protoplanetary disk, as a result of its gravitational interaction with the matter contained in the disk.

Planetesimal: small body a few kilometers in size, formed by the aggregation of dust in the protosolar or protoplanetary disk.

Planetoid: body formed by the agglomeration of several planetesimals following their mutual collisions.

Plasmasphere: toroidal region of the inner magnetosphere consisting of low energy plasma.

Plate tectonics: model explaining the global dynamics of the Earth's lithosphere. It is composed of about fifteen plates with relative movements with respect to each other. It is at the boundary between these plates that most geological phenomena occur (volcanic activity, earthquakes, formation of mountain ranges or oceanic trenches).

Poynting-Robertson (effect): effect by which solar radiation causes a particle of dust to spiral down towards the Sun. It is effective on micrometer-sized grains.

Precession: movement of the axis of a rotating body that describes a cone under the effect of external forces. In the case of the Earth, we speak of the precession of the equinoxes. This term can also refer to the rotation of the orbit of an object.

Protoplanetary disk: circumstellar disk formed of gas and dust, created by the collapse of a nebula rotating on itself. The protostar is formed in the central part, while the planets will form within the protoplanetary disk.

Protosolar (or primitive) nebula: mass of interstellar gas and dust from which the Solar System was formed, following its collapse into a disk.

Pulsar: a neutron star that rotates very rapidly, emitting a beam of radio waves like a lighthouse. It is the ultimate stage in the evolution of some massive stars.

Pyroxene: silicate of formula $XY(SiO_3)_2$, in which X and Y are two cations (examples: Ca, Mg). Pyroxene is a common compound in igneous and metamorphic rocks.

Radiation belts: the inner region of a planetary magnetosphere, in which high-energy particles are trapped by the magnetic field and can emit synchrotron radiation. In the case of the Earth, they are called Van Allen belts.

Radiation pressure: pressure exerted on a surface by the exchange of momentum between the object and the photons emitted by the Sun. It is effective for particles smaller than 0.1 micrometer in size, and tends to push them out of the Solar System.

Red dwarf: star of low mass (between 0.08 and 0.4 solar mass) and relatively faint. Red dwarfs are the most common stars in the Universe.

Reference system: system of celestial coordinates materialized by the position of very distant celestial objects (stars, quasars or galaxy nuclei).

Regolith: layer of fine dust covering the surface of objects without atmosphere or with a tenuous atmosphere, resulting from the pulverization of rocks following meteorite impacts.

Roche limit: distance from a planet below which a satellite begins to break up under the effect of tidal forces induced by the planet. It was established by the French astronomer Édouard Roche (1820–1883).

Scale height: the altitude at which one must rise, in an atmosphere, for the pressure to be decreased by a factor e. The scale height is about ten kilometers on the ground in the atmospheres of telluric planets.

Shooting star: light trail associated with the fall of an interplanetary particle (meteoroid) into the atmosphere. It can be associated with the passage of the Earth in the orbit of a comet; it is then called a swarm.

Small planet: see asteroid.

SNC (meteorites): Shergottites, Nakhlites and Chassignites meteorites (named according to the site where they fell) are considered to be of Martian origin, based on the analogy between their chemical, isotopic and petrological characteristics, as well as the observations of the Martian atmosphere made by the Viking probes.

Solar corona: region of the Sun located above the chromosphere that extends over several million kilometers. It is observed with coronagraphs or during total eclipses of the Sun.

Solar wind: plasma flow consisting essentially of electrons, protons and helium nuclei, ejected from the upper solar atmosphere and propagating radially through the Solar System to the confines of the heliosphere.

Stellar parallax: the angle under which one would see, from a star, the half-major axis of the Earth's orbit. Its measurement is used to obtain the distance to the stars.

Stratosphere: atmospheric region above the troposphere and below the mesosphere, in which solar ultraviolet radiation is absorbed, causing the temperature to increase with altitude. The Earth and the giant planets have a stratosphere, but Mars and Venus do not.

Supernova: late evolutionary stage of a massive star, during which all the heavy elements, up to iron, can be synthesized.

Synchrotron (emission): electromagnetic radiation emitted by a highly relativistic charged particle moving in a magnetic field, at the electronic cyclotron frequency and its harmonics.

T-Tauri (phase): phase of intense activity at the beginning of the evolution of a star (around 10 million years). The name refers to the first star on which the phenomenon was observed.

Thermosphere: the upper part of the atmosphere, located above the mesosphere and below the exosphere, with which it forms the heterosphere. The atoms are ionized there under the action of solar ultraviolet radiation.

Tidal effect: see *tide*.

Tide (effect of the): deformation of a star subjected to the gravitational forces of a nearby massive star.

Trans-Neptunian object (TNO): Solar System object in orbit around the Sun beyond the orbit of Neptune.

Trojans (asteroids): families of asteroids located on the orbit of Jupiter near the Lagrange points L4 and L5.

Tropopause: in an atmosphere, the upper limit of the troposphere. It is surmounted by the stratosphere or directly by the mesosphere.

Troposphere: the lower part of an atmosphere, characterized by a transfer of energy by convection and a decrease in temperature as altitude increases.

Van Allen (belts): see *radiation belts*.

Vernal (point): point of intersection (also called gamma point) between the plane of the equator and that of the ecliptic corresponding to the ascending node (passage of the Sun from the southern hemisphere to the northern hemisphere).

Virtual observatory: set of observational data, accompanied by exploitation programs, accessible to any researcher through the Internet.

White dwarf: very dense and not very luminous star, representing the ultimate stage in the evolution of stars with masses less than about eight times that of the Sun.

Zodiacal (cloud): disk of small interplanetary dust, close to the ecliptic, fed by dust from collisions between small bodies (comets or asteroids).

List of Authors

Thérèse ENCRENAZ
LESIA
Paris Observatory
PSL University
Paris
France

Marcello FULCHIGNONI
LESIA
Paris Observatory
PSL University
Paris
France

Laurent LAMY
LESIA
Paris Observatory
PSL University
Paris
France

James LEQUEUX
LERMA
Paris Observatory
PSL University
Paris
France

Françoise ROQUES
LESIA
Paris Observatory
PSL University
Paris
France

Index

J, K, L

M

N, O

P

R

S

T

U, V, W